Mathematics
for Engineers and Scientists

Mathematics
for Engineers and Scientists

K. WELTNER
J. GROSJEAN
P. SCHUSTER
W. J. WEBER

Stanley Thornes (Publishers) Ltd

First published in 1986 by:
Stanley Thornes (Publishers) Ltd
Ellenborough House
Wellington Street
CHELTENHAM GL50 1YW
United Kingdom

99   00   /   10   9   8   7

British Library Cataloguing in Publication Data

Mathematics for engineers and scientists
  1. Mathematics—1961–
  I. Weltner, K   II. Mathematik für Physiker.
  English
  510'.245   QA36

ISBN 0–85950–120–5

Typeset by Tech-Set, Gateshead, Tyne & Wear.
Printed and bound in Great Britain by
T.J. International Ltd, Padstow, Cornwall.

Contents

Acknowledgement

Originally published in the Federal Republic of Germany under the title

Mathematik für Physiker

by the authors
K. Weltner, H. Wiesner, P.-B. Heinrich, P. Englehardt and H. Schmidt.

The work has been translated by J. Grosjean and P. Schuster and adapted to the needs of engineering and science students in English speaking countries by J. Grosjean, P. Schuster, W.J. Weber and K. Weltner.

Preface

Mathematics is an essential tool for the engineer and the scientist which students must use from the beginning of their studies. This textbook aims to develop as rapidly as possible the students' ability to use and understand those parts of mathematics which they will most frequently encounter. Thus, functions, vectors, calculus, differential equations and functions of several variables are presented in a very accessible way. In addition, other chapters in the book provide the basic knowledge on various topics in applied mathematics.

From their extensive experience as lecturers, each of the authors has acquired a close awareness of the needs of first- and second-year students, and one of their special aims has been to help users to tackle successfully the difficulties with mathematics which are commonly met. A special feature which extends the supportive value of the main textbook is the separately published 'study guide'. The guide aims to satisfy two objectives simultaneously: it enables students to make more effective use of the main textbook, and it offers advice on the improvement of techniques on the study of textbooks generally.

Use of programmed and individualised instruction techniques in the study guide not only make it possible for students to maximise the value of the main textbook, but also helps them to understand important concepts. The learning process is further enhanced by the provision of graded exercises and problems designed to develop competence in the application of theory and in problem-solving.

Originally published in the Federal Republic of Germany in the academic year 1975/76 under the title *Mathematik für Physiker*, the whole work — textbook and study guide — has proved its worth in actual use. This new version has been modified and adapted to suit the needs of students studying in the English language in the fields of engineering and science, and to this end chapters on Numerical Methods, Applications of Integration, Eigenvalues and Laplace Transforms have been added.

Both the textbook and the study guide in the original edition and in this new version have resulted from teamwork. The authors of the original book were K. Weltner, P.-B. Heinrich, H. Wiesner, P. Engelhardt and H. Schmidt. The translation was undertaken by J. Grosjean and P. Schuster and the adaptation by J. Grosjean, P. Schuster, W.J. Weber and K. Weltner. The original authors acknowledge J. Grosjean's added contribution for his chapters on Applications of Integration and Laplace Transforms, and for modifications and additional material in various chapters.

K. Weltner
J. Grosjean
P. Schuster
W.J. Weber

Parts of this text marked with an asterisk may be omitted by the student on first reading.

CHAPTER 1

Functions

1.1 THE MATHEMATICAL CONCEPT OF FUNCTIONS AND ITS MEANING IN SCIENCE AND ENGINEERING

1.1.1 INTRODUCTION

The velocity of a body falling freely to Earth increases with time, i.e. the velocity of fall depends on the time. The pressure of a gas maintained at a constant temperature depends on its volume. The periodic time of a simple pendulum depends on its length. Such dependences between observed quantities are frequently encountered in science and engineering and they lead to the formulation of natural laws.

Two quantities are measured with the help of suitable instruments such as clocks, rulers, balances, ammeters, voltmeters etc.; one quantity is varied and the change in the second quantity observed. The former is called the *independent quantity*, or argument, and the latter the *dependent quantity*, all other conditions being carefully kep constant. The procedure to determine experimentally the relationships between physical quantities is called an *empirical* method. Such a method can be extended to the determination of the relationships between more than two quantities; thus the pressure of a gas depends on its volume and its temperature when both volume and temperature vary.

Relationships obtained experimentally may be tabulated or a graph drawn showing the variation at a glance. Such representations are useful but in practice we prefer to express the relationships mathematically.

A mathematical formulation has many advantages:

It is shorter and often clearer than a description in words.

It is unambiguous. Relationships described in such a way are easy to communicate and misunderstanding is out of the question.

It enables us to predict the behaviour of physical quantities in regions not yet verified experimentally; this is known as extrapolation.

The mathematical description of the relationship between physical quantities may give rise to a mathematical model.

1

1.1.2 THE CONCEPT OF A FUNCTION

We now investigate the exact mathematical description of the dependence of two quantities.

EXAMPLE   Consider a spring fixed at one end and stretched at the other end, as shown in figure 1.1. This results in a force which opposes the stretching or displacement. Two quantities can be measured:

the displacement x in metres (m); the force F in newtons (N).

Measurements are carried out for several values of x. Thus we obtain a series of paired values for x and F associated with each other.

(1) The paired values are tabulated as shown in figure 1.1.

Displacement (m)	Force (N)
0	0
0.1	1.2
0.2	2.4
0.3	3.6
0.4	4.8
0.5	6.0
0.6	7.2

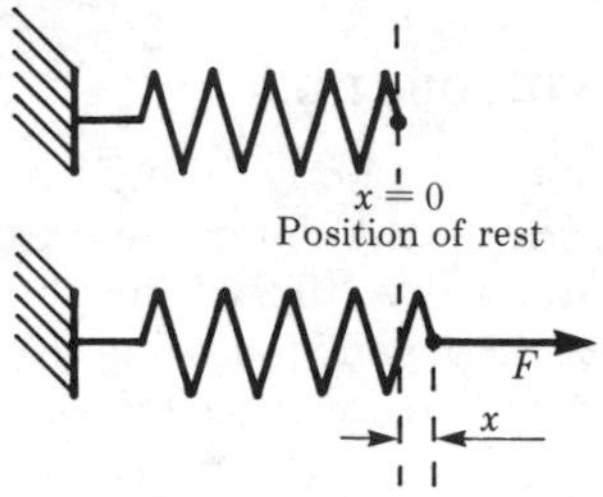

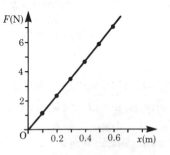

Figure 1.1

Such a table is called a table of values for all displacements x of the spring for which it is not permanently deformed or destroyed. The range of x is called the *range* or *domain of definition*. The corresponding range of the functional values is called the *range of values* (sometimes referred to as the *codomain*).

(2) We plot each paired value on a graph and draw a curve through the points. This enables us to obtain, approximately, intermediate values (figure 1.2).

Figure 1.2

(3) The relationship between x and F can be expressed by a formula which must be valid with the domain of definition. In this case the formula is

$$F = ax, \qquad \text{where} \qquad a = 12\,\text{N/m}$$

By substituting values of x we obtain the corresponding values of F. We notice that there is only one value of F for each value of x. The formula is unambiguous.

The letters x and y are frequently used in mathematics to represent paired values, so we can write $y = ax$.

Let us recapitulate. A function may be expressed in different ways:

by setting up a table of values;

graphically;

by means of a formula.

These three ways of representing a function are of course related. For example, we can draw up a table of values from the formula or from a graph.

If y depends on x then y is said to be a function of x; the relationship is expressed as

$$y = f(x)$$

It reads   "y equals f of x".

In order to define the function completely we must state the set of values of x for which it is valid, i.e. the domain of definition.

The quantity x is called the *argument* or *independent variable* and the quantity y the *dependent variable*.

Once the nature of the function is known, we can obtain the value of y for each value of the argument x within the domain of definition.

EXAMPLE $y = 3x^2$ The function in this case is $3x^2$. For a given value of the argument x, for example $x = 2$, we can calculate y:

$$y = 3 \times 2^2 = 12.$$

A function can be quite intricate, for example:

$$y = \frac{ax^2}{\sqrt{(1-x^2)^2 + bx^2}}$$

This is an expression found in the study of vibrations.

For the sake of clarity let us give a formal definition:

DEFINITION   Given two sets of real numbers, a domain (often referred to as the x-values) and a codomain (often referred to as the y-values), a *real function* assigns to each x-value a unique y-value.

In this book we will mainly be concerned with real functions, as opposed to more general functions like complex functions. Note that the concept of a function implies that the y-value is determined unambiguously. During the previous one or two decades the use of the term 'function' has changed. In the engineering literature, the term 'two-valued' or 'many-valued function' is still occasionally used. Strictly speaking, in modern terminology what is meant is not a function but a relationship.

EXAMPLE   Consider the equation $y^2 = x$; it is clear that x has only one value for two values of y, e.g. if $y_1 = +2$ and $y_2 = -2$ then $x = 4$ in each case. The equation may be rewritten thus:

$$y = \pm\sqrt{x}$$

This means that for every (positive) value of x, y has two possible values, $+\sqrt{x}$ and $-\sqrt{x}$. Hence y is a two-valued 'function'. Which root we assign to y will, in general, depend on the nature of the problem. For instance, the equation

$$y = \frac{ax^2}{\pm\sqrt{(1-x^2)^2 + bx^2}}$$

mentioned previously, is two-valued but the negative root has no physical meaning.

The ambiguity is removed by, e.g. restricting the value of y to the positive root; thus $y = +\sqrt{x}$ is unambiguous. This is a function; its range of values is $y \geqslant 0$. From now on, whenever we use the symbol $\sqrt{}$ for the square root, the positive root is to be understood.

1.2   GRAPHICAL REPRESENTATION OF FUNCTIONS

1.2.1   COORDINATE SYSTEM, POSITION VECTOR

Many functions can easily be represented graphically. Graphs are usually based on a rectangular coordinate system known as a *cartesian system* (after the French mathematician Descartes). The vertical axis is usually referred to as the *y-axis*, and the horizontal axis as the *x-axis* (figure 1.3). In certain applications they may bear different labels such as t, θ, etc. The axes intersect at the point O, called the *origin* of the coordinate system.

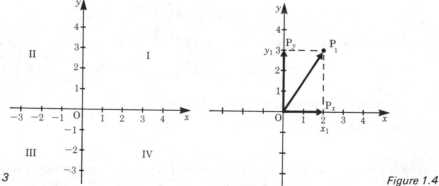

Figure 1.3   Figure 1.4

Associated with each axis is a scale and the choice of this scale depends on the range of values of the variables. The coordinate system divides a plane into four regions, known as *quadrants*, numbered counterclockwise. A point P_1 (figure 1.4) is uniquely defined in the coordinate system by two numerical values. If we drop a perpendicular from P_1, it meets the x-axis at P_x. P_x is called the *projection* of P_1 on to the x-axis and is related to a number x_1 on the x-axis, the x-coordinate or *abscissa*. In a similar way, P_y is the projection of P_1 on the y-axis, and we find a number y_1, the *y-coordinate* or *ordinate*. Thus, if we know both coordinates for the point P_1, then it is uniquely defined.

This is often written in the following way:

$$P_1 = (x_1, y_1)$$

The coordinates represent an ordered pair of numbers: x first and y second. The point P_1 in figure 1.4 is defined by $x_1 = 2$ and $y_1 = 3$, or $P_1 = (2, 3)$.

As an aside, note that the distance measured from the origin O of the coordinate system and the point P_1 as a directed distance is called the *position vector* and its projections on the axes are referred to as its *components*. These components are *directed line segments*. Vectors will be introduced in detail in Chapter 3.

1.2.2 THE LINEAR FUNCTION: THE STRAIGHT LINE

A straight line is defined by the equation

$$y = ax + b$$

We can obtain a picture of the line very quickly by giving x two particular values (figure 1.5):

$$\text{for} \quad x = 0, \quad y(0) = b$$

$$\text{for} \quad x = 1, \quad y(1) = a + b$$

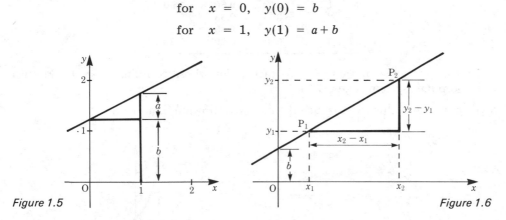

Figure 1.5   *Figure 1.6*

The constant a is the *slope of the line*. If two points of a straight line are known its slope can be calculated:

$$a = \frac{y_2 - y_1}{x_2 - x_1} \tag{1.1}$$

PROOF   Consider two arbitrary points on the line, $P_1 = (x_1, y_1)$ and $P_2 = (x_2, y_2)$. Substitution in the equation for the straight line gives

$$y_1 = ax_1 + b$$

$$y_2 = ax_2 + b$$

If these are now substituted in the right-hand side of the equation to find the slope, we have

$$\frac{(ax_2 + b) - (ax_1 + b)}{x_2 - x_1} = \frac{a(x_2 - x_1)}{x_2 - x_1} = a$$

The constant b is the intercept of the line on the y-axis (figure 1.6), i.e. the point of intersection of the line with the y-axis has the value b.

1.2.3 GRAPH PLOTTING

Consider the function

$$y = \frac{1}{x+1} + 1$$

and suppose we wish to plot its graph. There are three basic steps to follow:

(1) Set up a table of values. The best way to do this is to split up the function by taking a convenient number of simple terms as illustrated in the table below.

Table 1.7

x	$x+1$	$\dfrac{1}{x+1}$	y
−4	−3	−0.33	0.67
−3	−2	−0.50	0.50
−2	−1	−1.00	0
−1	0	∞	∞
0	1	1.00	2.00
1	2	0.50	1.50
2	3	0.33	1.33
3	4	0.25	1.25
4	5	0.20	1.20

(2) From the table of values, place each point whose coordinates are (x, y) onto the coordinate system (figure 1.7).

(3) Draw a smooth curve through the points as shown in figure 1.8.

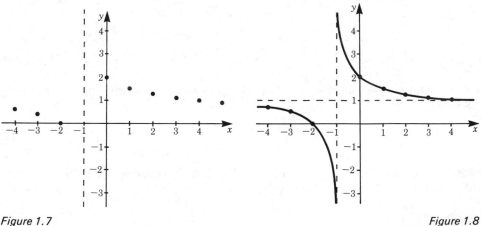

Figure 1.7   *Figure 1.8*

We observe that for this function there is a difficulty at $x = -1$. As x tends to the value −1 the function grows beyond limit: it tends to infinity. In order to obtain a better picture of the behaviour of the function it is advisable to take smaller steps and hence calculate additional values in the neighbourhood of $x = -1$, e.g. −1.01, −1.001, −0.95, −0.99, etc. This means that we increase the density of the points to be taken close to $x = -1$, whereas for other values of x where the graph changes less dramatically we can increase the distance of x between the points.

Scientists and engineers often require a picture of the way a function behaves rather than to know its exact behaviour. The process of obtaining such a picture is called *curve sketching*. For this purpose it is important to be able to identify the salient features of a function and these we will now investigate.

Poles

These are the points where the function grows beyond limit, i.e. tends to infinity $(+\infty$ or $-\infty)$. In the above discussion, such a point was found at $x = -1$. Poles are also referred to as *singularities*. The corresponding x-values are excluded from the domain of definition.

To determine where the poles occur we have to find the values of x for which the function $y = f(x)$ approaches infinity. In the case of fractions, this occurs when the denominator tends to zero, provided that the numerator is not zero. In our example we need to consider the fraction $\dfrac{1}{1+x}$.

We see that the denominator vanishes when $x = -1$; thus our function has a pole at the point $x_p = -1$.

Poles can also be found by taking the reciprocal of the function so that for y to tend to infinity the reciprocal must tend to zero.

Asymptotes

When a curve has a branch which extends to infinity, approaching a straight line, this line is called an *asymptote*. In our example such an asymptote is the line $y = 1$ which is parallel to the x-axis.

Zeros of a Function

These occur where the curve crosses the x-axis. To find their positions we simply have to equate the function to zero, i.e. $y = 0$, and solve for x.

In our example we have

$$\frac{1}{1+x} + 1 = 0$$

Solving for x gives $x = -2$.

Maxima and Minima and Points of Inflexion

These are other characteristic points of a function which are discussed in Chapter 5, section 5.7.

As a further example, consider the function

$$y = x^2 - 2x - 3$$

It is a parabola. It does not have poles or asymptotes. The zeros are found by equating it to zero and solving for x, which is shown in the following section:

$$x^2 - 2x - 3 = 0$$

The zeros are $x_1 = 3$ and $x_2 = -1$.

The graph of the function is shown in figure 1.9.

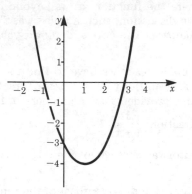

Figure 1.9

1.3 QUADRATIC EQUATIONS

Any equation in which the square, but no higher power of the unknown, occurs is called a *quadratic equation*. The simplest type, the *pure quadratic*, is $x^2 = 81$, for example. To solve for x we take the square root of both sides. Then $x = +9$ or $x = -9$ since $(+9)^2 = 81$ and $(-9)^2 = 81$; hence $x = \pm 9$. It is essential in practice to state both values or solutions, although in some situations only one value will have a physical significance.

The general expression for a quadratic takes the form

$$ax^2 + bx + c = 0$$

Because of the squared term this equation has two solutions or roots. It can be solved by 'completing the square'. We proceed as follows.

The terms containing the unknown are grouped on one side of the equation and the constants on the other side. The left-hand side is made into a perfect square by a suitable addition, the same amount being added to the right-hand side. Then the square root of both sides is taken.

Hence the roots are found as follows:

$$ax^2 + bx + c = 0$$

We subtract c on both sides, giving

$$ax^2 + bx = -c$$

We divide throughout by a: $\qquad x^2 + \dfrac{b}{a}x = -\dfrac{c}{a}$

To make the left-hand side into a perfect square we must add $\dfrac{b^2}{4a^2}$ to both sides:

$$x^2 + \frac{b}{a}x + \frac{b^2}{4a^2} = \frac{b^2}{4a^2} - \frac{c}{a} = \frac{b^2 - 4ac}{4a^2}$$

Hence
$$\left(x + \frac{b}{2a}\right)^2 = \frac{b^2 - 4ac}{4a^2}$$

Taking the square root of both sides gives

$$x + \frac{b}{2a} = \pm\sqrt{\frac{b^2 - 4ac}{4a^2}} = \pm\frac{1}{2a}\sqrt{b^2 - 4ac}$$

Hence $\quad x = -\dfrac{b}{2a} \pm \dfrac{1}{2a}\sqrt{b^2 - 4ac}$

The roots of the quadratic equation $ax^2 + bx + c = 0$ are

$$x_1 = \frac{-b + \sqrt{b^2 - 4ac}}{2a} \quad\text{and}\quad x_2 = \frac{-b - \sqrt{b^2 - 4ac}}{2a} \qquad (1.2a)$$

Quadratic equations occur frequently in science and engineering; the formulae for the two roots should be remembered. Often the quadratic equation is written in the form

$$x^2 + px + q = 0$$

The roots of the quadratic equation $x^2 + px + q = 0$ are

$$x_1 = \frac{-p + \sqrt{p^2 - 4q}}{2}, \quad x_2 = \frac{-p - \sqrt{p^2 - 4q}}{2} \qquad (1.2b)$$

As a check, it is easy to verify that with these values of x_1, x_2
$$x^2 + px + q = (x - x_1)(x - x_2)$$
i.e. the quadratic is expressed as the product of two linear expressions.

It should be noted that there are cases when no real solutions exist. The general quadratic equation $ax^2 + bx + c = 0$ has real solutions if the expression $b^2 - 4ac$, called the *discriminant*, is positive, i.e. the square root can be extracted. It is not hard to see that this corresponds to the function $f(x) = ax^2 + bx + c$ having zeros.

Conversely, if the discriminant $b^2 - 4ac$ is negative, then the function does not have a zero, i.e. its graph does not cut the x-axis.

We therefore have a criterion which allows us to decide whether a parabola, given by an algebraic expression, lies entirely above the x-axis or below the x-axis; the condition is $b^2 - 4ac < 0$.

1.4 **INVERSE FUNCTIONS**

Given a function $y = f(x)$, its inverse is obtained by interchanging the roles of x and y and then solving for y. The inverse function is denoted by $y = f^{-1}(x)$. For example, if $y = ax + b$ where a and b are constants, then the inverse function is $y = \dfrac{x}{a} - \dfrac{b}{a}$.

Geometrically, the formation of the inverse can be understood in two ways which are equivalent:

(1) Interchanging the roles of x and y is equivalent to interchanging the roles of the coordinate axes. In this case the graph remains unchanged but now we have a y-x coordinate system.

(2) If we keep the x-y coordinate system, as is generally done, the graph of the inverse function is obtained by reflecting the graph of $f(x)$ in the line $y = x$. This is shown in figure 1.10. AA' is the bisecting line.

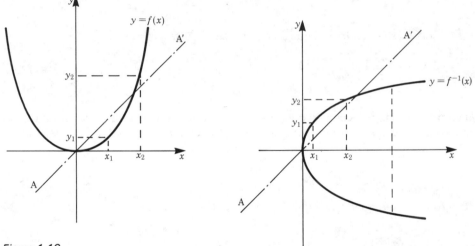

Figure 1.10

The figure shows the function $y = 2x^2$ and its inverse $y = \pm\sqrt{\dfrac{x}{2}}$, showing, in this case, that the inverse 'function' has two values for every x, i.e. it is not a *monotonic* function. The problem is solved by restricting the domain of the original function $y = 2x^2$ to $x \geqslant 0$. Then the inverse function is $y = \sqrt{\dfrac{x}{2}}$.

Multi-valued inverse functions cannot occur when $y = f(x)$ is a continuous and monotonic function. Such a function is defined as follows. A continuous function $y = f(x)$ is monotonic if, in the interval $x_1 < x < x_2$, it takes all the values between $f(x_1)$ and $f(x_2)$ only once. Such a function has a unique inverse, $y = f^{-1}(x)$, which is itself a continuous monotonic function in the corresponding interval. Thus the inverse is also single-valued. (The concept of continuity is treated in detail in Chapter 5, section 5.2. For the time being it will suffice to say that all functions commonly encountered, such as power functions, fractional rational functions and trigonometric functions, are continuous throughout their domain of definition.)

1.5   TRIGONOMETRIC OR CIRCULAR FUNCTIONS

1.5.1   UNIT CIRCLE

A circle having a radius equal to unity is called a unit circle and is used as a reference (figure 1.11).

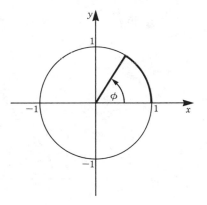

Figure 1.11

In geometry, angles are measured in degrees. A right angle has $90°$, whilst the angle around the four quadrants of a circle is $360°$ or the total angle at the centre is $360°$, this being subtended by the circumference.

In engineering and science, angles are usually measured in *radians* (abbreviated to rad). In radians an angle of $360°$ has the value of the circumference of the unit circle, namely 2π.

It follows that since

$$360° \cong 2\pi \text{ rad}$$

$$\text{then} \quad 1 \text{ rad} \cong \frac{360°}{2\pi} = 57.3°$$

$$\text{and} \quad 1° \cong \frac{1}{57.3} = 0.017\,45 \text{ rad}$$

To convert an angle from degrees to radians, we have

$$\phi_{\text{rad}} = \frac{\phi°}{57.3} \text{ rad}$$

and to convert an angle from radians to degrees we have

$$\phi° = \phi_{\text{rad}} \times 57.3$$

It is customary for angles to be considered positive when measured anticlockwise from the x-axis (figure 1.11) and negative when measured clockwise.

1.5.2 SINE FUNCTION

The sine function is frequently encountered in physical problems, e.g. in the study of vibrations. The sine of an angle is defined by means of a right-angled triangle, as shown in figure 1.12a; the sine of an angle is the quotient of the side opposite and the hypotenuse. Its magnitude is independent of the size of the triangle.

$$\sin \phi = \frac{a}{c} \qquad\qquad (1.3)$$

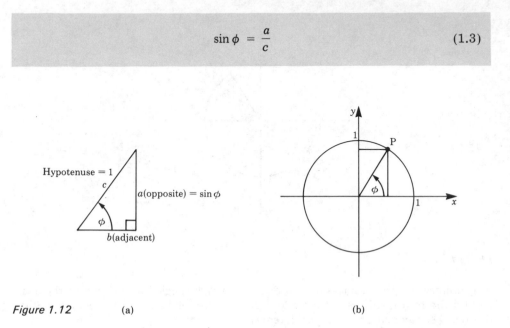

Figure 1.12   (a)   (b)

Considering now the unit circle shown in figure 1.12b, let P be a point on the circumference; the position vector of the point makes an angle ϕ with the x-axis. It follows that the y-coordinate of the point P is equal to the sine of the angle for, by definition,

$$\sin \phi = \frac{y}{r}$$

but since $r = 1$ we have $y = \sin \phi$.

This is true for all points on the circumference and therefore for all angles.

DEFINITION   The sine of an angle ϕ is the y-coordinate of the point P on the unit circle corresponding to ϕ.

A graphical representation of the sine function is obtained by plotting ϕ as the independent variable (the argument) and $\sin \phi$ (the dependent variable) as ordinate, as shown in figure 1.13 for ϕ between 0 and 2π radians. This corresponds to one complete revolution of the point P on the unit circle.

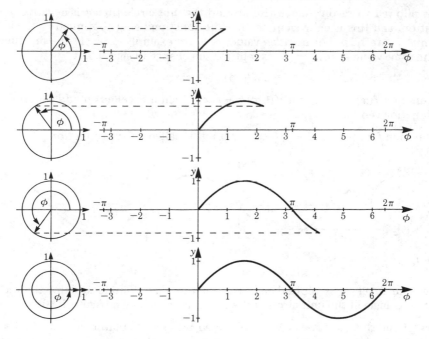

Figure 1.13

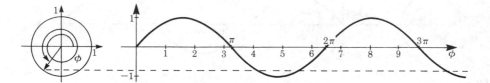

Figure 1.14

If P is allowed to move several times around the unit circle, then ϕ grows beyond 2π and takes on large values, as shown in figure 1.14. For each revolution of P the values of the sine function are repeated periodically.

DEFINITION   A function $y = f(x)$ is called *periodic* if for all x within the range of definition we have

$$f(x + p) = f(x)$$

where p is the smallest value for which this equation is valid. p is called the *period*.

The sine function has a period equal to 2π.

If P is allowed to revolve clockwise around the unit circle then ϕ is negative, by our definition, and this is equivalent to the sine function being continued to the left, as shown in figure 1.15. For negative values of ϕ, for example $\phi = -1$, the sine function has the same value as for $\phi = 1$ except for the change in sign, i.e.

$$\sin(-\phi) = -\sin\phi$$

If a function $f(x)$ is such that $f(-x) = -f(x)$, then it is called an *odd function*. The sine function is odd.

If a function $f(x)$ is such that $f(-x) = f(x)$ it is called an *even function*.

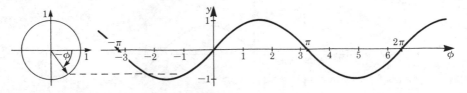

Figure 1.15

If the sine function is plotted in a cartesian x-y coordinate system, then x represents the angle in the unit circle and y the sine of the angle (figure 1.16).

Values of the sine functions can be obtained using a calculator or from tables, but the former is more convenient. They can also be calculated using power series, as will be shown in Chapter 8.

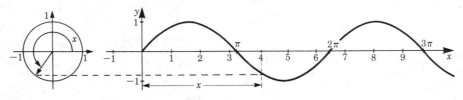

Figure 1.16

Amplitude

The function $y = \sin x$ has an amplitude of 1 unit; the range of values of the function is $-1 \leqslant y \leqslant 1$. If we multiply the sine function by a factor A we obtain a function having the same period but of amplitude A.

DEFINITION   The *amplitude* is the factor A of the function

$$y = A\sin x$$

Figure 1.17 shows sine functions with $A = 2, 1$ and 0.5 respectively.

$$y_1 = 2\sin x \quad \text{(dashed curve)}$$

$$y_2 = \sin x \quad \text{(full curve)}$$

$$y_3 = 0.7\sin x \quad \text{(dot–dash curve)}$$

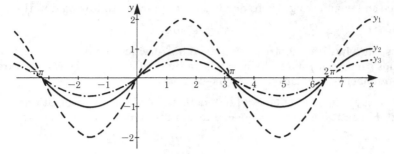

Figure 1.17

Period

Multiplying the argument of a sine function by a constant factor changes the period of the function. For example,

$$y = \sin 2x$$

where the argument is $2x$. Plotting this function as shown in figure 1.18 reveals that the period is π, i.e. the function oscillates at twice the frequency as does the function $y = \sin x$.

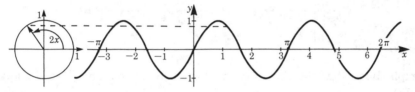

Figure 1.18

In general, the *period p* of the sine function

$$y = \sin bx$$

is given by

$$p = \frac{2\pi}{b} \tag{1.4}$$

Figure 1.19 shows the graphs of the function $y = \sin bx$ for a large and a small value of b.

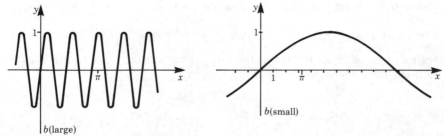

b(large)

b(small)

Figure 1.19

In science and engineering, we frequently encounter the following notation:

$$y = \sin \omega t$$

The constant b has been replaced by the symbol ω which stands for *circular frequency* (in radians per second) and t which stands for time (in seconds). The circular frequency ω is the number of oscillations in a time of 2π s.

The *frequency* f in cycles per second or hertz is the number of oscillations in a time of 1 second. Hence the circular frequency ω and the frequency f are related as follows:

$$\omega = 2\pi f$$

Phase

Consider the function

$$y = \sin(x + c)$$

The effect of adding a constant to the argument x is shown in figure 1.20 for a particular case where $c = \dfrac{\pi}{2}$ radians.

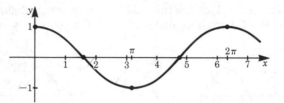

Figure 1.20

The graph shows that the sine curve is shifted by an amount $\dfrac{\pi}{2}$ to the left. This constant is called the phase.

> **DEFINITION**   The *phase* is the constant term added to the argument of a trigonometric function.

For a positive phase the curve is shifted to the left and for a negative phase it is shifted to the right. It is usual in science and engineering to use a Greek letter such as ϕ_0 to denote a phase.

The function $y = A \sin(\omega t + \phi_0)$ is a sine function of amplitude A, of the circular frequency ω and phase angle ϕ_0.

The function $y = A \sin(\omega t + \phi_0)$ is said to *lead* the function $y = B \sin \omega t$ by ϕ_0.

The function $y = A \sin(\omega t - \phi_0)$ is said to *lag* the function $y = B \sin \omega t$ by ϕ_0.

1.5.3   COSINE FUNCTION

The cosine of an angle is defined as the ratio of the adjacent side to the hypotenuse in a right-angled triangle (figure 1.21).

$$\cos \phi = \frac{b}{c} \tag{1.5}$$

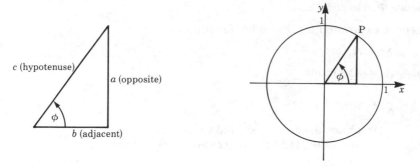

Figure 1.21   *Figure 1.22*

Consider a point P on the unit circle shown in figure 1.22. The cosine of the angle ϕ is equal to the length of the abscissa, the x component, which is the projection of P on the horizontal axis.

If, as shown in figure 1.23, x is the angle turned through by the radius of the unit circle and y the projection of the point P on the horizontal axis, then

$$y = \cos x$$

> **DEFINITION**   The cosine of an angle ϕ is the x-coordinate of the point P on the unit circle corresponding to ϕ.

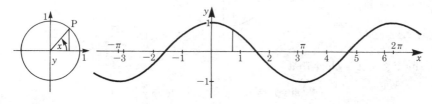

Figure 1.23

Figure 1.23 illustrates the graph of this function for positive and negative values of x. We observe that the cosine function is even. The cosine function can be obtained from the sine function by shifting the latter to the left by $\dfrac{\pi}{2}$ radians; hence, by inspection of the graphs,

$$\cos x = \sin\left(x + \frac{\pi}{2}\right) \tag{1.6}$$

It follows that the sine function can be obtained from the cosine function by shifting the latter by an amount $\dfrac{\pi}{2}$ radians to the right; hence

$$\sin x = \cos\left(x - \frac{\pi}{2}\right)$$

Whether one uses the sine or cosine function depends on the particular situation.

Amplitude, Period and Phase

The general expression for the cosine function is

$$y = A \cos(bx + c)$$

where A = amplitude

$\dfrac{2\pi}{b}$ = period

c = phase; the curve is shifted to the left if c is positive and to the right if c is negative.

1.5.4   RELATIONSHIPS BETWEEN THE SINE AND COSINE FUNCTIONS

(1) In figure 1.24, the position vector to the point P makes an angle ϕ with the x axis, while the point P_1 is obtained by subtracting a right angle from ϕ. We have

$$\phi_1 = \phi - \frac{\pi}{2}$$

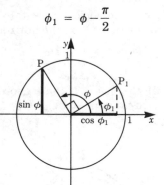

Figure 1.24

From the figure it is clear that

$$\sin \phi = \cos \phi_1 = \cos\left(\phi - \frac{\pi}{2}\right)$$

Similarly, as already mentioned in the previous section,

$$\cos \phi = \sin\left(\phi + \frac{\pi}{2}\right)$$

(2) Applying Pythagoras' theorem to the right-angled triangle in figure 1.25 gives

$$\sin^2 \phi + \cos^2 \phi = 1 \tag{1.7}$$

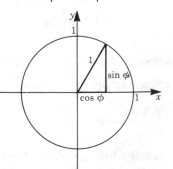

Figure 1.25

From this identity there follow two relationships which are frequently used:

$$\sin \phi = \sqrt{1 - \cos^2 \phi}$$
$$\cos \phi = \sqrt{1 - \sin^2 \phi}$$

They are valid for values of ϕ between 0 and $\frac{\pi}{2}$. For larger angles the sign of the root has to be chosen appropriately.

1.5.5   TANGENT AND COTANGENT

The tangent of the angle ϕ in figure 1.26 is defined as the ratio of the opposite side to the adjacent side.

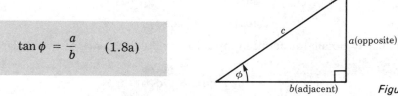

$$\tan \phi = \frac{a}{b} \qquad (1.8a)$$

Figure 1.26

From the definition of the sine and cosine functions it follows that

$$\tan \phi = \frac{\sin \phi}{\cos \phi} \qquad (1.8b)$$

As for the sine and cosine, the graph of the tangent function can be derived from the unit circle. In figure 1.27 we erect a tangent at A to the unit circle until it meets the radial line OP extended to P′. The point P′ has the value $\tan \phi$.

As ϕ approaches $\frac{\pi}{2}$ the value of $\tan \phi$ grows indefinitely. The function $y = \tan \phi$ is shown in figure 1.27. The period of $\tan \phi$ is π.

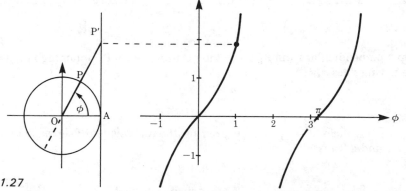

Figure 1.27

The cotangent is defined as the reciprocal of the tangent so that

$$\cot \phi = \frac{1}{\tan \phi} = \frac{\cos \phi}{\sin \phi} \qquad (1.9)$$

1.5.6 ADDITION FORMULAE

A trigonometric function of a sum or difference of two angles can be expressed in terms of the trigonometric values of the summands. These identities are called addition formulae.

$$\sin(\phi_1 + \phi_2) = \sin\phi_1 \cos\phi_2 + \cos\phi_1 \sin\phi_2$$
$$\cos(\phi_1 + \phi_2) = \cos\phi_1 \cos\phi_2 - \sin\phi_1 \sin\phi_2$$

(1.10)

If ϕ_2 is negative, then, noting that

$$\sin(-\phi_2) = -\sin\phi_2$$
$$\cos(-\phi_2) = \cos\phi_2$$

we immediately obtain the addition formulae for the difference of two angles:

$$\sin(\phi_1 - \phi_2) = \sin\phi_1 \cos\phi_2 - \cos\phi_1 \sin\phi_2$$
$$\cos(\phi_1 - \phi_2) = \cos\phi_1 \cos\phi_2 + \sin\phi_1 \sin\phi_2$$

(1.11)

The proof for $\sin(\phi_1 + \phi_2)$ is as follows. From figure 1.28 we have

$$\phi = \phi_1 + \phi_2$$

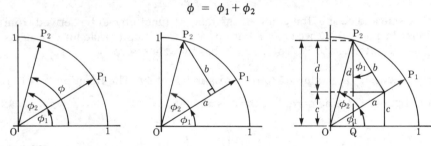

Figure 1.28

We drop a perpendicular from P_2 on to the position vector P_1 obtaining a right-angled triangle having sides

$$a = \cos\phi_2$$
$$b = \sin\phi_2$$

The sine of the angle $\phi = \phi_1 + \phi_2$ is given by the line segment P_2Q made up of two segments c and d; thus

$$\sin(\phi_1 + \phi_2) = c + d$$
$$= a \sin\phi_1 + b \cos\phi_1$$

Substituting for a and b gives

$$\sin(\phi_1 + \phi_2) = \sin\phi_1 \cos\phi_2 + \cos\phi_1 \sin\phi_2$$

The proof for $\cos(\phi_1 + \phi_2)$ can be developed geometrically in a similar fashion; alternatively it can now be given algebraically, since $\cos\phi = \sin\left(\phi + \dfrac{\pi}{2}\right)$.

Sum of a Sine and a Cosine Function with Equal Periods

The sine and cosine functions have the same period but the amplitudes may be different. Their sum should result in a single trigonometric function of the same period but with different amplitude and with a phase shift.

Superposition formula

$$A \sin \phi + B \cos \phi = C \sin (\phi + \phi_0)$$

where
$$C = \sqrt{A^2 + B^2}$$

and
$$\tan \phi_0 = \frac{B}{A} \qquad (1.12)$$

This relationship is important in the study of waves and vibrations.

Figure 1.29 illustrates the superposition of the two functions

$$y_1 = 1.2 \sin \phi \qquad \text{and} \qquad y_2 = 1.6 \cos \phi$$

with the resultant
$$y_3 = 2 \sin (\phi + 53°)$$

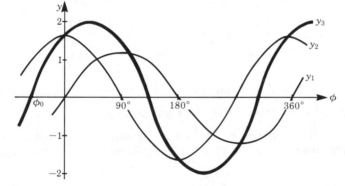

Figure 1.29

The proof of the superposition formula is as follows. From figure 1.30 we have

$$a = A \sin \phi$$
$$b = B \cos \phi$$

and
$$a + b = C \sin (\phi + \phi_0)$$

It follows that

$$a + b = A \sin \phi + B \cos \phi$$
$$= C \sin (\phi + \phi_0)$$

Furthermore, we see that

$$C = \sqrt{A^2 + B^2}$$

$$\tan \phi_0 = \frac{B}{A}$$

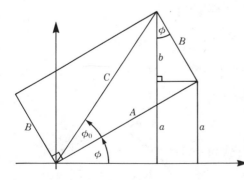

Figure 1.30

Further important relationships will be found in the appendix at the end of this chapter. All formulae follow from the addition formulae given and the known relationships between trigonometric functions.

1.6 INVERSE TRIGONOMETRIC FUNCTIONS

Trigonometric functions are periodic and therefore cannot be monotonic; consequently their inverses cannot be formed unless the domain of definition is restricted.

The restricted domains of definition are chosen as follows:

$$-\frac{\pi}{2} \leqslant x \leqslant \frac{\pi}{2} \qquad \text{for} \qquad y = \sin x$$

$$0 \leqslant x \leqslant \pi \qquad \text{for} \qquad y = \cos x$$

$$-\frac{\pi}{2} \leqslant x \leqslant \frac{\pi}{2} \qquad \text{for} \qquad y = \tan x$$

$$0 \leqslant x \leqslant \pi \qquad \text{for} \qquad y = \cot x$$

The symbol $\sin^{-1} x$ is used to denote the smallest angle, whether positive or negative, that has x for its sine. The symbol does not mean $\frac{1}{\sin x}$; it is understood as 'the angle whose sine is x'. It can be written as $\arcsin x$. The other inverse trigonometric functions may be written in a similar way, using either symbol, e.g. $\tan^{-1} x$ or $\arctan x$.

We know from our discussion of inverse functions in section 1.4 that the inverse function is the mirror image of the original function in the bisector of the first quadrant in a cartesian coordinate system. The graphs of the inverse trigonometric functions are shown in figure 1.31.

Inverse sine function: $y = \sin^{-1} x$

defined for $|x| \leqslant 1$ and $y \leqslant \frac{\pi}{2}$

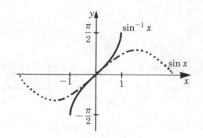

Figure 1.31

Inverse cosine function: $y = \cos^{-1}x$

defined for $|x| \leqslant 1$ and $0 \leqslant y \leqslant \pi$

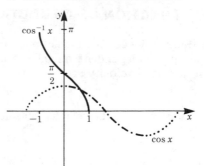

Inverse tangent function: $y = \tan^{-1}x$

defined for all real values
of x and $|y| \leqslant \dfrac{\pi}{2}$

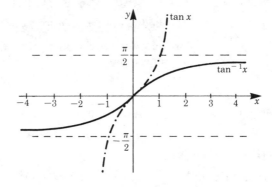

Inverse cotangent function: $y = \cot^{-1}x$

defined for all real values of x and $0 \leqslant y \leqslant \pi$

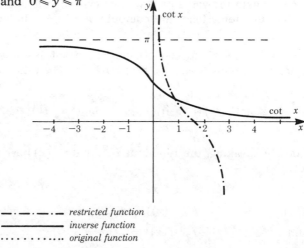

—·—·—·—·—   *restricted function*
——————   *inverse function*
· · · · · · · · · ·   *original function*

Figure 1.31 (continued)

1.7 FUNCTION OF A FUNCTION (COMPOSITION)

We frequently encounter functions where the independent variable is itself a function of another independent variable. For example, the kinetic energy T of a body is a function of its velocity v:

$$T = f(v)$$

But in many cases the velocity is itself a function of the time t, so that

$$v = g(t).$$

It is therefore evident that the kinetic energy can also be considered a function of time t; hence we have

$$T = f(g(t)).$$

> **DEFINITION** *A function of a function* is expressed in the following form:
>
> $$y = f[g(x)]$$
>
> f is called the *outer function*, g is called the *inner function*.
>
> The new function is also referred to as the *composition* of the functions f and g.

EXAMPLE
$$y = g^2$$
$$g = x + 1$$

We require $y = f(x)$

The solution is
$$y = (x + 1)^2$$

This is a rather simple example. To demonstrate that the use of the concept of composition may simplify calculations, consider the following function:

$$y = \sin(bx + c)$$

To calculate the value of y we first evaluate the inner function g and then apply the outer function. Hence we compute $g = bx + c$ independently and then obtain the sine of g.

As a very special, but nevertheless important example, consider the case where f and g are inverse functions, i.e. $g(x) = f^{-1}(x)$. Then the following identities hold true:

$$f(f^{-1}(x)) = x \quad \text{and} \quad f^{-1}(f(x)) = x \tag{1.13}$$

Loosely speaking, the functions $f(x)$ and $f^{-1}(x)$ have opposite effects. For example, let

$$f(x) = x^2 \quad (x \geqslant 0), \quad f^{-1}(x) = \sqrt{x}$$

Then
$$f(f^{-1}(x)) = (\sqrt{x})^2 = x$$
and
$$f^{-1}(f(x)) = \sqrt{x^2} = x$$

APPENDIX

Relationships between trigonometric functions

$$\sin(-\phi) = -\sin\phi \qquad \sin^2\phi + \cos^2\phi = 1$$

$$\cos(-\phi) = \cos\phi \qquad\qquad \tan\phi = \frac{\sin\phi}{\cos\phi}$$

$$\sin\left(\phi + \frac{\pi}{2}\right) = \cos\phi \qquad\qquad \cot\phi = \frac{\cos\phi}{\sin\phi}$$

$$\cos\left(\phi + \frac{\pi}{2}\right) = -\sin\phi$$

Addition theorems

$$\sin(\phi_1 + \phi_2) = \sin\phi_1\cos\phi_2 + \cos\phi_1\sin\phi_2$$

$$\sin(\phi_1 - \phi_2) = \sin\phi_1\cos\phi_2 - \cos\phi_1\sin\phi_2$$

$$\cos(\phi_1 + \phi_2) = \cos\phi_1\cos\phi_2 - \sin\phi_1\sin\phi_2$$

$$\cos(\phi_1 - \phi_2) = \cos\phi_1\cos\phi_2 + \sin\phi_1\sin\phi_2$$

$$\sin 2\phi = 2\sin\phi\cos\phi$$

$$\sin\frac{\phi}{2} = \sqrt{\frac{1}{2}(1 - \cos\phi)}$$

$$\sin\phi_1 + \sin\phi_2 = 2\left(\sin\frac{\phi_1 + \phi_2}{2}\cos\frac{\phi_1 - \phi_2}{2}\right)$$

TABLE OF PARTICULAR VALUES OF TRIGONOMETRIC FUNCTIONS

Radians	0	$\frac{\pi}{6}$	$\frac{\pi}{4}$	$\frac{\pi}{3}$	$\frac{\pi}{2}$
Degrees	$0°$	$30°$	$45°$	$60°$	$90°$
$\sin\phi$	0	$\frac{1}{2}$	$\frac{1}{2}\sqrt{2} \approx 0.707$	$\frac{1}{2}\sqrt{3} \approx 0.866$	1
$\cos\phi$	1	$\frac{1}{2}\sqrt{3} \approx 0.866$	$\frac{1}{2}\sqrt{2} \approx 0.707$	$\frac{1}{2}$	0
$\tan\phi$	0	$\frac{1}{3}\sqrt{3}$	1	$\sqrt{3} = 1.732$	$\pm\infty$

CHAPTER 2

Exponential, logarithmic and hyperbolic functions

2.1 POWERS, EXPONENTIAL FUNCTION

2.1.1 POWERS

Consider the multiplication of a number by itself, for example $a \times a \times a$. A simple way of expressing this is to write a^3. If we multiplied a by itself n times we would write a^n. We would say a to the power n; n is known as an *index* or *exponent*.

> **DEFINITION**   The power a^n is the product of n equal factors a.
>
> a is called the *base* and n the *index* or *exponent*.

This defines the powers for positive integral exponents only. What about negative exponents?

If a^n is reduced by the factor a, this is equivalent to dividing it by a, i.e. $\dfrac{a^n}{a}$. The number of factors is now $n-1$; hence we would write

$$\frac{a^n}{a} = a^n \frac{1}{a} = a^{n-1}$$

If we carry on dividing by a we obtain after n such divisions

$$a^{n-n} = a^0 = 1 \quad \text{since} \quad \frac{a^1}{a} = 1$$

By this process we have, in fact, given a meaning to a negative index. Hence

$$n > 0 \qquad a^2 = \frac{a^3}{a} = aa$$

$$a^1 = \frac{a^2}{a} = a$$

$$n = 0 \qquad a^0 = \frac{a^1}{a} = 1$$

$$n < 0 \qquad a^{-1} = \frac{1}{a}$$

$$a^{-2} = \frac{1}{a^2}$$

DEFINITION $a^{-n} = \dfrac{1}{a^n}$ (2.1)

$a^0 = 1$ is valid for any base a except for $a = 0$ because 0^0 is not defined.

2.1.2 LAWS OF INDICES OR EXPONENTS

Product $a^n \times a^m = a^{n+m}$ (2.2)

Quotient $\dfrac{a^n}{a^m} = a^{n-m}$ (2.3)

Power $(a^n)^m = a^{nm}$ (2.4)

Root $a^{\frac{1}{m}} = \sqrt[m]{a}$

$\qquad a^{n/m} = a^{n(\frac{1}{m})} = \sqrt[m]{a^n}$ (2.5)

This is only defined for $a > 0$.

These rules are valid for integral indices and also for arbitrary indices; the latter will be examined in Chapter 8.

There are three values of the base a which are in common use:

(1) **Base 10:** Order of magnitude of quantities can be easily expressed in terms of powers of 10.

Examples: Distance of the Earth from the Moon is given as 3.8×10^8 m,
average height of an adult as 1.8×10^0 m,
radius of a hydrogen atom as 0.5×10^{-10} m.

(2) **Base 2:** This is used in data processing and information theory (computers).

(3) **Base e:** e is a special number, known as Euler's number; its numerical value is

$$e = 2.718\,28\ldots$$

The importance of this number and powers of it will become clear in the Chapters 5, 6 and 10 on differential and integral calculus, and on differential equations. It is of paramount importance in higher mathematics.

2.1.3 BINOMIAL THEOREM

The following identities are well known:

$$(a+b)^2 = a^2 + 2ab + b^2$$

$$(a-b)^2 = a^2 - 2ab + b^2$$

The general formula for positive integral powers of a sum, $(a+b)^n$, is known as the *binomial theorem*; it states

$$(a+b)^n = a^n + \frac{n}{1} \times a^{n-1}b + \frac{n(n-1)}{1 \times 2} a^{n-2}b^2 + \frac{n(n-1)(n-2)}{1 \times 2 \times 3} a^{n-3}b^3 + \dots$$

$$+ \frac{n(n-1)(n-2) \dots \times 2 \times 1}{1 \times 2 \times \dots (n-2)(n-1)n} b^n \qquad (2.6)$$

The coefficients are known as *binomial coefficients*. The last coefficient is included for the sake of completeness; its value is 1.

2.1.4 EXPONENTIAL FUNCTION

The function a^x is called an *exponential function*; x, the index or exponent, is the independent variable.

EXAMPLE   Let $y = 2^x$

By giving x positive and negative integral values, the table on the right is easily produced.

x	2^x
-3	0.125
-2	0.25
-1	0.5
0	1
1	2
2	4
3	8

Figure 2.1 shows the graphs of the functions $y = 2^x$, $y = e^x$, $y = 10^x$.

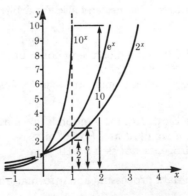

Figure 2.1

All exponential functions go through the point corresponding to $x = 0$, since all are equal when $y = 1$. Exponential functions grow very rapidly, as can be seen from the figure, even for small x-values; hence they are not easily represented graphically for large x-values. Exponential functions grow much faster than most other functions (unless $a < 1$).

Rates of growth in nature are described by exponential functions. An example is the increase in the number of bacteria in plants. Suppose that cell division doubles the number of bacteria every 10 hours. Let A be the number of bacteria at the start of an experiment; the table below gives the growth of the bacteria and this growth is represented graphically in figure 2.2.

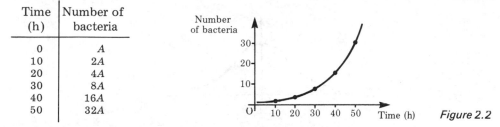

Time (h)	Number of bacteria
0	A
10	$2A$
20	$4A$
30	$8A$
40	$16A$
50	$32A$

Figure 2.2

The relationship between the growth of the bacteria and the time is expressed by means of the exponential function $y = A \times 2^{0.1t}$. The coefficient 0.1 is used because after exactly 10 time units (hours in this instance) the number of bacteria has doubled. In general this coefficient is the reciprocal of the time T required to double the bacterial population. Hence we write

$$y = A \times 2^{t/T}$$

Exponential functions, in particular decreasing exponential functions (decay and damping), are frequently encountered.

EXAMPLE   Radium is an element which decays without any external influence owing to the emission of α, β and γ radiation. Measurements show that the quantity of radium decays to half its original value in 1580 years. Unlike the bacteria example, the quantity of radium decreases with time. In this case we can write

$$y = A \times 2^{-ax}$$

The time required for the quantity of radium to decay to half its original value is referred to as the half-life t_h.

Hence the law for radioactive decay is given by

$$y = A \times 2^{-t/t_h}$$

Figure 2.3 shows the decreasing exponential function for the decay of radium.

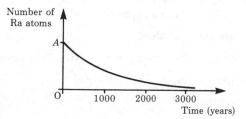

Figure 2.3

The decreasing exponential function describes damped vibrations, the discharge of capacitors, Newton's law of cooling and many other cases.

Finally, we will mention in passing another exponential function, namely

$$y = e^{-x^2}$$

for positive and negative values of x. The graph of the function is bell-shaped, as shown in figure 2.4. This function can play an important role, e.g. in statistics (the Gaussian or normal distribution).

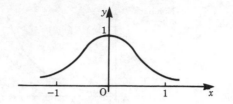

Figure 2.4

2.2 LOGARITHM, LOGARITHMIC FUNCTION

2.2.1 LOGARITHM

In section 2.1.4 we considered the exponential function

$$y = a^x$$

We calculated y for various integral values of x. For fractional values of x we need to use either tables or a calculator. We will now consider the inverse problem, i.e. given y, what is x?

EXAMPLE   If $10^x = 1000$, what is the value of x?

The solution is easy in this case, for we know that $10^3 = 1000$; hence $x = 3$. The required process is to transform the equation in such a way that both sides have powers to the same base; thus

$$10^x = 10^3$$

Hence, by comparing exponents, we have $x = 3$.

EXAMPLE   If $10^x = 100\,000$, what is x?

Step 1: We write both sides as powers to the same base, i.e. $10^x = 10^5$.

Step 2: We compare exponents: $x = 5$.

In both examples we required the exponent to the base 10 which yields a given value. This exponent has been given a name; it is called a *logarithm*.

The following statements are equivalent:

x is the exponent to the base 10 which gives the number $100\,000$.

x is the logarithm of the number $100\,000$.

This last statement is written as

$$x = \log 100\,000 = 5$$

In order to avoid any doubts about the base, it is common practice to specify it by a subscript. Thus

$$x = \log_{10} 100\,000 = 5$$

LOGARITHM, LOGARITHMIC FUNCTION

EXAMPLE   If $2^x = 64$, what is x?

We write both sides as powers to the same base, 2 in this case.

$$2^x = 2^6 \qquad \text{Hence} \quad x = 6$$

The exponent which raises the base 2 to 64 is 6. This result can be expressed as follows:

$$x = \log_2 64 = 6$$

DEFINITION   Logarithm:

The logarithm of a number c to a base a is the exponent x of the power to which the base must be raised to equal the number c.

As an equation, this definition is expressed as

$$a^{(\log_a c)} = c \qquad\qquad (2.7)$$

We have to remember that the logarithm is an exponent. To resolve the equation $a^x = c$ we should proceed in two steps:

Step 1: Write both sides as powers to the same base, i.e.

$$a^x = a^{(\log_a c)}$$

Step 2: Compare exponents

$$x = \log_a c$$

The examples we have just considered had integral exponents but in many cases where exponents are not integral, expressions cannot be resolved in such a simple manner. The case of fractional exponents will be explained in Chapter 8.

Equations in which exponents appear are more easily dealt with by taking logarithms. Taking logarithms is the transformation which consists of the two steps described above.

To avoid having to write the base below the log as a subscript, which is not very convenient, the following notation has been adopted:

Base 10: This base is mainly used in numerical calculations and it is written as

$$\log_{10} = \lg \text{ or } \log.$$

Logarithms to base 10 are also known as *common logarithms*.

Base 2: This base is mainly used in data processing and information theory. Logarithms to the base 2 are written

$$\log_2 = \text{ld}$$

Base e: Logarithms to the base e are called *natural logarithms* or Napierian logarithms; they are frequently used in calculations relating to physical problems. Logarithms to the base e are written

$$\log_e = \ln$$

2.2.2 OPERATIONS WITH LOGARITHMS

Operations with logarithms follow the power rules since logarithms are exponents. Thus, for example, the rule for multiplication is simplified to the addition of exponents, while the rule for division is simplified to subtraction of exponents, provided that the base is the same.

The rules are:

Multiplication $\log_a AB = \log_a A + \log_a B$ (2.8)

The logarithm of a product is equal to the sum of the logarithms of the factors.

Division $\log_a \dfrac{A}{B} = \log_a A - \log_a B$ (2.9)

The logarithm of a quotient is equal to the difference of the logarithms of the numerator and denominator.

Power $\log_a A^m = m \log_a A$ (2.10)

The logarithm of a number raised to a power is equal to the logarithm of that number multiplied by the exponent.

Root $\log_a \sqrt[m]{A} = \log_a A^{\frac{1}{m}} = \dfrac{1}{m} \log_a A$ (2.11)

The logarithm of the mth root of a number is equal to the logarithm of that number divided by m.

Conversion of a logarithm to base a into a logarithm to another base b is a fairly straightforward operation.

If $\qquad x = \log_a c$

then $\qquad c = a^x$

Taking log to the base b on both sides gives

$$\log_b c = \log_b a^x = x \log_b a$$

Since $x = \log_a c$, it follows that

$$\log_b c = \log_a c \times \log_b a$$

or

$$\log_a c = \frac{1}{\log_b a} \log_b c$$

EXAMPLE   We sometimes need to convert from base e to base 10 and vice versa.

$$\log c = 0.4343 \ln c$$
$$\ln c = 2.3026 \log c$$

2.2.3 LOGARITHMIC FUNCTIONS

The function $y = \log_a x$ is called a logarithmic function; it is equivalent to

$$x = a^y \qquad \text{for} \qquad a > 0$$

EXAMPLE $y = \log_2 x$ or $2^y = x$

Numerical values are given in the table below and the graph of the function is shown in figure 2.5.

x	y
0.25	-2
0.5	-1
1	0
2	1
4	2

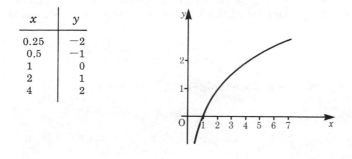

Figure 2.5

Figure 2.6 shows the logarithmic functions for three different bases, i.e. 10, 2 and e.

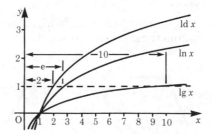

Figure 2.6

All logarithmic functions tend to minus infinity as x tends to zero, and they are all equal to zero at $x = 1$. Logarithmic functions are monotonic as they increase, and they tend to infinity as x goes to infinity. The reader should observe that the logarithmic function is the inverse of the exponential function. This is shown in figure 2.7 for bases 2 and e.

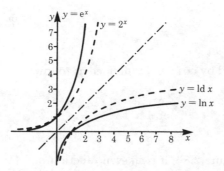

Figure 2.7

2.3 HYPERBOLIC FUNCTIONS AND INVERSE HYPERBOLIC FUNCTIONS

2.3.1 HYPERBOLIC FUNCTIONS

These functions play an important role in integration and in the solution of differential equations. They are simple combinations of the exponential function e^x and e^{-x} and are related to the hyperbola just as trigonometric (circular) functions are related to the circle. They are denoted by adding an 'h' to the abbreviations for the corresponding trigonometric functions. Graphs of the hyperbolic functions are shown in figure 2.8.

Hyperbolic Sine Function

This function is denoted by sinh (pronounced shine) and is defined as follows:

$$\sinh x = \frac{e^x - e^{-x}}{2} \tag{2.12}$$

Figure 2.8a shows $\sinh x$ and also the functions $\frac{1}{2}e^x$ and $-\frac{1}{2}e^{-x}$; $\sinh x$ is obtained by adding them together.

We observe that the hyperbolic sine is an odd function: it changes sign when x changes sign, i.e. it is symmetrical about the origin.

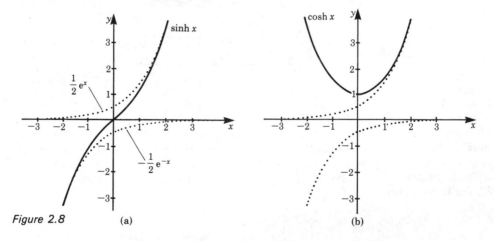

Figure 2.8   (a)   (b)

Hyperbolic Cosine Function

This function is denoted by cosh and is defined as follows:

$$\cosh x = \frac{e^x + e^{-x}}{2} \tag{2.13}$$

Its graph is shown in figure 2.8b; it is an even function.

A chain or cable which hangs under gravity sags in accordance with the cosh function. The curve is called a *catenary*.

Hyperbolic Tangent

This function is denoted by tanh and is defined as follows:

$$\tanh x \;=\; \frac{\sinh x}{\cosh x} \;=\; \frac{e^x - e^{-x}}{e^x + e^{-x}} \;=\; \frac{1 - e^{-2x}}{1 + e^{-2x}} \tag{2.14}$$

It is an odd function, i.e. its graph is symmetrical with respect to the origin. It is defined for all real values of x, and its range is $|y| < 1$. Its graph is shown in figure 2.9a. There are two asymptotes: $y = 1$ and $y = -1$.

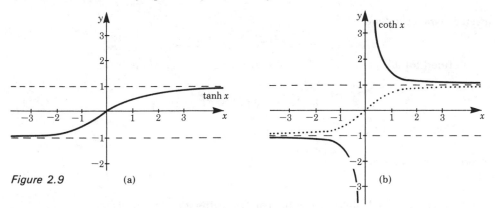

Figure 2.9   (a)   (b)

Hyperbolic Cotangent

This function is denoted by coth and is defined as follows:

$$\coth x \;=\; \frac{\cosh x}{\sinh x} \;=\; \frac{e^x + e^{-x}}{e^x - e^{-x}} \;=\; \frac{1 + e^{-2x}}{1 - e^{-2x}} \;=\; \frac{1}{\tanh x} \tag{2.15}$$

It is an odd function. It is defined for all real values of x, except $x = 0$. Its graph is shown in figure 2.9b; and it lies above 1 and below -1. It is asymptotic to $y = 1$ and $y = -1$.

An examination of the graphs of the hyperbolic functions reveals that they are not periodic, unlike the trigonometric functions.

There are a number of relationships between the hyperbolic functions which we will not discuss here. We will, however, derive an important one because of its similarity with the corresponding identity for trigonometric functions, i.e.

$$\sin^2 x + \cos^2 x \;=\; 1$$

Now consider the corresponding hyperbolic functions. We have

$$\sinh^2 x \;=\; \tfrac{1}{4}(e^x - e^{-x})^2 \;=\; \tfrac{1}{4}(e^{2x} - 2 + e^{-2x})$$

$$\cosh^2 x \;=\; \tfrac{1}{4}(e^x + e^{-x})^2 \;=\; \tfrac{1}{4}(e^{2x} + 2 + e^{-2x})$$

By subtraction we find

$$\cosh^2 x - \sinh^2 x = 1 \qquad (2.16)$$

2.3.2 INVERSE HYPERBOLIC FUNCTIONS

The hyperbolic functions are monotonic except for $\cosh x$. Hence we can form the inverse functions, except that for the inverse $\cosh x$ the range will be restricted to positive values of x. Figures 2.10 and 2.11 show the graphs of the inverse functions. They are formed by reflection in the bisectrix in the first quadrant.

Inverse Hyperbolic Sine

$$y = \sinh^{-1} x \quad \text{(figure 2.10a)}$$

It is defined for all real values of x. The following identity holds true:

$$\sinh^{-1} x = \ln(x + \sqrt{x^2 + 1}) \qquad (2.17)$$

Inverse Hyperbolic Cosine

$$y = \cosh^{-1} x \quad \text{(figure 2.10b)}$$

It is defined for $x \geqslant 1$; hence $y \geqslant 0$. The following identity holds true:

$$\cosh^{-1} x = \ln(x + \sqrt{x^2 - 1}) \qquad (2.18)$$

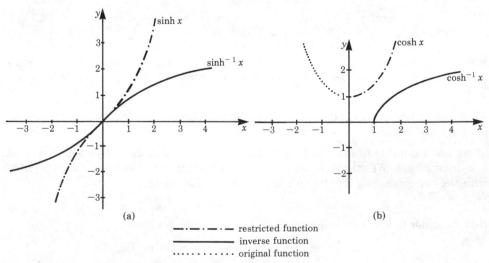

(a)   (b)

—·—·—·— restricted function
———————— inverse function
············· original function

Figure 2.10

Inverse Hyperbolic Tangent

$$y = \tanh^{-1}x \quad \text{(figure 2.11)}$$

It is defined for $|x| < 1$. The following identity holds true:

$$\tanh^{-1}x = \frac{1}{2}\ln\frac{1+x}{1-x} \qquad (2.19)$$

Inverse Hyperbolic Cotangent

$$y = \coth^{-1}x \quad \text{(figure 2.11)}$$

It is defined for $|x| > 1$; its range consists of all real numbers except $y = 0$. The following identity holds true:

$$\coth^{-1}x = \frac{1}{2}\ln\frac{1+x}{x-1} \qquad (2.20)$$

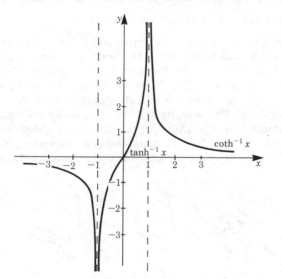

Figure 2.11

CHAPTER 3

Vector algebra I:
Scalars and vectors

3.1 SCALARS AND VECTORS

Mathematics is used in science and engineering to describe natural events in which
quantities are specified by numerical values and units of measurement. Such a
description does not always lead the scientist or the engineer to a successful conclusion.

Consider, for example, the following statement from a weather forecast:

'There is a force 4 westerly wind over the North Sea.'

This statement contains two pieces of information about the air movement, namely
the wind force which would be measured in physics as a wind velocity in metres per
second (m/s) and its direction. If the direction was not known, the movement of the
air would not be completely specified. Weather charts indicate the wind direction by
means of arrows, as shown in figure 3.1. It is evident that this is of considerable
importance to navigation. The velocity is thus completely defined only when both
its direction and magnitude are given. In science and engineering there are many
quantities which must be specified by magnitude and direction. Such quantities, of
which velocity is one, are called *vector quantities* or, more simply, *vectors*.

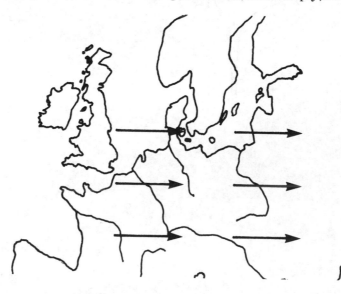

Figure 3.1

38

As an example from mathematics, consider the shift in position of a point from P_1 to P_2, as shown in figure 3.2a. This shift in position has a magnitude as well as a direction and it can be represented by an arrow. The magnitude is the length of the arrow and its direction is specified by reference to a suitable coordinate system. It follows that the shift of the point to a position P_3 is also a vector quantity (figure 3.2b).

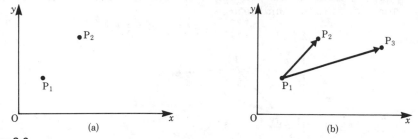

Figure 3.2

A figure in a plane or in space can be shifted parallel to itself; in such shifts the direction of all lines of the figure are preserved. Figure 3.3 shows a rectangle shifted from position A to position B where each point of the rectangle has been shifted by the same amount and in the same direction. Shifts which take place in the same direction and are equal in magnitude are considered to be equal shifts. A shift is uniquely defined by one representative vector, such as **a** in figure 3.3. Two vectors are considered to be equal if they have the same magnitude and direction.

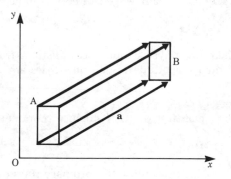

Figure 3.3

Furthermore, vectors may be shifted parallel to themselves, as shown in figure 3.4a, if the magnitudes and directions are preserved.

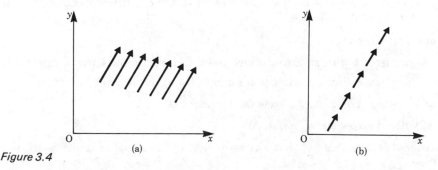

Figure 3.4

A vector may also be shifted along its line of action, as shown in figure 3.4b.

Vectors can be combined in various ways. Let us consider the *addition* of vectors. Consider the point P_1 in figure 3.5 shifted to P_2, and then shifted again to P_3. Each shift is represented by a vector, i.e. $\overrightarrow{P_1P_2}$ and $\overrightarrow{P_2P_3}$, and the result of the two shifts by the vector $\overrightarrow{P_1P_3}$. Hence we can interpret the succession of the two shafts as the sum of two vectors giving rise to a third vector.

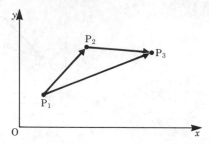

Figure 3.5

The length of the vector representing a physical quantity must be related to the unit of measurement.

> **DEFINITION**   Vectors are quantities defined by magnitude and direction. The geometrical representation of a vector is by means of an arrow whose length, to some scale, represents the magnitude of the physical quantity and whose direction indicates the direction of the vector.

On the other hand, there are physical quantities, distinct from vectors, which are completely defined by their magnitudes. Such quantities are called *scalar quantities* or, more simply, *scalars*.

> **DEFINITION**   A scalar quantity is one which is completely defined by its magnitude.

Calculations with scalar quantities follow the ordinary rules of algebra with positive and negative numbers. Calculations with vectors would appear, in the first instance, to be more difficult. However, the pictorial geometrical representation of vector quantities facilitates this task. With vectors it is possible to describe physical situations concisely. A clear notation is needed to represent vector quantities and there are, in fact, a number of notations in use.

Vectors are represented by

(1) two capital letters with an arrow above them to indicate the sense of direction, e.g. $\overrightarrow{P_1P_2}$ where P_1 is the starting point and P_2 the end point of the vector;

(2) bold-face letters, e.g. **a**, **A** (the style used in this text);

(3) letters with an arrow above, e.g. $\vec{a}$, $\vec{A}$;

(4) underlined letters, e.g. a͟, and occasionally, by a squiggle underneath the letter, e.g. a̰.

To distinguish the *magnitude* of a vector **a** from its direction we use the mathematical notation

$$|\mathbf{a}| = a$$

The quantity $|\mathbf{a}|$ is a scalar quantity.

3.2   ADDITION OF VECTORS

Geometrically, vectors may be combined by defining easy rules.

To scientists and engineers it is important that the results (sum, difference) should correspond exactly to the way actual physical quantities behave.

3.2.1   SUM OF TWO VECTORS: GEOMETRICAL ADDITION

Previously the sum of two vectors was shown to be made up of two shifts, i.e. the result of two successive shifts was represented by means of another shift. If two vectors **a** and **b** are to be added so that their sum is a third vector **c** then we write

$$\mathbf{c} = \mathbf{a} + \mathbf{b}.$$

Consider the two vectors **a** and **b**, shown in figure 3.6a, with a common origin at A. We can shift vector **b** parallel to itself until its starting point coincides with the end point of vector **a** (see figure 3.6b). As a result of this shift we define the vector **c** as a vector starting at A and ending at the end of vector **b** (see figure 3.6c). Then **c** is the vector sum of the two vectors **a** and **b** and is called the *resultant*. The triangle law of addition of vectors is expressed by the vector equation

$$\mathbf{c} = \mathbf{a} + \mathbf{b}$$

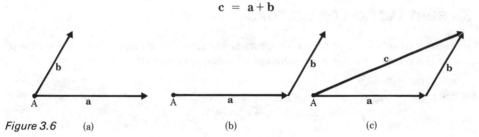

Figure 3.6   (a)   (b)   (c)

The sum of several vectors is obtained by successive application of the triangle law; a polygon is formed as illustrated in figure 3.7.

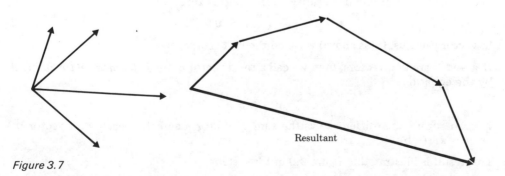

Resultant

Figure 3.7

The order in which the vectors are added is immaterial. This is known as the *commutative law*, i.e. $\mathbf{a} + \mathbf{b} = \mathbf{b} + \mathbf{a}$. Furthermore, the law for addition of vectors is *associative*, i.e. if $\mathbf{a}$, $\mathbf{b}$ and $\mathbf{c}$ are three vectors then their sum is

$$\mathbf{a} + (\mathbf{b} + \mathbf{c}) = (\mathbf{a} + \mathbf{b}) + \mathbf{c} \tag{3.1}$$

This means that we could add to $\mathbf{a}$ the sum of $\mathbf{b}$ and $\mathbf{c}$ or find the sum of $\mathbf{a}$ and $\mathbf{b}$ and add it to $\mathbf{c}$ and still obtain the same *resultant*.

Vector addition also follows Newton's parallelogram law of forces which applies to two forces acting at a point, as shown in figure 3.8. The vector sum of two such vectors $\mathbf{a}$ and $\mathbf{b}$ is obtained by drawing two lines parallel to the vectors $\mathbf{a}$ and $\mathbf{b}$, respectively, to form a parallelogram. The vector sum is then represented by the diagonal $\overrightarrow{AB}$; hence $\mathbf{c} = \mathbf{a} + \mathbf{b}$. A study of the figure shows that it is equivalent to the triangle law and that $\mathbf{c} = \overrightarrow{AB}$ is obtained by either adding $\mathbf{b}$ to $\mathbf{a}$ or $\mathbf{a}$ to $\mathbf{b}$, i.e.

$$\mathbf{c} = \mathbf{a} + \mathbf{b} = \mathbf{b} + \mathbf{c} \tag{3.2}$$

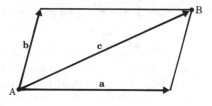

Figure 3.8

3.3 SUBTRACTION OF VECTORS

The method of subtraction for two vectors is obtained by an extension of the rule of addition if we first introduce the concept of a negative vector.

> **DEFINITION**   The negative of a vector $\mathbf{a}$ is a vector having the same magnitude but opposite direction. We write it as $-\mathbf{a}$.

If the vector $\mathbf{a}$ starts at A and ends at B, such that $\mathbf{a} = \overrightarrow{AB}$, then it follows that $-\mathbf{a} = \overrightarrow{BA}$.

The sum of a vector and its negative counterpart is zero, for

$$\mathbf{a} + (-\mathbf{a}) = \mathbf{0}$$

In vector calculus, $\mathbf{0}$ (as above) is called the *null vector*.

If $\mathbf{a}$ and $\mathbf{b}$ are two vectors, then we call a third vector $\mathbf{c}$ the *difference vector* defined by the equation

$$\mathbf{c} = \mathbf{a} - \mathbf{b}$$

We can regard this difference as the sum of vector $\mathbf{a}$ and the negative of vector $\mathbf{b}$, i.e. $\mathbf{c} = \mathbf{a} + (-\mathbf{b})$.

This result is illustrated in figure 3.9 in three steps.

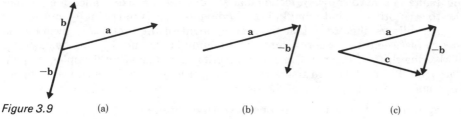

Figure 3.9   (a)   (b)   (c)

Firstly, we draw the negative vector $-\mathbf{b}$ (figure 3.9a); secondly, we shift this negative vector so that its end is at the tip of vector $\mathbf{a}$ (figure 3.9b); and thirdly, we form the sum of $\mathbf{a}$ and $(-\mathbf{b})$ in accordance with the triangle law and obtain the difference vector $\mathbf{c} = \mathbf{a} + (-\mathbf{b})$ (figure 3.9c).

To add and subtract vectors we proceed using the rules for addition and subtraction.

The difference vector $\mathbf{c} = \mathbf{a} - \mathbf{b}$ can also be constructed using the parallelogram rule. Figure 3.10a shows two vectors $\mathbf{a}$ and $\mathbf{b}$; in figure 3.10b the parallelogram is completed. The difference vector $\mathbf{c} = \mathbf{a} - \mathbf{b}$ is then given by the diagonal $\overrightarrow{BA}$ (figure 3.10c).

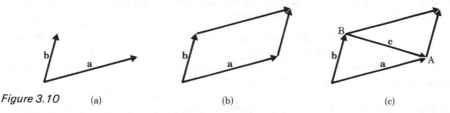

Figure 3.10   (a)   (b)   (c)

It is easy to see that both constructions lead to the same result, and that the latter construction shows clearly that the difference vector can be regarded geometrically by the line joining the end points of the two vectors.

3.4 COMPONENTS AND PROJECTION OF A VECTOR

Let us consider the shift of a point from position P_1 to position P_2 by the vector $\mathbf{a}$, as shown in figure 3.11a, and then find out by how much the point has shifted in the x-direction. To ascertain this shift we drop perpendiculars on to the x-axis from the points P_1 and P_2 respectively, cutting the axis at x_1 and x_2. The distance between these two points is the *projection* of the vector $\mathbf{a}$ on to the x-axis. This projection is also called the *x-component* of the vector.

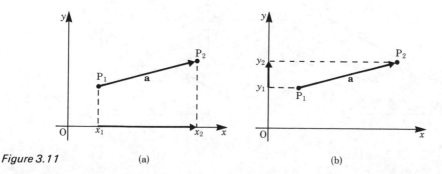

Figure 3.11   (a)   (b)

In the figure, we have shown a rectangular set of axes, i.e. axes which are perpendicular to each other so that point P_1 has coordinates (x_1, y_1) and point P_2 coordinates (x_2, y_2). It follows, therefore, that the x-component of the vector **a** is given by the difference between the x-coordinates of the points P_1 and P_2, i.e. by $x_2 - x_1$. Similarly, the shift of the point in the y-direction is obtained by dropping perpendiculars from P_1 and P_2 on to the y-axis, cutting it at y_1 and y_2, as shown in figure 3.11b. Hence the y-component of the vector **a** in the y-direction is given by $y_2 - y_1$.

The components of the vector **a** are usually written as follows:

$$a_x = x_2 - x_1$$
$$a_y = y_2 - y_1$$

If the coordinate axes are not perpendicular to each other the coordinate system is called *oblique*. Projections in such a coordinate system are obtained by using lines parallel to the axes instead of perpendiculars. We will not use this type of coordinate system in this book, even though oblique coordinates are very useful in certain cases, e.g. in crystallography.

Generalisation of the concept of projection. So far we have considered the projection of a vector on to a set of rectangular coordinates. We can generalise this concept by projecting a vector **a** on a vector **b** as follows.

We drop normals from the starting and end points of the vectors, as shown in figure 3.12a, on to the line of action of vector **b**. The line of action of the vector is the straight line determined by the direction of the vector. It extends on either side of the vector, as shown in figure 3.12a. The distance between the two normals is the component of vector **a** along vector **b**; this is written as $\mathbf{a}_b$. We can simplify the construction by shifting the vector **a** parallel to itself until its starting point meets the line of action of vector **b** and then dropping a normal on **b** from the tip of vector **a**, as shown in figure 3.12b. This results in a triangle and the projection or component of **a** in the **b** direction is $|\mathbf{a}_b| = |\mathbf{a}| \cos \alpha$

or
$$a_b = a \cos \alpha$$

Similarly, we can project vector **b** on to **a**, giving

$$b_a = b \cos \alpha$$

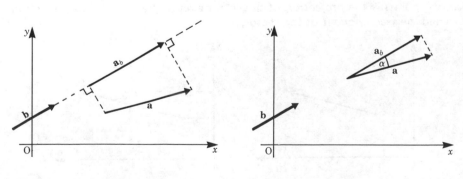

Figure 3.12   (a)   (b)

3.5 COMPONENT REPRESENTATION IN COORDINATE SYSTEMS

The graphical addition and subtraction of vectors can easily be carried out in a plane surface, e.g. in the x–y plane of a rectangular coordinate system. However, we frequently have to cope with spatial problems. These can be solved if the vector components in the direction of the coordinate axes are known. We can then treat the components in each of the axes as scalars obeying the ordinary rules of algebra.

3.5.1 POSITION VECTOR

The *position vector* of a point in space is a vector from the origin of the coordinate system to the point. Thus, to each point P in space there corresponds a unique vector. Such vectors are not movable and they are often referred to as *bound vectors*.

The addition of two position vectors is not possible. But in contrast subtracting them gives a sensible meaning. If P_1 and P_2 are two spatial points, i.e. two positions in space, and O is the origin of a coordinate system, then the position vectors are $\overrightarrow{OP_1}$ and $\overrightarrow{OP_2}$ and their difference, $\overrightarrow{OP_2} - \overrightarrow{OP_1}$, is a vector starting at P_1 and ending at P_2, as shown in figure 3.13.

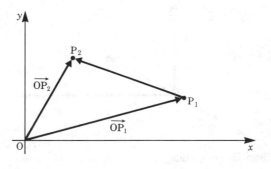

Figure 3.13

3.5.2 UNIT VECTORS

Vectors have magnitude and direction, but if we wish to indicate the direction *only* we define a *unit vector*. A unit vector has a magnitude of 1 unit; consequently it defined the direction only.

Figure 3.14 shows three such unit vectors (bold arrows) belonging to vectors **a**, **b** and **c** respectively.

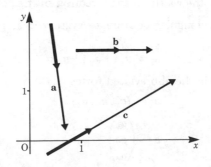

Figure 3.14

Of special significance are unit vectors along a cartesian or rectangular coordinate system (see figure 3.15). In such a three-dimensional system, these unit vectors are denoted by the letters **i**, **j**, **k** or $\mathbf{e}_x$, $\mathbf{e}_y$, $\mathbf{e}_z$ or $\mathbf{e}_1$, $\mathbf{e}_2$, $\mathbf{e}_3$. Here we shall adopt the **i**, **j**, **k** notation.

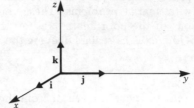

Figure 3.15

Figure 3.16 shows a point P with coordinates P_x, P_y, P_z. The position vector $\overrightarrow{OP}$ = P has three *components*:

a component P_x **i** along the x-axis;

a component P_y **j** along the y-axis;

a component P_z **k** along the z-axis.

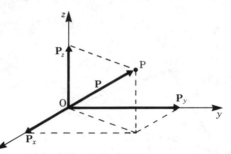

Figure 3.16

From the figure it then follows that

$$\overrightarrow{OP} = \mathbf{P} = P_x\,\mathbf{i} + P_y\,\mathbf{j} + P_z\,\mathbf{k}$$

3.5.3 COMPONENT REPRESENTATION OF A VECTOR

A vector can be constructed if its components along the axes of a coordinate system are known. Hence the following information is sufficient to fix a vector:

a defined coordinate system;

the components of the vector in the direction of the coordinate axes.

Figure 3.17 shows a rectangular coordinate system xyz. If the vector **a** has components a_x, a_y, a_z then

$$\mathbf{a} = a_x\,\mathbf{i} + a_y\,\mathbf{j} + a_z\,\mathbf{k}$$

It can also be expressed in the abbreviated forms

$$\mathbf{a} = (a_x, a_y, a_z)$$

or

$$\mathbf{a} = \begin{pmatrix} a_x \\ a_y \\ a_z \end{pmatrix}$$

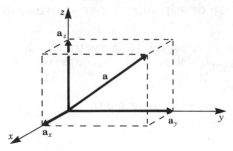

Figure 3.17

Thus the vector **a** is defined by the three numbers a_x, a_y, a_z. To obtain the vector **a** we simply multiply these numbers by the appropriate unit vectors. a_x, a_y and a_z are the 'coordinates' of the vector; they are scalar quantities.

DEFINITION   We may express a vector $\mathbf{a} = a_x\mathbf{i} + a_y\mathbf{j} + a_z\mathbf{k}$ thus:

$$\mathbf{a} = (a_x, a_y, a_z) = \begin{pmatrix} a_x \\ a_y \\ a_z \end{pmatrix}$$

These are called the **component representations** of the vector **a**.

EXAMPLE   The vector shown in figure 3.18 is given by $\mathbf{a} = (1, 3, 3)$.

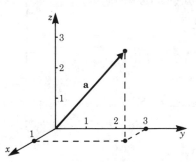

Figure 3.18

Two vectors are equal if and only if their components are equal. Hence if $\mathbf{a} = \mathbf{b}$, then

$$a_x = b_x$$
$$a_y = b_y$$
$$a_z = b_z$$

3.5.4   REPRESENTATION OF THE SUM OF TWO VECTORS IN TERMS OF THEIR COMPONENTS

We will now show that the result of the geometrical addition of two vectors can be obtained by adding separately the components of the vectors in given directions.

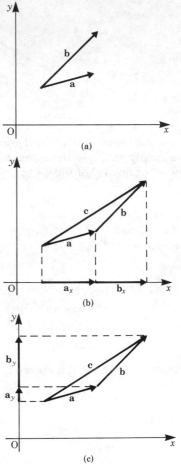

Two vectors **a** and **b** in the x–y plane, as shown in figure 3.19a, can be expressed in terms of unit vectors thus:

$$\mathbf{a} = a_x\mathbf{i} + a_y\mathbf{j}$$

and

$$\mathbf{b} = b_x\mathbf{i} + b_y\mathbf{j}$$

(a)

We now add **a** and **b** to give the resultant vector **c**, as shown in figure 3.19b:

$$\mathbf{c} = \mathbf{a} + \mathbf{b}$$

The x-component of **c** is $c_x\mathbf{i} = a_x\mathbf{i} + b_x\mathbf{i}$ or $c_x\mathbf{i} = (a_x + b_x)\mathbf{i}$. Hence the x-component of the resultant vector is equal to the algebraic sum of the x-components of the original vectors.

(b)

Similarly, the y-component (see figure 3.19c) is

$$c_y\mathbf{j} = (a_y + b_y)\mathbf{j}$$

(c)

Figure 3.19

It then follows that the vector **c**, the resultant of vectors **a** and **b**, is given by

$$\mathbf{c} = (a_x + b_x)\mathbf{i} + (a_y + b_y)\mathbf{j}$$

whose coordinates are $(a_x + b_y)$, $(a_y + b_y)$.

The same procedure can be adopted in the case of three-dimensional vectors. If **a** and **b** are two such vectors so that

$$\mathbf{a} = (a_x, a_y, a_z) \qquad \text{and} \qquad \mathbf{b} = (b_x, b_y, b_z)$$

then it follows that

$$\mathbf{a} + \mathbf{b} = (a_x + b_x, \quad a_y + b_y, \quad a_z + b_z) \tag{3.3a}$$

Generally, the sum of two or more vectors is found by adding separately their components in the directions of the axes.

3.5.5   SUBTRACTION OF VECTORS IN TERMS OF THEIR COMPONENTS

The task of finding the difference $\mathbf{a} - \mathbf{b}$ between two vectors $\mathbf{a}$ and $\mathbf{b}$ can be reduced to that of adding vector $\mathbf{a}$ and the negative of vector $\mathbf{b}$.

It therefore follows that, for the two-dimensional case

$$\mathbf{a} - \mathbf{b} = (a_x - b_x, \quad a_y - b_y)$$

and for the three-dimensional case

$$\mathbf{a} - \mathbf{b} = (a_x - b_x, \quad a_y - b_y, \quad a_z - b_z) \tag{3.3b}$$

EXAMPLE   Let the vectors be $\mathbf{a} = (2, 5, 1)$ and $\mathbf{b} = (3, -7, 4)$ then, in terms of the components, we have

$$\mathbf{a} - \mathbf{b} = (2 - 3, \quad 5 + 7, \quad 1 - 4) = (-1, 12, -3)$$

Of special significance is the difference vector of two position vectors. This is given by the vector which joins the end points of the two position vectors.

Figure 3.20 shows that vector $\mathbf{c}$ is obtained by joining the two points P_1 and P_2 so that $\mathbf{c} = P_1 - P_2$.

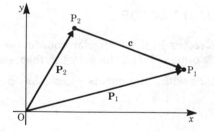

Figure 3.20

In terms of the components of P_1 and P_2,

$$\mathbf{c} = (P_{1x} - P_{2x}, \quad P_{1y} - P_{2y})$$

EXAMPLE   If $P_1 = (3, -1, 0)$ and $P_2 = (-2, 3, -1)$ are two points in space then the difference vector $\mathbf{c}$ given by $\mathbf{c} = P_2 - P_1$ is

$$\mathbf{c} = (-2 - 3, \quad 3 + 1, \quad -1 - 0) = (-5, 4, -1)$$

3.6 MULTIPLICATION OF A VECTOR BY A SCALAR

Multiplication of a vector by a scalar quantity results in a vector whose magnitude is that of the original vector multiplied by the scalar and whose direction is that of the original vector or reversed if the scalar is negative.

> **DEFINITION**   Multiplication of a vector **a** by a scalar λ gives the vector λ**a** having length λa and the same direction as **a** when $\lambda > 0$. If $\lambda < 0$ it has the opposite direction.

In terms of the components of the vector **a**, the new vector λ**a** is given by

$$\lambda \mathbf{a} = (\lambda a_x, \quad \lambda a_y, \quad \lambda a_z) \tag{3.4}$$

If $\lambda = 0$, then the vector λ**a** is the *null vector* $(0, 0, 0)$.

EXAMPLE   Given **a** $= (2, 5, 1)$, then when $\lambda = 3$ we have

$$\lambda \mathbf{a} = (6, 15, 3)$$

and when $\lambda = -3$ we have

$$\lambda \mathbf{a} = (-6, -15, -3)$$

3.7 MAGNITUDE OF A VECTOR

If the components of a vector in a rectangular coordinate system are known, the magnitude of the vector is obtained with the aid of Pythagoras' theorem.

Figure 3.21 shows a vector **a** with components a_x, a_y, i.e. **a** $= (a_x, a_y)$.

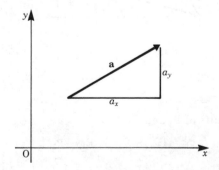

Figure 3.21

Since the vector and its components form a right-angled triangle, we have

$$a^2 = a_x{}^2 + a_y{}^2$$

and the magnitude of the vector is

$$|\mathbf{a}| = a = \sqrt{a_x{}^2 + a_y{}^2} \tag{3.5a}$$

The three-dimensional vector $\mathbf{a} = (a_x, a_y, a_z)$ shown in figure 3.22 has a magnitude given by

$$|\mathbf{a}| = a = \sqrt{a_x{}^2 + a_y{}^2 + a_z{}^2} \qquad (3.5\text{b})$$

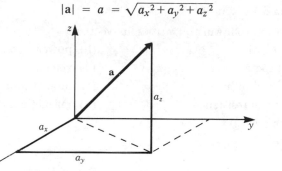

Figure 3.22

EXAMPLE   The magnitude of the vector $\mathbf{a} = (3, -7, 4)$ is

$$a = \sqrt{3^2 + 7^2 + 4^2} = \sqrt{74} \approx 8.60$$

The distance between two points in space is thus easily determined if the components are known.

EXAMPLE   Figure 3.23 shows two given points in the plane $P_1 = (x_1, y_1)$ and $P_2 = (x_2, y_2)$. It is required to find the distance between them.

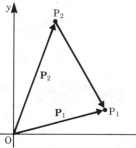

Figure 3.23

To find the distance we require the coordinates of the connecting vector $\overrightarrow{P_2P_1}$. These are

$$\overrightarrow{P_2P_1} = (x_1 - x_2, \quad y_1 - y_2)$$

and the magnitude is

$$|\overrightarrow{P_2P_1}| = \sqrt{(x_1 - x_2)^2 + (y_1 - y_2)^2}$$

If P_1 and P_2 are two points in space, then the distance between them is

$$|\overrightarrow{P_2P_1}| = \sqrt{(x_1 - x_2)^2 + (y_1 - y_2)^2 + (z_1 - z_2)^2}$$

Any vector may be expressed in terms of a unit vector. If $\mathbf{a} = (a_x, a_y, a_z)$ is any vector, its magnitude is

$$|\mathbf{a}| = a = \sqrt{a_x{}^2 + a_y{}^2 + a_z{}^2}$$

If the unit vector in the direction of $\mathbf{a}$ is denoted by $\mathbf{e}_a$, then

$$\mathbf{e}_a = \frac{a_x \mathbf{i} + a_y \mathbf{j} + a_z \mathbf{k}}{\sqrt{a_x{}^2 + a_y{}^2 + a_z{}^2}}$$

or

$$\mathbf{e}_a = \lambda \mathbf{a} = \frac{1}{|\mathbf{a}|} \mathbf{a} = \left(\frac{a_x}{|\mathbf{a}|}, \quad \frac{a_y}{|\mathbf{a}|}, \quad \frac{a_z}{|\mathbf{a}|} \right)$$

Hence

$$\mathbf{a} = a \mathbf{e}_a$$

EXERCISES

3.1   Scalars and Vectors

1.   Which of the following quantities are vectors?

(a) acceleration

(b) power

(c) centripetal force

(d) velocity

(e) quantity of heat

(f) momentum

(g) electrical resistance

(h) magnetic intensity

(i) atomic weight

3.2   Addition of Vectors;   3.3   Subtraction of Vectors

2.   Given the vectors **a**, **b** and **c**, draw the vector sum $S = a + b + c$ in each case.

(a)   (b)

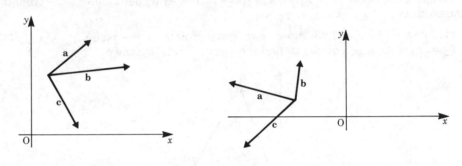

Figure 3.24   Figure 3.25

3.   Draw the vector sum $a_1 + a_2 + \ldots + a_n$.

(a)   (b)

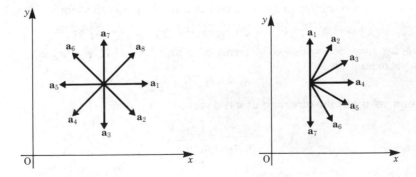

Figure 3.26   Figure 3.27

4. Draw the vector $\mathbf{c} = \mathbf{a} - \mathbf{b}$.

    (a)   (b)

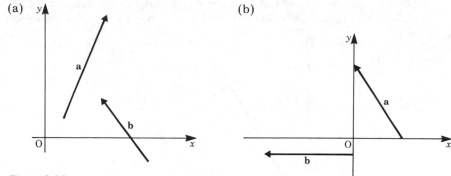

Figure 3.28   Figure 3.29

3.4 Components and Projection of a Vector

5. Project vector $\mathbf{a}$ on to vector $\mathbf{b}$.

    (a)   (b)

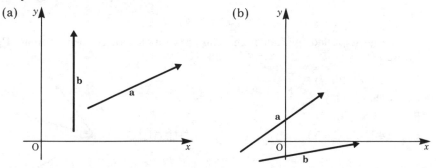

Figure 3.30   Figure 3.31

6. Calculate the magnitude of the projection of $\mathbf{a}$ on to $\mathbf{b}$.

    (a) $|\mathbf{a}| = 5$, $\angle\,(\mathbf{a}, \mathbf{b}) = \dfrac{\pi}{3}$ (b) $|\mathbf{a}| = 2$, $\angle\,(\mathbf{a}, \mathbf{b}) = \dfrac{\pi}{2}$

    (c) $|\mathbf{a}| = 4$, $\angle\,(\mathbf{a}, \mathbf{b}) = 0$ (d) $|\mathbf{a}| = \dfrac{3}{2}$, $\angle\,(\mathbf{a}, \mathbf{b}) = \dfrac{2}{3}\pi$

3.5 Component Representation

7. Given the points $P_1 = (2, 1)$, $P_2 = (7, 3)$ and $P_3 = (5, -4)$, calculate the coordinates of the fourth corner P_4 of the parallelogram $P_1 P_2 P_3 P_4$ formed by the vectors $\mathbf{a} = \overrightarrow{P_1 P_2}$ and $\mathbf{b} = \overrightarrow{P_1 P_3}$.

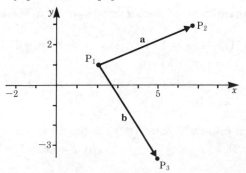

Figure 3.32

8. If $P_1 = (x_1, y_1)$, $P_2 = (x_2, y_2)$, $P_3 = (x_3, y_3)$ and $P_4 = (x_4, y_4)$ are four arbitrary points in the x-y plane and if $a = \overrightarrow{P_1 P_2}$, $b = \overrightarrow{P_2 P_3}$, $c = \overrightarrow{P_3 P_4}$, $d = \overrightarrow{P_4 P_1}$, calculate the components of the resultant vector $S = a + b + c + d$ and show that $S = 0$.

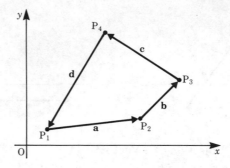

Figure 3.33

9. A carriage is pulled by four men. The components of the four forces F_1, F_2, F_3, F_4 are

$F_1 = (20\,\text{N}, \quad 25\,\text{N})$
$F_2 = (15\,\text{N}, \quad 5\,\text{N})$
$F_3 = (25\,\text{N}, \, -5\,\text{N})$
$F_4 = (30\,\text{N}, \, -15\,\text{N})$

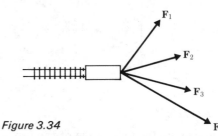

Figure 3.34

Calculate the resultant force.

10. If $a = (3, 2, 1)$, $b = (1, 1, 1)$, $c = (0, 0, 2)$, calculate
(a) $a + b - c$ (b) $2a - b + 3c$

3.6 Multiplication of a Vector by a Scalar

11. Calculate the magnitude of vector $a = \lambda_1 a_1 + \lambda_2 a_2 - \lambda_3 a_3$ for the following cases:
(a) $a_1 = (2, -3, 1)$, $a_2 = (-1, 4, 2)$, $a_3 = (6, -1, 1)$,
  $\lambda_1 = 2$, $\lambda_2 = \frac{1}{2}$, $\lambda_3 = 3$
(b) $a_1 = (-4, 2, 3)$, $a_2 = (-5, -4, 3)$, $a_3 = (2, -4, 3)$,
  $\lambda_1 = -1$, $\lambda_2 = 3$, $\lambda_3 = -2$

12. Calculate in each case the unit vector e_a in the direction of a:
(a) $a = (3, -1, 2)$ (b) $a = (2, -1, -2)$

3.7 Magnitude of a Vector

13. Calculate the distance **a** between the points P_1 and P_2 in each case:

    (a) $P_1 = (3, 2, 0)$
    $P_2 = (-1, 4, 2)$

    (b) $P_1 = (-2, -1, 3)$
    $P_2 = (4, -2, -1)$

14. An aircraft is flying on a northerly course and its velocity relative to the air is

    $$V_1 = (0\,\text{km/h}, \quad 300\,\text{km/h})$$

    Calculate the velocity of the aircraft relative to the ground for the following three different air velocities:

    (a) $V_2 = (0, -50)\,\text{km/h}, \quad$ headwind
    (b) $V_3 = (50, 0)\,\text{km/h}, \quad$ crosswind
    (c) $V_4 = (0, 50)\,\text{km/h}, \quad$ tailwind

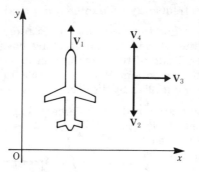

Figure 3.35

Calculate the magnitude of the absolute velocity relative to the ground for the three cases:

    (d) $|V_1 + V_2|$ (e) $|V_1 + V_3|$ (f) $|V_1 + V_4|$

CHAPTER 4

Vector algebra II: Scalar and vector products

We saw in the previous chapter how vector quantities may be added and subtracted. In this chapter we consider the products of vectors and define rules for them. First we will examine two cases frequently encountered in practice.

1) In applied science we define the work done by a force as the magnitude of the force multiplied by the distance it moves along its line of action, or by the component of the magnitude of the force in a given direction multiplied by the distance moved in that direction. Work is a scalar quantity and the product obtained when force is multiplied by displacement is called the *scalar product*.

2) The torque on a body produced by a force F (figure 4.1) is defined as the product of the force and the length of the lever arm OA, the line of action of the force being perpendicular to the lever arm. Such a product is called a *vector product* or *cross product* and the result is a vector in the direction of the axis of rotation, i.e. perpendicular to both the force and the lever arm.

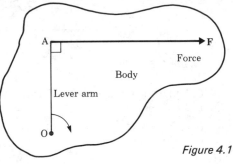

Figure 4.1

4.1 SCALAR PRODUCT

Consider a carriage running on rails. It moves in the x-direction (figure 4.2) under the application of a force F which acts at an angle α to the direction of travel. We require the work done by the force when the carriage moves through a distance s in the x-direction (figure 4.3).

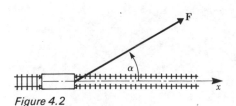

Figure 4.2

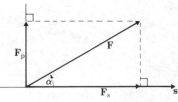

Figure 4.3

In order to study the action of the force F on the carriage we resolve it into two components: one along the rails (in the x-direction), and one perpendicular to the rails, i.e. $\mathbf{F_s}$ and $\mathbf{F_p}$ respectively. $\mathbf{F_s}$, $\mathbf{F_p}$ and $\mathbf{s}$ are vector quantities; the work is, by definition, the product of the force along the direction of motion and the distance moved. In this case, it is the product of $\mathbf{F_s}$ and $\mathbf{s}$. It follows also from the definition that the work done by $\mathbf{F_p}$ is zero since there is no displacement in that direction. Furthermore, if the rails are horizontal then the motion of the carriage and the work done is not influenced by gravity, since it acts in a direction perpendicular to the rails.

If W is the work done then $W = F \cos \alpha s$ or $Fs \cos \alpha$ in magnitude.

Since work is a scalar quantity the product of the two vectors is called a *scalar product* or *dot product*, because one way of writing it is with a dot between the two vectors:

$$W = \mathbf{F_s \cdot s}$$

where $$|\mathbf{F_s}| = |\mathbf{F}| \cos \alpha$$

It is also referred to as the *inner product* of two vectors. Generally, if $\mathbf{a}$ and $\mathbf{b}$ are two vectors their inner product is written $\mathbf{a \cdot b}$.

> **DEFINITION** The *inner* or *scalar product* of two vectors is equal to the product of their magnitude and the cosine of the angle between their directions:
>
> $$\mathbf{a \cdot b} = ab \cos \alpha \qquad (4.1)$$

Geometrical interpretation. The scalar product of two vectors $\mathbf{a}$ and $\mathbf{b}$ is equal to the product of the magnitude of vector $\mathbf{a}$ with the projection of $\mathbf{b}$ on $\mathbf{a}$ (figure 4.4a):

$$\mathbf{a \cdot b} = ab \cos \alpha$$

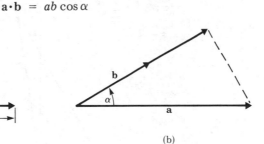

Figure 4.4 (a) (b)

Or it is the product of the magnitude of $\mathbf{b}$ with the projection of $\mathbf{a}$ on $\mathbf{b}$ (figure 4.4b):

$$\mathbf{a \cdot b} = ba \cos \alpha$$

In the case of the carriage, we can also evaluate the work done by the product of the magnitude of the force and the component of the displacement along the direction of the force (figure 4.5).

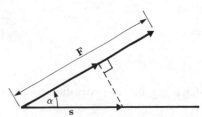

Figure 4.5

EXAMPLE   A force of 5 N is applied to a body. The body is moved through a distance of 10 m in a direction which subtends an angle of 60° with the line of action of the force.

The mechanical work done is

$$U = \mathbf{F} \cdot \mathbf{s} = Fs \cos\alpha = 5 \times 10 \times \cos 60° = 25 \, \text{N m}$$

The unit of work should be noted: it is newtons × metres = N m or joules (J). This example could be considered to represent the force of gravity acting on a body which slides down a chute through a distance s; the force $\mathbf{F} = mg$ where m is the mass of the body and g the acceleration due to gravity (figure 4.6).

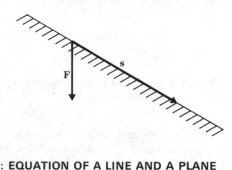

Figure 4.6

4.1.1   APPLICATION: EQUATION OF A LINE AND A PLANE

The scalar product can be used to obtain the *equation of a line* in an x–y plane if the normal from the origin to the line is given (figure 4.7). In this case the scalar product of $\mathbf{n}$ with any position vector $\mathbf{r}$ to a point on the line is constant and equal to n^2.

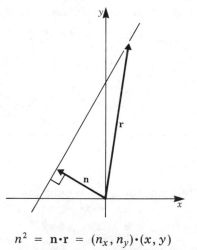

Figure 4.7

Thus

$$n^2 = \mathbf{n} \cdot \mathbf{r} = (n_x, n_y) \cdot (x, y)$$

$$n^2 = xn_x + yn_y \tag{4.2a}$$

$$y = \frac{n_x}{n_y}x + \frac{n^2}{n_y}$$

If we extend the procedure to three dimensions we obtain the *equation of a plane* in an x–y–z coordinate system:

$$n^2 = xn_x + yn_y + zn_z \tag{4.2b}$$

4.1.2   SPECIAL CASES

Scalar Product of Perpendicular Vectors

If two vectors **a** and **b** are perpendicular to each other so that $\alpha = \dfrac{\pi}{2}$ and hence $\cos\alpha = 0$, it follows that the scalar product is zero, i.e. $\mathbf{a} \cdot \mathbf{b} = 0$.

The converse of this statement is important. If it is known that the scalar product of two vectors **a** and **b** vanishes, then it follows that the two vectors are perpendicular to each other, provided that $\mathbf{a} \neq \mathbf{0}$ and $\mathbf{b} \neq \mathbf{0}$.

Scaler Product of Parallel Vectors

If two vectors **a** and **b** are parallel to each other so that $\alpha = 0$ and hence $\cos\alpha = 1$, it follows that their scalar product $\mathbf{a} \cdot \mathbf{b} = ab$.

4.1.3   COMMUTATIVE AND DISTRIBUTIVE LAWS

The scalar product obeys the commutative and distributive laws. These are given without proof.

Commutative law	$\mathbf{a} \cdot \mathbf{b} = \mathbf{b} \cdot \mathbf{a}$	(4.3)
Distributive law	$\mathbf{a} \cdot (\mathbf{b} + \mathbf{c}) = \mathbf{a} \cdot \mathbf{b} + \mathbf{a} \cdot \mathbf{c}$	(4.4)

As an example of the scalar product let us derive the *cosine rule*. Figure 4.8 shows three vectors; α is the angle between the vectors **a** and **b**.

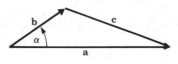

Figure 4.8

We have
$$\mathbf{b} + \mathbf{c} = \mathbf{a}$$
$$\mathbf{c} = \mathbf{a} - \mathbf{b}$$

We now form the scalar product of the vectors with themselves, giving
$$\mathbf{c} \cdot \mathbf{c} = c^2 = (\mathbf{a} - \mathbf{b})^2$$
$$c^2 = \mathbf{a} \cdot \mathbf{a} + \mathbf{b} \cdot \mathbf{b} - 2\mathbf{a} \cdot \mathbf{b}$$
$$c^2 = a^2 + b^2 - 2ab\cos\alpha \qquad (4.5)$$

If $\alpha = \dfrac{\pi}{2}$, we have Pythagoras' theorem for a right-angled triangle.

4.1.4 **SCALAR PRODUCT IN TERMS OF THE COMPONENTS OF THE VECTORS**

If the components of two vectors are known, their scalar product can be evaluated. It is useful to consider the scalar product of the unit vectors i along the x-axis and j along the y-axis, as shown in figure 4.9.

From the definition of the scalar product we deduce the following:

$$\mathbf{i} \cdot \mathbf{i} = 1$$
$$\mathbf{j} \cdot \mathbf{j} = 1$$
$$\mathbf{i} \cdot \mathbf{j} = 0$$
$$\mathbf{j} \cdot \mathbf{i} = 0$$

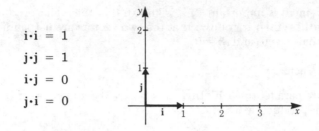

Figure 4.9

Figure 4.10 shows two vectors **a** and **b** that issue from the origin of a cartesian coordinate system. If a_x, b_x, a_y and b_y are the components of these vectors along the x-axis and y-axis, respectively, then

$$\mathbf{a} = a_x \mathbf{i} + a_y \mathbf{j}$$
$$\mathbf{b} = b_x \mathbf{i} + b_y \mathbf{j}$$

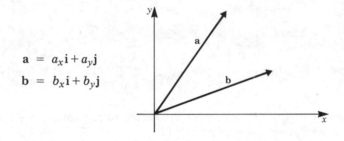

Figure 4.10

The scalar product is

$$\begin{aligned}
\mathbf{a} \cdot \mathbf{b} &= (a_x \mathbf{i} + a_y \mathbf{j}) \cdot (b_x \mathbf{i} + b_y \mathbf{j}) \\
&= a_x b_x \mathbf{i} \cdot \mathbf{i} + a_x b_y \mathbf{i} \cdot \mathbf{j} + a_y b_x \mathbf{j} \cdot \mathbf{i} + a_y b_y \mathbf{j} \cdot \mathbf{j} \\
\mathbf{a} \cdot \mathbf{b} &= a_x b_x + a_y b_y
\end{aligned}$$

Thus the scalar product is obtained by adding the products of the components of the vectors along each axis (figure 4.11).

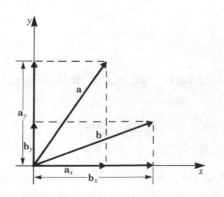

Figure 4.11

In the case of three-dimensional vectors it is easily demonstrated that the following rule holds true:

$$\mathbf{a} \cdot \mathbf{b} = a_x b_x + a_y b_y + a_z b_z \quad \text{(scalar product)} \tag{4.6}$$

It is also an easy matter to calculate the magnitude of a vector in terms of its components. Thus

$$a^2 = \mathbf{a} \cdot \mathbf{a}$$
$$= a_x a_x + a_y a_y + a_z a_z$$
$$= a_x^2 + a_y^2 + a_z^2$$
$$a = |\mathbf{a}| = \sqrt{a_x^2 + a_y^2 + a_z^2}$$

(In section 3.7, equation (3.5b))

EXAMPLE   Given that $\mathbf{a} = (2, 3, 1)$, $\mathbf{b} = (-1, 0, 4)$, calculate the scalar product.

$$\mathbf{a} \cdot \mathbf{b} = a_x b_x + a_y b_y + a_z b_z$$
$$= 2 \times (-1) + 3 \times 0 + 1 \times 4 = 2$$

The magnitude of each vector is

$$a = \sqrt{2^2 + 3^2 + 1} = \sqrt{14} \approx 3.74$$
$$b = \sqrt{1 + 4^2} = \sqrt{17} \approx 4.12$$

4.2   VECTOR PRODUCT

4.2.1   TORQUE

At the beginning of this chapter we defined the torque C, resulting from a force $\mathbf{F}$ applied to a body at a point P (figure 4.12), to be the product of that force and the position vector $\mathbf{r}$ from the axis of rotation O to the point P, the directions of the force and the position vector being perpendicular.

The magnitude of the torque is therefore $C = |\mathbf{r}||\mathbf{F}|$ or, more simply, $C = rF$.

This is known as the *lever law*.

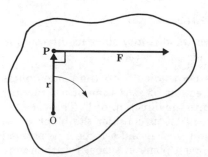

Figure 4.12

A special case is illustrated in figure 4.13 where the line of action of the force **F** is in line with the axis (the angle between force and position vector **r** is zero). In this situation, the force cannot produce a turning effect on the body and consequently $C = 0$.

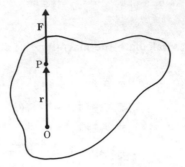

Figure 4.13

The general case is when the force **F** and the radius **r** are inclined to each other at an angle α, as shown in figure 4.14. To calculate the torque C applied to the body we resolve the force into two components: one perpendicular to **r**, $\mathbf{F}_\perp$, and one in the direction of **r**, $\mathbf{F}_\parallel$.

The first component is the only one that will produce a turning effect on the body. Now $\mathbf{F}_\perp = F \sin \alpha$ in magnitude; hence $C = rF \sin \alpha$.

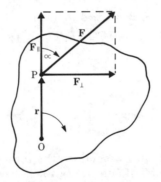

Figure 4.14

DEFINITION   Magnitude of torque C

$$C = rF \sin \alpha$$

4.2.2 TORQUE AS A VECTOR

Physically, torque is a vector quantity since its direction is taken onto account. The following convention is generally accepted.

The torque vector **C** is perpendicular to the plane containing the force **F** and the radius vector **r**. The direction of **C** is that of a screw turned in a way that brings **r** by the shortest route into the direction of **F**. This is called the right-hand rule.

To illustrate this statement let us consider the block of wood shown in figure 4.15 where the axis of rotation is at A and a force **F** is applied at P at a distance **r**. The two vectors **r** and **F** define a plane in space. **F** is then moved parallel to itself to act at A; as the screw is turned it rotates the radius vector **r** towards **F** through an angle α. Hence the direction of the torque **C** coincides with the penetration of the screw.

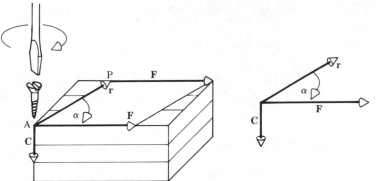

Figure 4.15

4.2.3 DEFINITION OF THE VECTOR PRODUCT

The *vector product* of two vectors **a** and **b** (figure 4.16) is defined as a vector **c** of magnitude $ab \sin \alpha$, where α is the angle between the two vectors. It acts in a direction perpendicular to the plane of the vectors **a** and **b** in accordance with the right-hand rule.

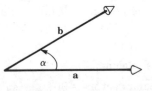

Figure 4.16

This product, sometimes referred to as the *outer product* or *cross product*, is written

$$c = a \times b \qquad \text{or} \qquad c = a \wedge b \qquad (4.7)$$

It is pronounced 'a cross b' or 'a wedge b'. Its magnitude is $c = ab \sin \alpha$.
Note that **a** X **b** = −**b** X **a**.

This definition is quite independent of any physical interpretation. It has geometrical significance in that the vector **c** represents the area of a parallelogram having sides **a** and **b**, as shown in figure 4.17. **c** is perpendicular to the plane containing **a** and **b**, direction given by the right-hand rule.

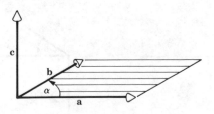

Figure 4.17

The *distributive laws* for vector products are given here without proof.

$$a \times (b+c) = a \times b + a \times c \qquad (4.8)$$

and

$$(a+b) \times c = a \times c + b \times c \qquad (4.9)$$

Further, we note with respect to a scalar λ that

$$\lambda a \times b = a \times \lambda b = \lambda(a \times b) \qquad (4.10)$$

EXAMPLE   Given two vectors **a** and **b** of magnitudes $a = 4$ and $b = 3$ and with an angle $\alpha = \dfrac{\pi}{6} = 30°$ between them, determine the magnitude of $c = a \times b$.

$$c = ab \sin 30° = 4 \times 3 \times 0.5 = 6$$

4.2.4   SPECIAL CASES

Vector Product of Parallel Vectors

The angle between two parallel vectors is zero. Hence the vector product is **0** and the parallelogram degenerates into a line. In particular

$$a \times a = 0$$

It is important to note that the converse of this statement is also true. Thus, if the vector product of two vectors is zero, we can conclude that they are parallel, provided that $a \neq 0$ and $b \neq 0$.

Vector Product of Perpendicular Vectors

The angle between perpendicular vectors is $90°$, i.e. $\sin\alpha = 1$. Hence

$$|a \times b| = ab$$

4.2.5   ANTI-COMMUTATIVE LAW FOR VECTOR PRODUCTS

If **a** and **b** are two vectors then

$$a \times b = -b \times a \qquad (4.11)$$

PROOF   Figure 4.18 shows the formation of the vector product. The vector product is $c = a \times b$ and **c** points upwards. In figure 4.19, **c** is now obtained by turning **b** towards **a**, then, by our definition, the vector **b** $\times$ **a** points downwards. It follows therefore that $a \times b = -b \times a$. The magnitude is the same, i.e. $ab \sin\alpha$.

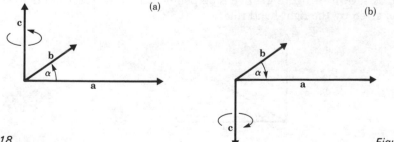

(a)   (b)

Figure 4.18   *Figure 4.19*

4.2.6 COMPONENTS OF THE VECTOR PRODUCT

Let us first consider the vector products of the unit vectors **i**, **j** and **k** (figure 4.20).

According to our definition the following relationships hold:

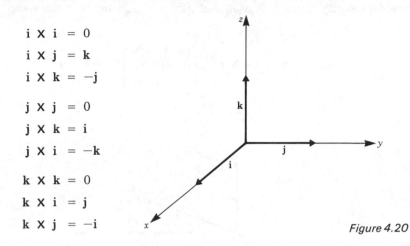

$$\mathbf{i} \times \mathbf{i} = 0$$
$$\mathbf{i} \times \mathbf{j} = \mathbf{k}$$
$$\mathbf{i} \times \mathbf{k} = -\mathbf{j}$$

$$\mathbf{j} \times \mathbf{j} = 0$$
$$\mathbf{j} \times \mathbf{k} = \mathbf{i}$$
$$\mathbf{j} \times \mathbf{i} = -\mathbf{k}$$

$$\mathbf{k} \times \mathbf{k} = 0$$
$$\mathbf{k} \times \mathbf{i} = \mathbf{j}$$
$$\mathbf{k} \times \mathbf{j} = -\mathbf{i}$$

Figure 4.20

Let us now try to express the vector product in terms of components. The vectors **a** and **b** expressed in terms of their components are

$$\mathbf{a} = a_x\mathbf{i} + a_y\mathbf{j} + a_z\mathbf{k}$$

$$\mathbf{b} = b_x\mathbf{i} + b_y\mathbf{j} + b_z\mathbf{k}$$

The vector product is

$$\mathbf{a} \times \mathbf{b} = (a_x\mathbf{i} + a_y\mathbf{j} + a_z\mathbf{k}) \times (b_x\mathbf{i} + b_y\mathbf{j} + b_z\mathbf{k})$$

Expanding in accordance with the distributive law gives

$$\mathbf{a} \times \mathbf{b} = (a_x b_x\mathbf{i} \times \mathbf{i}) + (a_x b_y\mathbf{i} \times \mathbf{j}) + (a_x b_z\mathbf{i} \times \mathbf{k})$$

$$+ (a_y b_x\mathbf{j} \times \mathbf{i}) + (a_y b_y\mathbf{j} \times \mathbf{j}) + (a_y b_z\mathbf{j} \times \mathbf{k})$$

$$+ (a_z b_x\mathbf{k} \times \mathbf{i}) + (a_z b_y\mathbf{k} \times \mathbf{j}) + (a_z b_z\mathbf{k} \times \mathbf{k})$$

Using the relationships for the vector products of unit vectors we obtain

$$\mathbf{a} \times \mathbf{b} = (a_y b_z - a_z b_y)\mathbf{i} + (a_z b_x - a_x b_z)\mathbf{j} + (a_x b_y - a_y b_x)\mathbf{k} \qquad (4.12a)$$

The vector product may conveniently be written in determinant form. A detailed treatment of determinants can be found in Chapter 15.

$$\mathbf{a} \times \mathbf{b} = \begin{vmatrix} \mathbf{i} & \mathbf{j} & \mathbf{k} \\ a_x & a_y & a_z \\ b_x & b_y & b_z \end{vmatrix} \qquad (4.12b)$$

EXAMPLE   The velocity of a point P on a rotating body is given by the vector product of the angular velocity and the position vector of the point from the axis of rotation. In figure 4.21, if the z-axis is the axis of rotation, the angular velocity ω is a vector along this axis. If the position vector of a point P is $\mathbf{r} = (0, r_y, r_z)$ and the angular velocity $\omega = (0, 0, \omega_z)$, as shown in the figure, then the velocity $\mathbf{v}$ of P is

$$\mathbf{v} = \omega \times \mathbf{r} = \begin{vmatrix} \mathbf{i} & \mathbf{j} & \mathbf{k} \\ 0 & 0 & \omega_z \\ 0 & r_y & r_z \end{vmatrix} = -r_y \omega_z \mathbf{i}$$

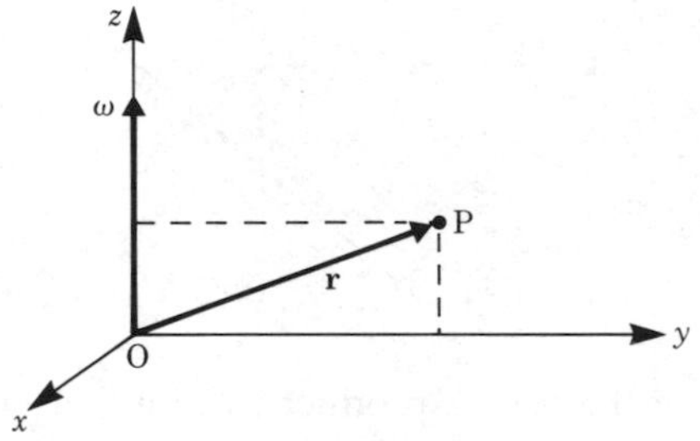

Figure 4.21

EXERCISES

4.1   Scalar Product

1.   Calculate the scalar products of the vectors **a** and **b** given below:

    (a) $\mathbf{a} = 3$ $\mathbf{b} = 2$ $\alpha = \pi/3$ (b) $\mathbf{a} = 2$ $\mathbf{b} = 5$ $\alpha = 0$

    (c) $\mathbf{a} = 1$ $\mathbf{b} = 4$ $\alpha = \pi/4$ (d) $\mathbf{a} = 2.5$ $\mathbf{b} = 3$ $\alpha = 120°$

2.   Considering the scalar products, what can you say about the angle between the vectors **a** and **b**?

    (a) $\mathbf{a} \cdot \mathbf{b} = 0$ (b) $\mathbf{a} \cdot \mathbf{b} = ab$

    (c) $\mathbf{a} \cdot \mathbf{b} = \dfrac{ab}{2}$ (d) $\mathbf{a} \cdot \mathbf{b} < 0$

3.   Calculate the scalar product of the following vectors:

    (a) $\mathbf{a} = (3, -1, 4)$ (b) $\mathbf{a} = (3/2, 1/4, -1/3)$
         $\mathbf{b} = (-1, 2, 5)$ $\mathbf{b} = (1/6, -2, 3)$

    (c) $\mathbf{a} = (-1/4, 2, -1)$ (d) $\mathbf{a} = (1, -6, 1)$
         $\mathbf{b} = (1, 1/2, 5/3)$ $\mathbf{b} = (-1, -1, -1)$

4. Which of the following vectors **a** and **b** are perpendicular?

(a) $\mathbf{a} = (0, -1, 1)$
   $\mathbf{b} = (1, 0, 0)$

(b) $\mathbf{a} = (2, -3, 1)$
   $\mathbf{b} = (-1, 4, 2)$

(c) $\mathbf{a} = (-1, 2, -5)$
   $\mathbf{b} = (-8, 1, 2)$

(d) $\mathbf{a} = (4, -3, 1)$
   $\mathbf{b} = (-1, -2, -2)$

(e) $\mathbf{a} = (2, 1, 1)$
   $\mathbf{b} = (-1, 3, -2)$

(f) $\mathbf{a} = (4, 2, 2)$
   $\mathbf{b} = (1, -4, 2)$

5. Calculate the angle between the two vectors **a** and **b**:

(a) $\mathbf{a} = (1, -1, 1)$
   $\mathbf{b} = (-1, 1, -1)$

(b) $\mathbf{a} = (-2, 2, -1)$
   $\mathbf{b} = (0, 3, 0)$

6. A force $\mathbf{F} = (0, 5\,\mathrm{N})$ is applied to a body and moves it through a distance s. Calculate the work done by the force.

(a) $\mathbf{s}_1 = (3\,\mathrm{m}, 3\,\mathrm{m})$ (b) $\mathbf{s}_2 = (2\,\mathrm{m}, 1\,\mathrm{m})$ (c) $\mathbf{s}_3 = (2\,\mathrm{m}, 0)$

4.2 Vector Product

7. Indicate in figures 4.22 and 4.23 the direction of the vector **c** if $\mathbf{c} = \mathbf{a} \times \mathbf{b}$
   (a) when **a** and **b** lie in the x–y plane
   (b) when **a** and **b** lie in the y–z plane

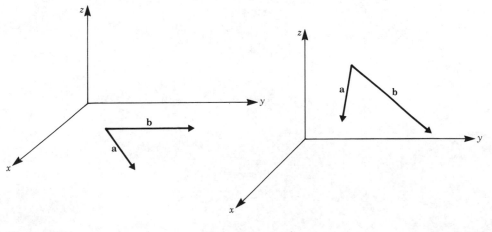

Figure 4.22   Figure 4.23

8. Calculate the magnitude of the vector product of the following vectors:

(a) $a = 2$ $b = 3$ $\alpha = 60°$ (b) $a = 1/2$ $b = 4$ $\alpha = 0°$

(c) $a = 8$ $b = 3/4$ $\alpha = 90°$

9. In figure 4.24 $\mathbf{a} = 2\mathbf{i}$, $\mathbf{b} = 4\mathbf{j}$, $\mathbf{c} = -3\mathbf{k}$ (i, j and k are the unit vectors along the x-, y- and z-axes, respectively). Calculate

(a) $\mathbf{a} \times \mathbf{b}$ (b) $\mathbf{a} \times \mathbf{c}$ (c) $\mathbf{c} \times \mathbf{a}$

(d) $\mathbf{b} \times \mathbf{c}$ (e) $\mathbf{b} \times \mathbf{b}$ (f) $\mathbf{c} \times \mathbf{b}$

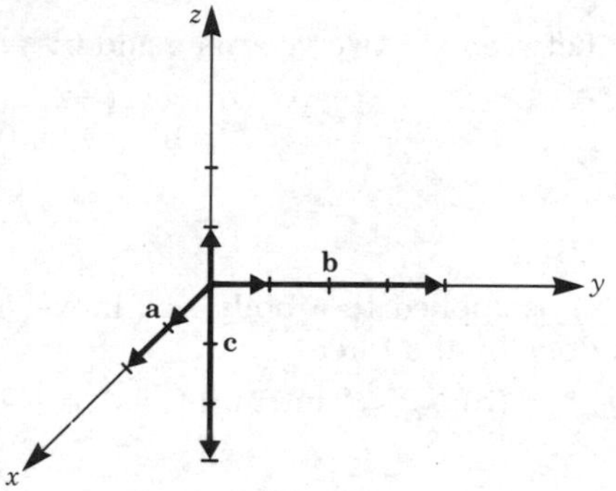

Figure 4.24

10. Calculate $\mathbf{c} = \mathbf{a} \times \mathbf{b}$ when

(a) $\mathbf{a} = (2, 3, 1)$ (b) $\mathbf{a} = (-2, 1, 0)$
    $\mathbf{b} = (-1, 2, 4)$ $\mathbf{b} = (1, 4, 3)$

CHAPTER 5

Differential calculus

5.1 SEQUENCES AND LIMITS

5.1.1 THE CONCEPT OF SEQUENCE

As a preliminary example, consider the fraction $\dfrac{1}{n}$.

By giving n the values of the natural numbers $1, 2, 3, 4, 5 \ldots$ successively, we obtain the following sequence:

$$1, \frac{1}{2}, \frac{1}{3}, \frac{1}{4}, \frac{1}{5}, \ldots$$

These values are illustrated graphically in figure 5.1.

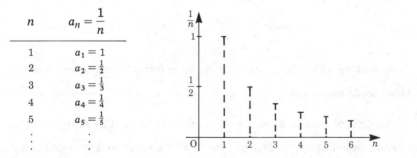

Figure 5.1

In this example, $\dfrac{1}{n}$ defined the form of the sequence and to n we assigned the values of the natural numbers. The functional representation of the terms of a sequence is usually denoted by a_n, so that the *sequence* becomes

$$a_1, a_2, a_3, \ldots, a_n, a_{n+1}, \ldots$$

This can be abbreviated thus: $\quad\quad \{a_n\}$

a_n is the nth term of the sequence, sometimes referred to as the *general term*.

EXAMPLE Let a_n be given by $\quad a_n = \dfrac{1}{n(n+1)}$

It then gives rise to the following sequence:

$$\frac{1}{1 \times 2}, \quad \frac{1}{2 \times 3}, \quad \frac{1}{3 \times 4}, \quad \cdots$$

69

EXAMPLE   Let a_n be given by

$$a_n = (-1)^n n$$

It then gives rise to the following sequence

$$-1, 2, -3, 4, -5, 6, \ldots$$

EXAMPLE   Let a_n be given by $a_n = aq^n$, where a and q are real numbers. By giving n the values $0, 1, 2, 3, \ldots$ the following sequence is obtained: $a, aq, aq^2, aq^3, \ldots$ This type of sequence is called a *geometric progression* (GP), and q is known as the *common ratio*.

A sequence may be finite or infinite. In the case of a finite sequence, the range of n is limited, i.e. the sequence terminates after a certain number of terms. An infinite sequence has an unlimited number of terms.

5.1.2   LIMIT OF A SEQUENCE

Consider the sequence formed by $a_n = \dfrac{1}{n}$. If we let n grow indefinitely then it follows that $\dfrac{1}{n}$ converges to zero or tends to zero. This is expressed in the following way:

$$\frac{1}{n} \to 0 \qquad \text{as} \qquad n \to \infty$$

or

$$\lim_{n \to \infty} \frac{1}{n} = 0$$

Zero denotes the *limiting value* of $\dfrac{1}{n}$ as n tends to infinity. Such a sequence is referred to as a *null sequence*.

The sequence whose general term is $a_n = 1 + \dfrac{1}{n}$, on the other hand, converges to the value 1 as n increases indefinitely. In general, the limiting value of a sequence may be any number g.

DEFINITION   If a sequence whose general term is a_n converges towards a finite value g as $n \to \infty$, then g is called the limit of the sequence. This is written as $\lim\limits_{n \to \infty} a_n = g$.

A precise mathematical definition is as follows:

The sequence formed by the general term a_n is said to converge towards the constant value g if for any preassigned positive number ϵ, however small, it is possible to find a positive integer M such that

$$|a_n - g| < \epsilon \qquad \text{for all} \qquad n > M$$

If a sequence converges towards the value g it is said to be *convergent*. A sequence that does not converge is said to be *divergent*.

The following examples are given without proof to illustrate convergent and divergent sequences.

Convergent Sequences

EXAMPLE   The sequence defined by $a_n = \dfrac{n}{n+1}$ has the limiting value 1 as $n \to \infty$.

This is because the number 1 in the denominator becomes less and less important as n becomes larger and larger. This sequence is illustrated in figure 5.2.

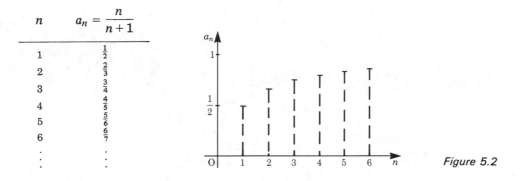

Figure 5.2

EXAMPLE   The sequence defined by $a_n = 2 + (-\tfrac{1}{2})^n$ has the limiting value 2 as $n \to \infty$. This sequence is illustrated in figure 5.3.

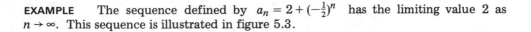

n	$a_n = 2 + (-\tfrac{1}{2})^n$
1	1.5
2	2.25
3	1.875
4	2.0625
⋮	⋮

Figure 5.3

EXAMPLE   A sequence of great significance is defined by $a_n = \left(1 + \dfrac{1}{n}\right)^n$. This sequence has a limiting value which is denoted by the letter e, named after Euler who discovered it.

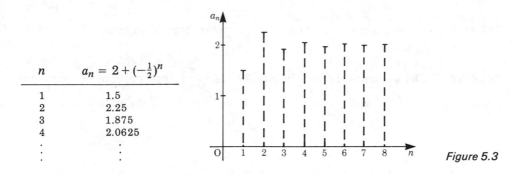

DEFINITION   Euler's number:

$$e = \lim_{n \to \infty} \left(1 + \frac{1}{n}\right) = 2.718\,281\,828\ldots$$

EXAMPLE   The following limit, given here without proof, will be used in section 5.5.3:

$$\lim_{n \to \infty} (e^{\frac{1}{n}} - 1)n = 1 \qquad (5.1)$$

Divergent Sequences

EXAMPLE   The sequence defined by $a_n = n^2$ grows beyond all bounds as $n \to \infty$. This sequence is illustrated in figure 5.4.

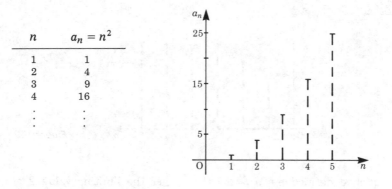

n	$a_n = n^2$
1	1
2	4
3	9
4	16
⋮	⋮

Figure 5.4

EXAMPLE   The sequence defined by $a_n = \dfrac{2^n}{n^2}$ grows beyond all bounds as $n \to \infty$.

EXAMPLE   The sequence defined by $a_n = (-1)^n \dfrac{n}{n+1}$ has no limit. It oscillates between the values $+1$ and -1 as $n \to \infty$. This sequence is illustrated in figure 5.5.

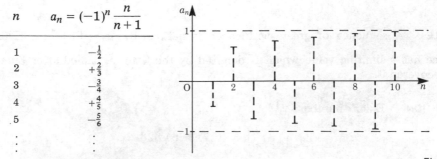

n	$a_n = (-1)^n \dfrac{n}{n+1}$
1	$-\frac{1}{2}$
2	$+\frac{2}{3}$
3	$-\frac{3}{4}$
4	$+\frac{4}{5}$
5	$-\frac{5}{6}$
⋮	⋮

Figure 5.5

5.1.3 LIMIT OF A FUNCTION

The concept of the limit of a sequence can be extended without difficulty to functions. Consider the function $y = f(x)$. The independent variable x can take the values $x_1, x_2, \ldots$ If these values do not exceed the domain of definition of the function $f(x)$, then the corresponding values $y_n = f(x_n)$ form a sequence of values for y.

> **DEFINITION**   Within the domain of definition of the function $y = f(x)$ we take out all possible sequences $\{x_n\}$ which converge towards a determined fixed value x_0. If $\{y_n\} = \{f(x_n)\}$ tends to one single value, g, for all $\{x_n\}$, then we call g the *limit* of the function $f(x)$ as $x \to x_0$.

We say that the function $f(x)$ *converges* and write

$$\lim_{x \to x_0} f(x) = g \qquad \text{if } x \text{ tends to the finite value } x_0$$

$$\lim_{x \to \infty} f(x) = g \qquad \text{if } x \text{ tends to infinity } (\infty)$$

EXAMPLE $y = \dfrac{1}{x}$ for $x \to \infty$

Let us assume that x takes the values $1, 2, 3, \ldots$ successively. We then have a sequence whose general term is $a_n = \dfrac{1}{n}$ and which tends to zero as n increases beyond limit. But we could equally let x run through sequences such as $1, 3, 9, 27, \ldots$ or $\frac{3}{7}, \frac{6}{7}, \frac{9}{7}, \frac{12}{7}, \ldots$, or indeed many other sequences of real numbers. In each case we shall find that y tends to zero, i.e. $y = \dfrac{1}{x}$ has the limit $g = 0$ as $x \to \infty$.

Hence
$$\lim_{x \to \infty} \frac{1}{x} = 0$$

5.1.4 EXAMPLES FOR THE PRACTICAL DETERMINATION OF LIMITS

Up to now we have not given a clear and precise procedure for obtaining limits. In fact, such a procedure is not readily available; but to some extent the successes in obtaining the limits in certain cases give rise to a procedure for achieving our objective in other cases. This is illustrated by the following examples.

EXAMPLE $\lim\limits_{x \to \infty} \dfrac{x^2}{x^2 + x + 1}$

We know that $\lim\limits_{x \to \infty} \dfrac{1}{x} = 0$; hence, if we divide each summand of the fraction by the highest power of x, we get terms in the denominator which vanish if $x \to \infty$.

$$\frac{x^2}{x^2 + x + 1} = \frac{1}{1 + 1/x + 1/x^2} \qquad (\text{if } x \neq 0)$$

As $x \to \infty$, $\dfrac{1}{x} \to 0$ and $\dfrac{1}{x^2} \to 0$ \qquad\qquad Hence $\lim\limits_{x \to \infty} \dfrac{1}{1 + 1/x + 1/x^2} = 1$

EXAMPLE $\lim\limits_{x \to 0} \dfrac{\sin x}{x}$ (x in radians)

This is an important limit as we shall see later when we calculate the differential coefficient of the sine function (section 5.5.3).

Figure 5.6a shows a circular sector OAB of unit radius. If x is the angle between the radii OA and OB, then it follows from the definitions of the trigonometric functions that $BD = \sin x$, $OD = \cos x$ and $AC = \tan x$. Also, x is the length of the arc AB.

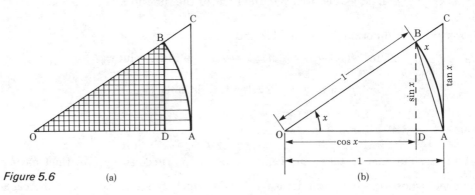

Figure 5.6   (a)   (b)

Now consider the areas of the triangles ODB and OAC and the area of the sector OAB shown in figure 5.6b. We see that the area of the sector OAB is greater than the area of the triangle ODB and smaller than the area of the triangle OAC, i.e.

$$\text{Area } \triangle \text{ODB} < \text{Area sector OAB} < \text{Area } \triangle \text{OAC}$$

$$\frac{\sin x \cos x}{2} < \frac{x}{2} < \frac{\tan x}{2}$$

Dividing by $\dfrac{\sin x}{2}$ gives

$$\cos x < \frac{x}{\sin x} < \frac{1}{\cos x}$$

We take the reciprocal

$$\frac{1}{\cos x} > \frac{\sin x}{x} > \cos x$$

As $x \to 0$, $\cos x \to 1$ and hence $\dfrac{\sin x}{x}$ lies between two expressions which both tend to the limit 1. It follows, therefore, that

$$\lim_{x \to 0} \frac{\sin x}{x} = 1 \tag{5.2}$$

5.2 CONTINUITY

If a function takes a sudden jump at $x = x_0$ it is said to be *discontinuous*; if, on the other hand, no such jump occurs then the function is said to be *continuous*.

This is illustrated in figure 5.7a and b respectively.

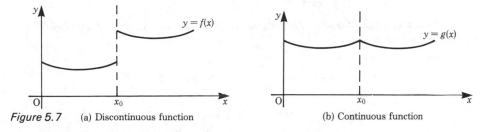

Figure 5.7   (a) Discontinuous function   (b) Continuous function

DEFINITION   The function $y = f(x)$ is continuous at the point $x = x_0$ if the following conditions are satisfied:

$f(x)$ has the same limit g as $x \to x_0$ whether x_0 is approached from the left or from the right on the x-axis.

This limit g agrees with the value $f(x_0)$ at $x = x_0$.

A notation for the limit approached from the right-hand side is

$$\lim_{x \to x_0 + 0} f(x)$$

When the limit is approached from the left-hand side we write

$$\lim_{x \to x_0 - 0} f(x)$$

Hence the function $f(x)$ is continuous at $x = x_0$ if

$$\lim_{x \to x_0 - 0} f(x) = \lim_{x \to x_0 + 0} f(x) = \lim_{x \to x_0} f(x) = f(\lim_{x \to x_0} x) = f(x_0)$$

5.3 SERIES

A *series* is formed by adding the terms of a sequence or progression.

As an example, consider the sequence

$$1, \frac{1}{2}, \frac{1}{3}, \frac{1}{4}, \ldots, \frac{1}{n}, \ldots, \frac{1}{r}$$

By adding the terms we obtain the series

$$1 + \frac{1}{2} + \frac{1}{3} + \frac{1}{4} + \ldots + \frac{1}{n} + \ldots \frac{1}{r}$$

We should note the following:

Sequence: $a_1, a_2, a_3, \ldots, a_n, \ldots, a_r$

Series: $a_1 + a_2 + a_3 + \ldots + a_n + \ldots a_r$

a_1 is referred to as the *leading term*,

a_n the *general term*,

a_r the *last* or *end term*,

n is a *variable number* and assumes all values between 1 and r. Other letters such as i, j, k are often used to denote the variable.

To indicate a series, the Greek letter sigma (Σ) is used to avoid writing down long and complicated expressions.

$$a_1 + a_2 + a_3 + \ldots + a_r = \sum_{n=1}^{r} a_n = S_r$$

S_r denotes the sum of r terms.

The notation $\displaystyle\sum_{n=1}^{r} a_n$ means that n takes on all the values between 1 and r, e.g.

$$\sum_{n=1}^{2} a_n = a_1 + a_2 = S_2$$

$$\sum_{n=1}^{4} a_n = a_1 + a_2 + a_3 + a_4 = S_4$$

The variable is completely determined by its limits (1 to r), and it does not matter which letter is used to represent it.

As an example, consider the series formed by adding the squares of the natural numbers

$$1^2 + 2^2 + 3^2 + 4^2 + \ldots + n^2 + \ldots + r^2.$$

Here the general term is

$$a_n = n^2$$

By using the summation sign, the series is expressed as follows:

$$\sum_{n=1}^{r} n^2 = 1^2 + 2^2 + 3^2 + \ldots + r^2$$

If a series has an infinite number of terms, such that $r \to \infty$, then we write

$$\lim_{r \to \infty} S_r = \lim_{r \to \infty} \sum_{n=1}^{r} a_n$$

This is usually written

$$S = \sum_{n=1}^{\infty} a_n$$

Such a series is referred to as an *infinite series*.

We should note, however, that, strictly speaking, this summation indicates a limiting process. $r \to \infty$ means that we take r as large as we please. S for $r \to \infty$ is a limiting value of $\{S_r\}$, provided it exists.

5.3.1 GEOMETRIC SERIES

The following series is called a *geometric series*, the sum of the geometric progression (GP)

$$a + aq + aq^2 + aq^3 + \ldots + aq^n + \ldots$$

The sum of the first r terms of this series is

$$S_r = \sum_{n=0}^{n=r-1} aq^n$$

To obtain an expression for this sum we multiply the original series by the common ratio q and then subtract the original series from the new one:

$$S_r q = \quad aq + aq^2 + \ldots + aq^{r-1} + aq^r$$
$$S_r = a + aq + aq^2 + \ldots + aq^{r-1}$$

Subtracting gives

$$S_r q - S_r = -a + aq^r$$

or

$$S_r(q-1) = a(q^r - 1)$$

Geometric series: $\quad S_r = a\dfrac{q^r - 1}{q - 1} = a\dfrac{1 - q^r}{1 - q} \quad$ for $\quad q \neq 1 \qquad$ (5.3)

This is the sum of the first r terms of the GP.

To obtain the sum for an infinite number of terms we need to find the limit of S_r as $r \to \infty$. We have to distinguish between the following two cases:

Case 1: $|q| < 1$

Here $\lim\limits_{r \to \infty} q^r = 0$

Hence $S = \lim\limits_{r \to \infty} S_r = \lim\limits_{r \to \infty} a \dfrac{1 - q^r}{1 - q} = \dfrac{a}{1 - q}$

Case 2: $|q| > 1$

In this case q^r grows beyond all bounds as $r \to \infty$ and the geometric series has no finite limit.

5.4 DIFFERENTIATION OF A FUNCTION

5.4.1   GRADIENT OR SLOPE OF A LINE

DEFINITION   The *gradient of a line* is the ratio of the rise Δy to the base line Δx from which this rise is achieved.

The symbol Δ is the Greek letter 'delta' and is used here to mean the 'difference between'. Hence Δx does not mean Δ multiplied by x but the difference between two values of x such as x_1 and x_2, i.e. $\Delta x = x_2 - x_1$.

The gradient or slope is also given by the tangent of the angle of elevation α, as shown in figure 5.8.

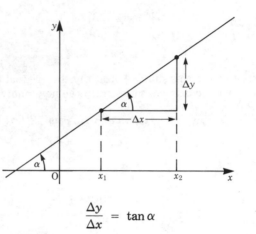

Figure 5.8

$$\frac{\Delta y}{\Delta x} = \tan \alpha$$

5.4.2   GRADIENT OF AN ARBITRARY CURVE

The gradient of an arbitrary curve, unlike the gradient of a line, varies from point to point, as can be seen by examining the curve shown in figure 5.9. If we could find the gradient at some fixed point P on the curve, then the line through P with the same slope is called *the tangent to the curve at P*, and we can use as synonyms the expressions 'gradient of the curve' and 'slope or gradient of the tangent'.

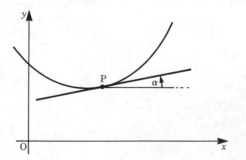

Figure 5.9

The problem now is to find an expression for the gradient of any curve at a given point P.

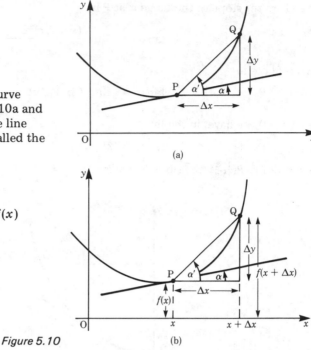

Consider a point P on the curve $y = f(x)$ shown in figure 5.10a and a neighbouring point Q. The line drawn through P and Q is called the *secant*, whose slope is

$$\tan \alpha' = \frac{\Delta y}{\Delta x}$$

We have $\Delta y = f(x + \Delta x) - f(x)$ (see figure 5.10b).

Figure 5.10

With P fixed for the moment, let the point Q move towards P. It follows that in the limit, when Q coincides with P, the angle α' is equal to the angle α, the slope of the tangent to the curve at P. As Q gets nearer to P we notice that Δx tends to zero and, as a consequence, Δy tends to zero also; but the ratio $\dfrac{\Delta y}{\Delta x}$ tends to a definite limit since the secant PQ becomes the tangent to the curve at P. Hence

$$\tan \alpha = \lim_{\alpha' \to \alpha} \tan \alpha' = \lim_{\Delta x \to 0} \frac{\Delta y}{\Delta x} = \lim_{\Delta x \to 0} \frac{f(x + \Delta x) - f(x)}{\Delta x}$$

This is the slope of the tangent at P.

DEFINITION   The fraction $\dfrac{f(x + \Delta x) - f(x)}{\Delta x} = \dfrac{\Delta y}{\Delta x}$ is called

the *difference quotient*.   (5.4)

EXAMPLE   Calculate the slope of the parabola $y = x^2$ at the point $P = (\frac{1}{2}, \frac{1}{4})$.

From figure 5.11, the slope of the secant PQ, where Q is any other point, is

$$\tan \alpha' = \frac{f(x + \Delta x) - f(x)}{\Delta x}$$

We wish to obtain the slope of the tangent at P. We know that

$$f(x + \Delta x) = (x + \Delta x)^2$$

Therefore the slope of the tangent at P is

$$\tan \alpha = \lim_{\Delta x \to 0} \frac{(x + \Delta x)^2 - x^2}{\Delta x}$$

This reduces to

$$\tan \alpha = \lim_{\Delta x \to 0} (2x + \Delta x)$$

As $\Delta x \to 0$ we have, in the limit,

$$\tan \alpha = 2x$$

At the point $P(\frac{1}{2}, \frac{1}{4})$ the slope is $\tan \alpha = 2 \times \frac{1}{2} = 1$, giving $\alpha = 45°$.

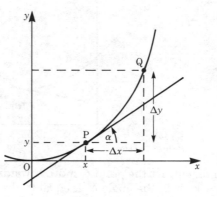

Figure 5.11

It is not always true that the difference quotient has a limit as $\Delta x \to 0$ because not every curve $f(x)$ has a well-defined slope at a particular point. For example, consider the point P on the curve shown in figure 5.12.

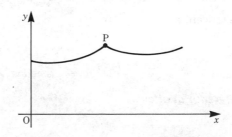

Figure 5.12

5.4.3   DERIVATIVE OF A FUNCTION

Moving from the geometrical concept above to the general case, we consider the difference quotient of a function $f(x)$, namely

$$\frac{\Delta y}{\Delta x} = \frac{f(x + \Delta x) - f(x)}{\Delta x}$$

DEFINITION   If the difference quotient $\Delta y/\Delta x$ has a limit as $\Delta x \to 0$, this limit is called the *derivative* or *differential coefficient* of the function $y = f(x)$ with respect to x and we write

$$\frac{dy}{dx} = \lim_{\Delta x \to 0} \frac{\Delta y}{\Delta x} \qquad (5.5)$$

This differential coefficient is denoted by y', $f'(x)$ or $\dfrac{dy}{dx}$. It must be clearly understood that the d does not multiply y or x but is the symbol for the differential of y or x; $\dfrac{dy}{dx}$ is read as 'dy by dx'.

Using the above notations, we have

$$y' = f'(x) = \frac{dy}{dx} = \frac{d}{dx} f(x) = \lim_{\Delta x \to 0} \frac{\Delta y}{\Delta x} = \lim_{\Delta x \to 0} \frac{f(x + \Delta x) - f(x)}{\Delta x}$$

We have thus defined analytically the *first* derivative by a limiting process which we can also interpret geometrically as the slope of the tangent to the curve at a point x, as shown in figure 5.13.

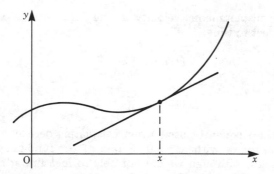

Figure 5.13

By this process we have obtained substantially more than we had postulated; instead of obtaining the slope at some fixed point P, we have, in fact, obtained the slope as a function of the independent variable x.

The importance of the differential calculus lies in the fact that it describes relationships between variable entities. The differential coefficient y' gives the rate of change of y with respect to x. In the next section we will look at an example taken from the physics of motion.

5.4.4 PHYSICAL APPLICATION: VELOCITY

The vehicle shown in figure 5.14 is observed to cover a distance Δx in a time Δt, i.e. we start the clock at some time t and stop it at a time $t + \Delta t$. The magnitude of the average velocity, the rate of change of displacement with time, is given by

$$v_0 = \frac{\Delta x}{\Delta t}$$

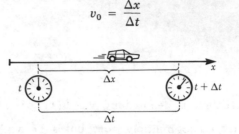

Figure 5.14

This expression gives us an average value only: it does not tell us how fast the vehicle is moving at a particular instant in time, i.e. we do not know its *instantaneous velocity* $v(t)$.

The smaller we take Δt, and hence Δx, the closer we get to the value of the instantaneous velocity at a particular time. Figure 5.15 shows the vehicle travelling a shorter distance Δx which it will cover in a shorter interval of time Δt.

Figure 5.15

We now define the instantaneous velocity as the first derivate of the position co-ordinate x with respect to time:

$$v(t) = \lim_{\Delta t \to 0} \frac{\Delta x}{\Delta t}$$

or

$$v(t) = \frac{\mathrm{d}x}{\mathrm{d}t} = \dot{x}$$

The 'dot' notation is frequently used when calculating derivatives with respect to time. This limiting process with $\Delta t \to 0$ is one of the fundamental mathematical abstractions of physics. Although we are not able to measure arbitrary small times, we are nevertheless justified in taking limits with respect to time because we can draw conclusions which can be verified experimentally. The expression for the velocity at any instant of time t has been written $v(t)$ to imply that the velocity itself may vary with time. This variation, i.e. the rate of change of velocity with respect to time, is referred to as *acceleration*. This is a quantity which has a relationship with another measurable physical quantity — the force as shown in treatises on mechanics.

Hence the acceleration $a(t)$ is defined by

$$a(t) = \lim_{\Delta t \to 0} \frac{\Delta v}{\Delta t} = \frac{\mathrm{d}v}{\mathrm{d}t} = \dot{v}$$

5.4.5 **THE DIFFERENTIAL**

We have defined the derivative or differential coefficient as

$$\frac{dy}{dx} = \lim_{\Delta x \to 0} \frac{\Delta y}{\Delta x}$$

where $\dfrac{dy}{dx}$ was not to be regarded as dy divided by dx but as the limit of the quotient $\Delta y / \Delta x$ as $\Delta x \to 0$. There are, however, situations where it is important to give separate meanings to dx and dy.

Let us arbitrarily assume that dx is a finite quantity! dx is called the *differential of x*. Consider two points P and Q on the curve $y = f(x)$ shown in figure 5.16.

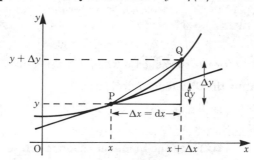

Figure 5.16

In going from P to Q along the curve, y changes by an amount Δy given by

$$\Delta y = f(x + \Delta x) - f(x)$$

The tangent to the curve at the point x changes by an amount

$$dy = f'(x)\,dx$$

during the same interval $\Delta x = dx$. dy is called the differential of the function $y = f(x)$.

We see quite clearly that, in general, the differential of the function is not equal to the functional change in y, i.e.

$$dy \neq \Delta y$$

Thus the differential dy is an approximation for the change Δy: the smaller the interval Δx, the better the approximation. Hence, as soon as we are able to calculate the derivative y' of a function we are also able to calculate its differential.

The differential is used extensively as a *first approximation* for the change in the function. Geometrically, it means that the function is replaced by its tangent at a particular point.

Notation

$$x = \text{independent variable}$$
$$y = \text{dependent variable}$$
$$dx = \textit{differential of the independent variable } x$$
$$dy = \textit{differential of the dependent variable } y,$$
$$\text{i.e. } dy = f'(x)\,dx$$

dy is often replaced by df

5.5 CALCULATING DIFFERENTIAL COEFFICIENTS

We first demonstrate the calculation of differential coefficients for power functions. The calculation of the difference quotient and the limiting process is easy.

In the following sections, we will derive some general rules for calculating differential coefficients. Using these rules, we will be able to treat the functions most often used in practical applications.

5.5.1 DERIVATIVES OF POWER FUNCTIONS; CONSTANT FACTORS

First we state the formula for obtaining the derivatives of power functions.

If $\qquad y = f(x) = x^n \qquad$ where n is any rational number

then $\qquad y' = nx^{n-1}$

(5.6)

The proof is given only for the special case when n is a positive integer.

We start by investigating the difference quotient:

$$\frac{\Delta y}{\Delta x} = \frac{(x + \Delta x)^n - x^n}{\Delta x}$$

Expanding the term $(x + \Delta x)^n$ by the binomial theorem (cf. section 2.1.3) gives

$$\frac{\Delta y}{\Delta x} = \frac{x^n + nx^{n-1}\Delta x + \ldots + (\Delta x)^n - x^n}{\Delta x}$$

$$= \frac{nx^{n-1}\Delta x + \ldots + (\Delta x)^n}{\Delta x}$$

Factorising Δx gives

$$\frac{\Delta y}{\Delta x} = nx^{n-1} + \frac{n(n-1)}{2}x^{n-2}\Delta x + \ldots + (\Delta x)^{n-1}$$

We proceed with the limiting process. Making $\Delta x \to 0$ results in all terms vanishing except the first one:

$$y' = nx^{n-1}$$

EXAMPLE If $y = x^3$, then $n = 3$.

Applying the above rule gives

$$y' = 3x^{3-1} = 3x^2$$

It can be shown that if n is a negative integer, $n = -\alpha$, then $y' = -\alpha x^{-(\alpha-1)}$

It can also be shown that if $y = x^{p/q}$, where p and q are both integers, then

$$y' = \frac{p}{q}x^{(p/q)-1}$$

Hence the rule applies whether n is positive, negative or a fraction.

EXAMPLE If $y = \dfrac{1}{\sqrt{x}} = x^{-1/2}$, i.e. $n = -\frac{1}{2}$, then $y' = -\frac{1}{2}x^{-1/2-1} = -\frac{1}{2}x^{-3/2}$

The derivative of a *constant* vanishes:

$$y(x) \;=\; c, \quad c \;=\; \text{constant}$$
$$y'(x) \;=\; 0$$

The graph of this function is shown in
figure 5.17. It is parallel to the x-axis.
The slope is zero. This obvious result
is also obtained by systematic
calculation:

$$y' \;=\; \lim_{\Delta x \to 0} \frac{f(x+\Delta x)-f(x)}{\Delta x}$$

$$=\; \lim_{\Delta x \to 0} \frac{c-c}{\Delta x} \;=\; 0$$

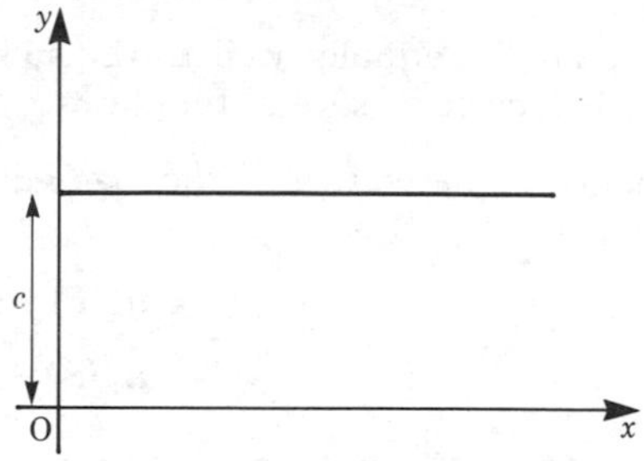

Figure 5.17

5.5.2 RULES FOR DIFFERENTIATION

Constant Factor

A constant factor is preserved during differentiation:

$$y \;=\; cf(x) \qquad \text{where } c \text{ is a constant}$$
$$y' \;=\; cf'(x) \tag{5.7}$$

PROOF   We can take out the constant c and place it in front of the limit sign, since
it is not affected by the limiting process

$$y' \;=\; \lim_{\Delta x \to 0} \frac{cf(x+\Delta x)-cf(x)}{\Delta x}$$

$$=\; c \lim_{\Delta x \to 0} \frac{f(x+\Delta x)-f(x)}{\Delta x}$$

Hence $y' = cf'(x).$

Differentiation of a Sum:Sum Rule

The derivative of the sum of several functions is the sum of the individual derivatives:

$$y \;=\; u(x)+v(x)$$
$$y' \;=\; u'(x)+v'(x) \tag{5.8}$$

PROOF   We separate the limit into a sum of limits.

By definition, $y' \;=\; \lim_{\Delta x \to 0} \dfrac{u(x+\Delta x)+v(x+\Delta x)-u(x)-v(x)}{\Delta x}$

Provided that the limit of each function exists we can separate the two functions, so that

$$y' = \lim_{\Delta x \to 0} \frac{u(x + \Delta x) - u(x)}{\Delta x} + \lim_{\Delta x \to 0} \frac{v(x + \Delta x) - v(x)}{\Delta x}$$

Hence
$$y' = u'(x) + v'(x).$$

The rule applies equally well to the sum or difference of two functions and to the sum or difference of several functions.

Generally, the derivative of the algebraic sum of n functions is the algebraic sum of their derivatives:

If
$$y = u_1(x) + u_2(x) + \ldots + u_n(x)$$
then
$$y' = u_1'(x) + u_2'(x) + \ldots + u_n'(x)$$

Product of Two Functions: Product Rule

If $u(x)$ and $v(x)$ are two functions, the derivative of the product is given by the following expression:

$$y = u(x) v(x)$$
$$y' = u'(x)\ v(x) + u(x)\ v'(x) \tag{5.9}$$

PROOF   By definition,

$$y' = \lim_{\Delta x \to 0} \frac{u(x + \Delta x)\ v(x + \Delta x) - u(x)\ v(x)}{\Delta x}$$

Adding and subtracting $u(x)\ v(x + \Delta x)$ to the numerator gives

$$y' = \lim_{\Delta x \to 0} \frac{u(x + \Delta x)v(x + \Delta x) - u(x)v(x) + u(x)v(x + \Delta x) - u(x)v(x + \Delta x)}{\Delta x}$$

Collecting terms in such a way that difference quotients are formed gives

$$y' = \lim_{\Delta x \to 0} \frac{u(x + \Delta x) - u(x)}{\Delta x} v(x + \Delta x) + \lim_{\Delta x \to 0} \frac{v(x + \Delta x) - v(x)}{\Delta x} u(x)$$

Hence $y' = u'(x) v(x) + v'(x) u(x)$

Quotient of Two Functions: Quotient Rule

If $u(x)$ and $v(x)$ are two functions, the derivative of the quotient is given by the following expression:

$$y = \frac{u(x)}{v(x)}$$

$$y' = \frac{u'(x)\ v(x) - u(x)\ v'(x)}{[v(x)]^2} \tag{5.10}$$

The proof follows the pattern given above and is omitted here.

Derivative of a Function of a Function: Chain Rule

If $g(x)$ is a function of x and $f(g)$ is a function of g, let $y = f(g)$, i.e.

$$y = f(g(x))$$

y is said to be a function of a function. Its derivative is obtained by differentiating the outer function f with respect to g $\left(\text{written } \dfrac{df}{dg}\right)$ and the inner function with respect to x and multiplying the two derivatives.

$$y = f(g(x))$$

$$y' = \frac{df}{dg}\, g'(x)$$

(5.11)

PROOF   By definition,

$$\frac{dy}{dg} = \frac{df}{dg} = \lim_{\Delta g \to 0} \frac{f(g + \Delta g) - f(g)}{\Delta g} \qquad \text{and} \qquad \frac{dg}{dx} = \lim_{\Delta x \to 0} \frac{g(x + \Delta x) - g(x)}{\Delta x}$$

$$\frac{dy}{dg} = \lim_{\Delta g \to 0} \frac{\Delta f}{\Delta g} \qquad \text{and} \qquad \frac{dg}{dx} = \lim_{\Delta x \to 0} \frac{\Delta g}{\Delta x}$$

Before proceeding to the limit, consider the product $\dfrac{\Delta f}{\Delta g}\, \dfrac{\Delta g}{\Delta x}$

This is equal to $\dfrac{\Delta y}{\Delta x}$

Thus $\qquad \dfrac{dy}{dx} = \lim_{\Delta x \to 0} \dfrac{\Delta y}{\Delta x} = \lim_{\Delta g \to 0} \dfrac{\Delta f}{\Delta g} \lim_{\Delta x \to 0} \dfrac{\Delta g}{\Delta x}$

giving $\qquad \dfrac{dy}{dx} = \dfrac{df}{dg}\, \dfrac{dg}{dx}$

Hence $\qquad y' = \dfrac{df}{dg}\, g'(x)$

EXAMPLE $\qquad\qquad y = (1 + x^2)^3$

$$g(x) = 1 + x^2 \qquad \text{(inner function)}$$

$$f(g) = g^3 \qquad \text{(outer function)}$$

$$\frac{df}{dg} = 3g^2 \qquad \text{and} \qquad g'(x) = 2x$$

$$y' = 3g^2 \times 2x = 6(1 + x^2)^2 x$$

Derivative of the Inverse Function

If the function $f(x)$ is differentiable in a given interval where $f'(x) \neq 0$, then the inverse function $f^{-1}(x)$ possesses a derivative at all points in the corresponding interval. The following relationship holds true:

$$\frac{d}{dx} f^{-1}(x_0) = [f^{-1}(x_0)]' = \frac{1}{f'(y_0)}$$

To demonstrate this, consider the graphs of $f^{-1}(x)$ and $f(x)$, as shown in figure 5.18. It will be remembered from the geometrical correlation discussed in Chapter 1, section 1.4 that these graphs are symmetrical about the bisection line.

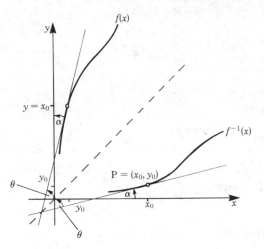

Figure 5.18

At a point $P = (x_0, y_0)$ the slope of the curve is the derivative of the function $f^{-1}(x_0)$. We will now determine the slope.

$$\tan \alpha = [f^{-1}(x_0)]'$$

We denote by α the angle made by the tangent to the curve at P and the x-axis. The slope of the tangent measured with respect to the y-axis is denoted by θ. Furthermore, the slope of the tangent to $f(y_0)$, at the position symmetrical to P, is also θ. (Note $y_0 = f^{-1}(x_0)$.)

$$\tan \theta = f'(y_0)$$

Since $\alpha + \theta = 90°$, it follows that $\tan \alpha = \cot \theta = \dfrac{1}{\tan \theta}$.

Hence
$$[f^{-1}(x)]' = \frac{1}{f'(y)} = \frac{1}{f'[f^{-1}(x)]}$$

This formula can be rewritten as follows:

$f'(y)$ is the derivative of f at the point y; remember that $y = f^{-1}(x)$ means $x = f(y)$. Thus $\dfrac{dx}{dy} = f'(y)$. If we insert $\dfrac{dx}{dy}$ in the formula we obtain

$$\frac{dy}{dx} = \frac{1}{dx/dy}$$

Derivative of the inverse function $y = f^{-1}(x)$:

$$\frac{d}{dx} f^{-1}(x) = \frac{dy}{dx} = \frac{1}{f'(f^{-1}(x))}$$

$$\frac{dy}{dx} = \frac{1}{dx/dy}$$

(5.12)

5.5.3 DIFFERENTIATION OF FUNDAMENTAL FUNCTIONS

We now evaluate the differential coefficients for functions which are frequently used. Fortunately, we do not, in each individual case, have to carry out the limiting process for the function $f(x)$ under consideration:

$$f'(x) = \lim_{\Delta x \to 0} \frac{f(x + \Delta x) - f(x)}{\Delta x}$$

The basic difficulty in obtaining this limit is that the numerator and the denominator of the difference quotient both become zero as $\Delta x \to 0$, giving the expression 0/0. To overcome this difficulty we try to transform the difference quotient in such a way that the denominator does not become zero during the limiting process; this can only be achieved with some fundamental functions like power functions. In some cases (for example with sine functions and exponential functions) we are forced to carry out the limiting process. But in most other cases we may reduce the differential coefficient to the known differential coefficients of other functions, using the differentiation rules derived in the previous section. The following brief proofs for a number of fundamental functions illustrate this.

Trigonometric Functions

$$
\begin{aligned}
y &= \sin x & y' &= \cos x \\
y &= \cos x & y' &= -\sin x \\
y &= \tan x & y' &= \frac{1}{\cos^2 x} = 1 + \tan^2 x \\
y &= \cot x & y' &= \frac{-1}{\sin^2 x} = -1 - \cot^2 x
\end{aligned}
$$

(5.13)

In the case of the last two derivatives we must exclude the values of x for which the denominator becomes zero.

PROOF   Sine function

We start with the difference quotient

$$\frac{\Delta y}{\Delta x} = \frac{\sin(x + \Delta x) - \sin x}{\Delta x}$$

$$\frac{\Delta y}{\Delta x} = \frac{2 \sin (\Delta x/2) \cos (x + \Delta x/2)^\dagger}{\Delta x}$$

$$= \frac{\sin (\Delta x/2)}{(\Delta x/2)} \cos (x + \Delta x/2)$$

We saw in section 5.1.4 (equation 5.2) that

$$\lim_{\Delta x \to 0} \frac{\sin \Delta x}{\Delta x} = 1$$

Hence
$$\frac{dy}{dx} = \lim_{\Delta x \to 0} \frac{\sin (\Delta x/2)}{\Delta x/2} \cos \left(x + \frac{\Delta x}{2} \right) = \cos x$$

PROOF   Cosine function

$$y = \cos x = \sin \left(x + \frac{\pi}{2} \right)$$

We apply the chain rule for a function of a function with

$$g(x) = x + \frac{\pi}{2} \qquad f(g) = \sin g$$

Differentiating gives

$$y' = \cos g \, g' = \cos \left(x + \frac{\pi}{2} \right)$$

$$= -\sin x$$

The derivatives of the tan and cot functions can be obtained by applying the quotient rule:

$$y = \tan x = \frac{\sin x}{\cos x}, \qquad y' = \frac{\cos^2 x - (-\sin^2 x)}{\cos^2 x} = \frac{1}{\cos^2 x}$$

and
$$y = \cot x = \frac{\cos x}{\sin x}, \qquad y' = \frac{-\sin^2 x - \cos^2 x}{\sin^2 x} = \frac{-1}{\sin^2 x}$$

---

†To obtain this transformation, we use the relationships from Chapter 1, page 25:

$$\sin \alpha - \sin \beta = 2 \left(\sin \frac{\alpha - \beta}{2} \cos \frac{\alpha + \beta}{2} \right)$$

In our case $\alpha = x + \Delta x$ and $\beta = x$

EXAMPLE   The vibration equation. A sine function with an arbitrary period has the form

$$y = \sin ax \qquad\qquad \text{The period is } \frac{2\pi}{a}$$

This can be treated as a function of a function with $g(x) = ax$ and $f(g) = \sin(g)$.

The derivative is obtained by means of the chain rule. First we differentiate f with respect to g and then g with respect to x.

$$f(g) = \sin(g) \qquad \frac{df}{dg} = \cos(g)$$

$$g(x) = ax \qquad \frac{dg}{dx} = g'(x) = a$$

Hence, by the chain rule,

$$y' = a \cos ax$$

In science and engineering, we often have to deal with quantities which depend on time. Mechanical and electrical vibrations are typical examples.

A vibration with an amplitude A and a frequency ω (also referred to as circular frequency) is described by the equation

$$x = A \sin(\omega t)$$

(When there is no possible confusion, the bracket around ωt may be omitted, so that $x = A \sin \omega t$.)

To obtain the velocity of the vibration we have to differentiate this equation with respect to the time t:

$$v(t) = \frac{dx}{dt} = \dot{x}$$

Remember that the 'dot' above the x indicates differentiation with respect to time t. Hence

$$v(t) = \dot{x} = \omega A \cos(\omega t)$$

since A is a constant factor which remains unchanged during differentiation. The rest of the equation is identical to the equation $y = \sin ax$ where a replaces ω, x replaces t, y replaces x.

Inverse Trigonometric Functions

$$y = \sin^{-1} x \qquad y' = \frac{1}{\sqrt{1-x^2}}$$

$$y = \cos^{-1} x \qquad y' = \frac{-1}{\sqrt{1-x^2}}$$

$$y = \tan^{-1} x \qquad y' = \frac{1}{1+x^2} \qquad\qquad (5.14)$$

$$y = \cot^{-1} x \qquad y' = \frac{-1}{1+x^2}$$

To prove the derivatives of the inverse trigonometric functions we use the general equation derived in section 5.5.2, i.e.

$$\frac{dy}{dx} = \frac{1}{dx/dy}$$

PROOF   Derivative of the inverse sine function

$$y = \sin^{-1} x$$

$$x = \sin y$$

We differentiate with respect to y, obtaining

$$\frac{dx}{dy} = \cos y = \sqrt{1 - \sin^2 y} = \sqrt{1 - x^2}$$

Since

$$\frac{dy}{dx} = \frac{1}{dx/dy}$$

it follows that

$$y' = \frac{1}{\sqrt{1 - x^2}}$$

Other proofs follow the same pattern.

Exponential and Logarithmic Functions

$$y = e^x \qquad y' = e^x$$
$$y = \ln x \qquad y' = \frac{1}{x}$$

(5.15)

PROOF   Exponential function

$$\frac{\Delta y}{\Delta x} = \frac{e^{(x + \Delta x)} - e^x}{\Delta x} = \frac{e^x (e^{\Delta x} - 1)}{\Delta x}$$

According to equation 5.1, $\lim\limits_{n \to \infty} (e^{1/n} - 1)n = 1$.

This limit remains valid if we substitute any suitable sequence of numbers for n. If we substitute $\dfrac{1}{n} = \Delta x$ then, as $n \to \infty$, $\Delta x \to 0$. Hence

$$\lim_{\Delta x \to 0} \frac{e^{\Delta x} - 1}{\Delta x} = 1$$

Consequently

$$\frac{dy}{dx} = y' = e^x$$

PROOF   Logarithmic function

$$y = \ln x \qquad \text{(log to the base e)}$$

This function is equivalent to

$$e^y = e^{\ln x} = x$$

We now obtain the derivative of x with respect to y:

$$\frac{dx}{dy} = e^y$$

Remembering the equation 5.12

$$\frac{dy}{dx} = \frac{1}{dx/dy}$$

we find

$$y' = \frac{dy}{dx} = \frac{1}{e^y} = \frac{1}{x}$$

Comments on the Importance of the Exponential Function

We notice the exponential function, $y = e^x$, remains unchanged when differentiated, i.e. $y' = y$. According to our geometrical interpretation of the derivative (see section 5.4.3), y' indicates how y changes with x. Therefore this function will play an important role in all fields where the rate of change of a function is closely related to the function itself. This is, for example, the case with natural growth and decay processes.

The equation $y' = y$ is, by the way, the first 'differential equation' encountered in this book. It is called a differential equation because it involves not only y but also the derivative of y. We note that the function $y = e^x$ satisfies this differential equation; it is said to be a solution of $y' = y$. We shall use this also when we consider the solution of other differential equations (Chapter 10).

Hyperbolic Functions

The derivatives of hyperbolic functions and their inverses have a special significance in the evaluation of certain integrals.

$$
\begin{aligned}
y &= \sinh x & y' &= \cosh x \\
y &= \cosh x & y' &= \sinh x \\
y &= \tanh x & y' &= \frac{1}{\cosh^2 x} = 1 - \tanh^2 x \\
y &= \coth x & y' &= -\frac{1}{\sinh^2 x} = 1 - \coth^2 x
\end{aligned}
\qquad (5.16)
$$

The proofs of the derivatives are quite straightforward. We shall concentrate on the derivative of $\sinh x$.

The derivative of the hyperbolic cosine function can be obtained in a similar manner. The derivatives of the hyperbolic tangent and cotangent can be obtained using the quotient rule.

PROOF   Derivative of the hyperbolic sine

$$y = \sinh x = \tfrac{1}{2}(e^x - e^{-x})$$

We know the derivatives of the exponential functions,

$$(e^x)' = e^x, \qquad (e^{-x})' = -e^{-x}$$

Hence

$$y' = \tfrac{1}{2}(e^x + e^{-x}) = \cosh x$$

Inverse Hyperbolic Functions

$$
\begin{aligned}
y &= \sinh^{-1}x & y' &= \frac{1}{\sqrt{1+x^2}} \\[2mm]
y &= \cosh^{-1}x & y' &= \frac{1}{\sqrt{x^2-1}} & (x>1) \\[2mm]
y &= \tanh^{-1}x & y' &= \frac{1}{1-x^2} & (|x|<1) \\[2mm]
y &= \coth^{-1}x & y' &= -\frac{1}{x^2-1} & (|x|>1)
\end{aligned}
\qquad (5.17)
$$

The derivatives of $\tanh^{-1}x$ and $\coth^{-1}x$ look identical. But they do differ in their domain.

PROOF   Derivative of the inverse hyperbolic sine function

$$y = \sinh^{-1}x$$

We use the rule for inverse functions:

$$\frac{dy}{dx} = \frac{1}{dx/dy}$$

If

$$y = \sinh^{-1}x$$

then

$$x = \sinh y$$

Thus

$$\frac{dx}{dy} = \cosh y = \sqrt{1 + \sinh^2 y} = \sqrt{1+x^2}$$

It follows that

$$y' = \frac{1}{\sqrt{1+x^2}}$$

5.6 HIGHER DERIVATIVES

The differential coefficient of the function $y = f(x)$ not only gives the slope of the function at a particular point but also gives the slope at every other point within the range for which the function $f(x)$ is defined and for which the derivative exists. The differential coefficient is itself a function of x.

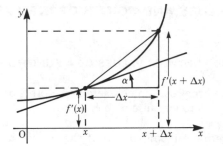

Figure 5.19

This, therefore, suggests that we can differentiate the derivative $f'(x)$ once more with respect to x. In this way the *second derivative* of $y = f(x)$ with respect to x is obtained (figure 5.19).

DEFINITION   The limiting value

$$\lim_{x \to 0} \frac{f'(x + \Delta x) - f'(x)}{\Delta x} = f''(x) = y''(x) \qquad (5.18)$$

is called the *second derivative* of $y = f(x)$ with respect to x.

It is denoted by $f''(x)$, $y''(x)$, $\dfrac{\mathrm{d}}{\mathrm{d}x}\dfrac{\mathrm{d}y}{\mathrm{d}x}$, $\dfrac{\mathrm{d}^2 y}{\mathrm{d}x^2}$, or $\dfrac{\mathrm{d}^2}{\mathrm{d}x^2} f(x)$.

($\dfrac{\mathrm{d}^2 y}{\mathrm{d}x^2}$ is read as 'd-two y by dx squared'!)

This second derivative will, in general, be a function of x and we can obtain the third derivative of $f'''(x) = \dfrac{\mathrm{d}^3 y}{\mathrm{d}x^3}$.

Hence, by repeated differentiation, we can obtain the 4th, 5th, ... nth *derivative*:

$$y^{(n)} = \frac{\mathrm{d}^{(n)} y}{\mathrm{d}x^n} = \frac{\mathrm{d}^n}{\mathrm{d}x^n} f(x) = f^{(n)}(x)$$

In the same way as the first derivative gave us information about the slope of a function $f(x)$, the second derivative gives us information about the slope of the function $f'(x)$, the third derivative about $f''(x)$, and so on.

EXAMPLE   If $y = x$, then $y' = 1$ and $y'' = 0$. All the higher derivatives will be zero.

EXAMPLE   Consider the equation for SHM (simple harmonic motion):

$$x = A \sin \omega t$$

$$\dot{x} = A \omega \cos \omega t \qquad \text{(velocity)}$$

$$\ddot{x} = \frac{\mathrm{d}^2 x}{\mathrm{d}t^2} = -A \omega^2 \sin \omega t \qquad \text{(acceleration)}$$

We note in passing that, since $x = A \sin \omega t$, $\ddot{x} = -\omega^2 x$, i.e. the acceleration is proportional to the displacement. (This is another example of a differential equation.)

5.7 EXTREME VALUES AND POINTS OF INFLEXION; CURVE SKETCHING

5.7.1 MAXIMUM AND MINIMUM VALUES OF A FUNCTION

In Chapter 1 we showed that certain characteristic points of a function (zeros, poles and asymptotes) helped us to visualise its behaviour.

We are now able to refine our knowledge of the behaviour of a function with the help of the first and second derivatives and to find points where the function has extreme values (referred to as local maxima and minima).

In what follows we will assume that the function possesses a second derivative.

> **DEFINITION**   A function $f(x)$ possesses a *local maximum* at a point x_0 if all the values of the function in the immediate neighbourhood of the point x_0 are less than $f(x_0)$ (see figure 5.20a).
>
> A function $f(x)$ possesses a *local minimum* at a point x_0 if all the values of the function in the immediate neighbourhood of the point x_0 are greater than $f(x_0)$ (see figure 5.20b).

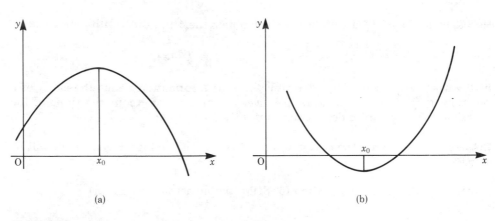

(a)   (b)

Figure 5.20

A *necessary* condition for a function $f(x)$ to have a maximum or a minimum at a point $x = x_0$ is that its first derivative $f'(x_0)$ should be zero. Conversely, is it possible to conclude that, if $f'(x_0) = 0$, the function has a minimum or a maximum value?

The answer to this question is 'No', as can be seen from figure 5.21. At $x = x_1$ the slope $f'(x_1) = 0$, but to the right of x_1 the value of the function is greater than at x_1 and to the left of x_1 it is less than x_1. Hence we have neither a maximum nor a minimum. Such a point is referred to as a *point of inflexion* with a horizontal tangent.

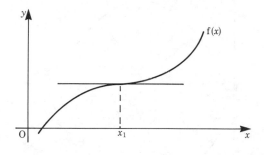

Figure 5.21

At the point of inflexion shown, the curvature changes. Figure 5.21 shows a point of inflexion where the tangent to the curve at x_1 is horizontal. Such a point is also called a *saddle point*. An examination of the derived curve y' (figure 5.22) shows that the derivative decreases left of the point of inflexion and increases right of that point. At the point of inflexion it happens to be zero (hence the horizontal tangent) and the curve of y' goes through a minimum at $x = x_1$; thus $f''(x_1) = 0$. This holds for any point of inflexion with a horizontal or non-horizontal tangent.

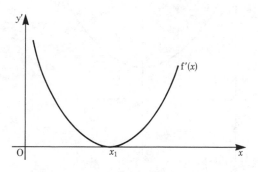

Figure 5.22

This example has shown that the condition $f'(x_0) = 0$, although necessary, is not *sufficient* to determine whether the function has a minimum or maximum value at the point $x = x_0$. The value of the second derivative, $f''(x_0)$, will give us the second condition for a minimum or a maximum.

Consider the slope of the function in the neighbourhood of a maximum, as shown in figure 5.23. On the left of x_0 it is positive and on the right of x_0 it is negative. Hence, in the immediate vicinity of x_0, the slope of the function $f(x)$ decreases monotonically and $y''(x_0) < 0$.

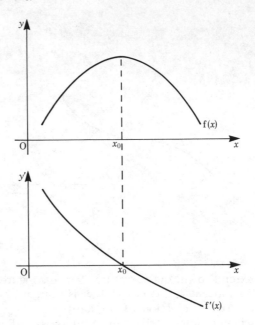

Figure 5.23

By a similar argument, if $y''(x_0) > 0$ and $y'(x_0) = 0$ then the function has a minimum at $x = x_0$ (figure 5.24).

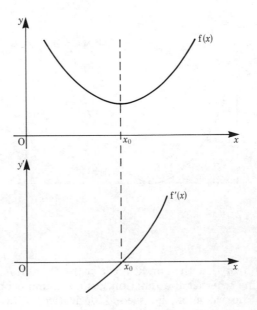

Figure 5.24

Maximum or minimum: If $f'(x_0) = 0$ (a necessary condition) and if, in addition, $f''(x_0) < 0$, then there exists a local maximum at $x = x_0$.

If $f''(x_0) > 0$ there is a local minimum.

Point of inflexion: The condition $f''(x_0) = 0$ is necessary for the existence of a point of inflexion and $f'''(x_0) \neq 0$ furnishes the sufficient criterion.

Procedure for the Determination of Maxima or Minima[†]

Step 1:   Calculate the first derivative $f'(x)$. Set $f'(x) = 0$. Solve this equation and obtain its roots $x_0, x_1, x_2 \ldots$, at which points the function may have a minimum or a maximum.

Step 2:   Calculate the second derivative $f''(x)$. If $f''(x_0) < 0$, there is a maximum at $x = x_0$. If $f''(x_0) > 0$, there is a minimum at $x = x_0$. Furthermore, if $f''(x_0) = 0$, then there may neither be a minimum nor a maximum, and, provided $f'''(x_0) \neq 0$, there is a point of inflexion at $x = x_0$.

Similar checks will have to be made for the points $x_1, x_2, \ldots$

EXAMPLE   Consider the function $y = x^2 - 1$.

Step 1 gives $y' = 2x$, so that $x_0 = 0$.

Step 2 gives $y'' = 2$ which is positive.

Hence the function has a minimum at $x = 0$.

EXAMPLE   Consider the function $y = x^3 + 6x^2 - 15x + 51$.

Step 1 gives $y' = 3x^2 + 12x - 15 = 0$

This is a quadratic equation whose roots are $x_0 = -5$, $x_1 = 1$.

Step 2 gives $y'' = 6x + 12$

For $x_0 = -5$, $y'' = -30 + 12 = -18$.

Hence the function has a maximum at this point.

For $x_1 = 1$, $y'' = 6 + 12 = 18$.

Hence the function has a minimum at this point.

Verify for yourself that this function also has a point of inflexion at $x = -2$.

5.7.2   FURTHER REMARKS ON POINTS OF INFLEXION (CONTRAFLEXURE)

Consider the curve shown in figure 5.25a, defined by the equation $y = f(x)$. As we trace the curve from N to N' its slope varies; it decreases from x_1 to x_2 and increases from x_2 to x_3. The curvature changes from concave downwards to concave upwards at N and N', respectively. The slope of the curve is shown in figure 5.25b. It can be

[†]This procedure is only valid if the maxima or minima are within the range of definition of the function. It is not valid if the maxima or minima coincide with the boundary of the range of definition. In order to identify such cases, it is helpful to sketch the graph of the function.

seen that at x_2 the slope has a minimum value. The second derivative can be understood as the 'slope' of the slope. Figure 5.25c shows that at x_1 we have $y'' < 0$. This means that in the direction of the x-axis the slope decreases, while at x_3 we have $y'' > 0$ and the slope increases.

It was mentioned earlier that a point such as P is called a point of inflexion or a point of *contraflexure*. The function $y = f(x)$ possesses a point of inflexion at $x = x_2$ if $y''(x_2) = 0$ and $y'''(x_2) \neq 0$.

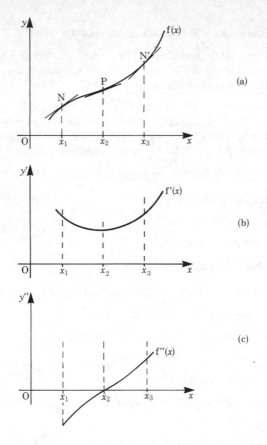

Figure 5.25

EXAMPLE   Suppose that the deflexion of a uniformly loaded beam, fixed at one end and simply supported at the other, is given by

$$y = K(3X^3 - 2X^4 - X) \qquad \text{where} \qquad X = x/L$$

x is the distance along the beam, K is a constant and L is the length, as shown in figure 5.26. (Note the directions of the axes.)

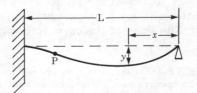

Figure 5.26

For such a beam there is a point of inflexion at P, and to locate it the condition is that $y'' = 0$.

Differentiating gives $y' = K(9X^2 - 8X^3 - 1)$ and $y'' = K(18X - 24X^2)$.

Putting $y'' = 0$ gives $X_1 = 0$ and $X_2 = 18/24 = 0.75$.

Of course, only the point X_2 is of interest to us. We now have to check that $y'''(X_2) \neq 0$:

$$y'''(X) = K(18 - 48X)$$

$$y'''(X_2) = K(18 - 36) < 0$$

Thus there is, in fact, a point of inflexion at $X_2 = 0.75$, i.e. at $x_2 = 0.75L$, three-quarters of the way along the length of the beam.

5.7.3 CURVE SKETCHING

Relationships between variables are frequently derived from physical laws leading to equations or functions. These are often difficult to visualise, so it is not easy to picture the way the function behaves. The difficulty can be overcome by sketching the curve. This is not a matter of plotting each point but of deriving a trend from particular points, such as zeros, poles and asymptotes, as has been shown in Chapter 1. We now have more precise methods which enable us to find further important features like extreme values and points of inflexion. An example of how a curve may be sketched is given below.

To sketch a curve given by $y = f(x)$, the following steps may be taken in any order:

(i)   find the intersections with the x-axis (see section 1.2.3);
(ii)   find the poles (see section 1.2.3);
(iii) examine the behaviour of the function as $x \to \pm\infty$ and find the asymptotes (see section 1.2.3);
(iv)   find the range of values for x and y;
(v)   find the extreme values (i.e. maxima, minima) and points of inflexion.

Also, in certain cases, it may be useful to look for symmetry.

EXAMPLE   Let us investigate the behaviour of the function

$$f(x) = \frac{(x-1)^2}{x^2 + 1}$$

By rearranging we find

$$f(x) = \frac{(x-1)^2}{x^2+1} = \frac{x^2+1-2x}{x^2+1} = 1 - \frac{2x}{x^2+1}$$

This shows that $f(x)$ is the result of shifting an odd function $\left(\text{namely } -\dfrac{2x}{x^2+1}\right)$ one unit along the positive y-axis. Remember that a function $g(x)$ is called odd if $g(-x) = -g(x)$. An odd function is symmetric with respect to the origin. $f(x)$ is, therefore, symmetric with respect to the point $(0,1)$.

Intersections with the x-axis

$$(x-1)^2 = 0 \qquad \text{(set numerator to zero)}$$

giving $\qquad\qquad x_0 = 1 \qquad$ (repeated)

Note: the denominator does not vanish at that point.

Pole Positions

$$x^2 + 1 = 0 \qquad \text{(set denominator to zero)}$$

There are no poles for any real value of x.

Asymptotes

$$f(x) = \frac{(x-1)^2}{x^2+1} = 1 - \frac{2x}{x^2+1}$$

In the limit we obtain

$$\lim_{x \to \pm\infty} f(x) = \lim_{x \to \pm\infty} \left(1 - \frac{2x}{x^2+1}\right) = 1$$

The proper fractional function $\left(\dfrac{2x}{x^2+1}\right)$ vanishes as $x \to \pm\infty$, and the line parallel to the x-axis $f(x) = 1$ is the asymptote as x tends to plus or minus infinity.

Range of Definition

$f(x)$ is a fractional rational function. It is defined for all values for which the denominator is different from zero. In this case it is the entire x-axis.

Maxima and Minima

The necessary condition is

$$f'(x) = 0 = 2\frac{(x-1)(x+1)}{(x^2+1)^2}$$

It is satisfied for $x_1 = +1$ and $x_2 = -1$.

The sufficient condition is

$$f''(x) \neq 0$$

$$f''(x) = 4\frac{3x - x^3}{(x^2 + 1)^3}$$

Since

$$f''(x_1) = 1 > 0 \qquad \text{we have a minimum at } x_1 = +1$$

and since

$$f''(x_2) = -1 < 0 \qquad \text{we have a maximum at } x_2 = -1$$

The coordinates of the extreme points are

$$\text{minimum } (+1, 0)$$

$$\text{maximum } (-1, 2)$$

Points of Inflexion

The necessary condition is

$$f''(x) = 0 = 4\frac{3x - x^3}{(x^2 + 1)^3}$$

It is satisfied for $x_3 = 0$, $x_4 = +\sqrt{3}$ and $x_5 = -\sqrt{3}$.

The sufficient condition is

$$f'''(x) \neq 0$$

$$f'''(x) = 4\frac{3x^4 - 18x^2 + 3}{(x^2 + 1)^4}$$

Since

$$f'''(x_3) = 12 \neq 0$$
$$f'''(x_4) = -3/8 \neq 0$$

and

$$f'''(x_5) = -3/8 \neq 0$$

there are three points of inflexion. Their coordinates are

$$(0, 1); (+\sqrt{3}, 1 - 1/2 \times \sqrt{3}) \qquad \text{and} \qquad (-\sqrt{3}, 1 + 1/2 \times \sqrt{3})$$

Figure 5.27 shows a sketch of the function, using the information obtained above.

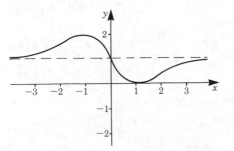

Figure 5.27

By considering symmetry with respect to a point, we could have shortened our calculation. We know that there is a minimum at $x = +1$; therefore there must be a maximum at $x = -1$. We know too that there is a point of inflexion at $x = +\sqrt{3}$; therefore there must also be another one at $x = -\sqrt{3}$.

5.8 APPLICATIONS OF DIFFERENTIAL CALCULUS

We have developed a number of rules for obtaining the derivatives of various functions. We are now in a position to apply them to the solution of practical problems.

5.8.1 EXTREME VALUES

Here we consider the application of the rule for calculating minimum and maximum values.

EXAMPLE   A cylindrical tank, flat at the top and the bottom, is to be made from thin sheet metal. The volume is to be 4 cubic metres. We wish to know the diameter D and the height H of the cylinder for which the total area A of sheet metal is a minimum.

The volume of the cylinder is

$$V = \frac{\pi}{4}D^2H = 4$$

Hence

$$H = \frac{16}{\pi D^2}$$

The area is

$$A = \pi DH + \frac{\pi}{2}D^2$$

Substituting for H gives

$$A = \frac{16}{D} + \frac{\pi}{2}D^2$$

For a minimum

$$\frac{\mathrm{d}A}{\mathrm{d}D} = 0, \quad \text{i.e.} \quad -\frac{16}{D^2} + \pi D = 0$$

Solving for D gives

$$D = \sqrt[3]{\frac{16}{\pi}} = 2 \times \sqrt[3]{\frac{2}{\pi}} = 1.721\,\text{m}$$

Note that $H = D$!

The other condition for a minimum is

$$\frac{\mathrm{d}^2A}{\mathrm{d}D^2} > 0$$

It is satisfied, since

$$\frac{d^2A}{dD^2} = \frac{32}{D^3} + \pi > 0$$

The required area of metal will be

$$A = \frac{3\pi}{2} D^2 = \frac{3\pi}{2} \times 1.721^2 = 13.949 \, \text{m}^2$$

5.8.2   INCREMENTS

A useful application of differential calculus is the calculation of small increments. When an experiment is carried out, readings are taken and results deduced from them. Normally there is the possibility of some error in the measurements and it is then required to calculate the incremental effect on the result. This effect may be calculated as follows. The experimental data may be denoted by x and the result by $y = f(x)$. Figure 5.28 shows a portion of a graph representing $y = f(x)$. Consider the function at P and let x increase by a small amount Δx (error); the corresponding increment in y is $\Delta y = f(x + \Delta x) - f(x)$.

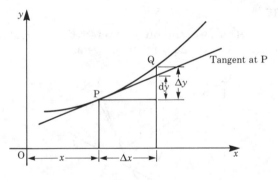

Figure 5.28

An approximate measure for the increment in the value of the function at Q is given by Δy. At P the slope is $\frac{dy}{dx}$, hence the approximate increment in the function is

$$\Delta y \approx \frac{dy}{dx} \Delta x = f'(x) \Delta x \quad \text{for small } \Delta x \qquad (5.19)$$

The expression is called the *absolute error*. The *relative error* is

$$\frac{\Delta y}{y} \approx \frac{f'(x)}{f(x)} \Delta x$$

(See also section 5.4.5.)

EXAMPLE   Suppose a cylindrical vessel of the type encountered in section 5.8.1 (i.e. height H = diameter D) is produced automatically. Supposing there is an error of 2% in the dimensions of H and D, what is the resulting error in the volume of the vessel?

$$V = \frac{\pi}{4}D^2H = \frac{\pi}{4}D^3$$

Differentiating gives

$$\frac{dV}{dD} = \frac{3}{4}\pi D^2$$

If ΔD is the error in D (and H) then

$$\Delta V \approx \frac{3}{4}\pi D^2 \,\Delta D$$

The relative error in V is

$$\frac{\Delta V}{V} \approx 3\frac{\Delta D}{D}$$

Thus, if the dimensions vary by 2%, the volumes of the vessel may vary by up to 6%.

5.8.3 CURVATURE

Given a function $y = f(x)$, we are often interested in calculating the radius of curvature of the function, e.g. in the bending of beams.

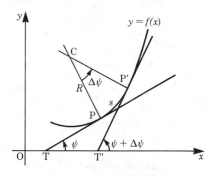

Figure 5.29

Figure 5.29 shows a portion of the graph of the function $y = f(x)$. P and P′ are two points close to each other. Draw tangents PT and P′T′, making angles ψ and $\psi + \Delta\psi$, respectively, with the x-axis. From P and P′ draw the normals to meet at the point C. In the limit, as P′ approaches P, this point is called the *centre of curvature*. The length of the normal to C is called the *radius of curvature*, denoted by R.

$\dfrac{1}{R}$ is called the *curvature*. Let us calculate R. We will consider the segment of the curve between P and P′, the length of which is denoted by s. It is approximately an arc of a circle of radius $CP = CP' = R$. From the diagram we have

$$R\Delta\psi = \Delta s \qquad \text{or} \qquad \frac{1}{R} = \frac{\Delta\psi}{\Delta s}$$

In the limit, as we take Δs smaller and smaller, we find

$$\frac{1}{R} = \frac{d\psi}{ds} \quad \text{or} \quad R = \frac{ds}{d\psi}$$

We wish to relate R to y and its derivatives and can now use a relationship derived in section 7.2 (lengths of curves) which is based on Pythagoras' theorem $ds^2 = dx^2 + dy^2$:

$$\frac{ds}{dx} = \sqrt{1 + (y')^2}$$

Using the chain rule we find

$$\frac{ds}{d\psi} = \frac{ds}{dx}\frac{dx}{d\psi} = \frac{dx}{d\psi}\sqrt{1 + (y')^2}$$

We also know that

$$\tan\psi = y'$$

Differentiating this expression with respect to x gives

$$(\tan\psi)' = y'' = \frac{1}{\cos^2\psi}\cdot\frac{d\psi}{dx}$$

$$y'' = (1 + \tan^2\psi)\frac{d\psi}{dx} = [1 + (y')^2]\frac{d\psi}{dx}$$

Substituting for $\dfrac{dx}{d\psi}$ it follows that

Radius of curvature	$R = \dfrac{ds}{d\psi} = \dfrac{[1 + (y')^2]^{3/2}}{y''}$	(5.20)

This is the desired expression for the radius of curvature at any point x in terms of the first and second derivatives of the given function.

EXAMPLE   Calculate the radius of curvature of the function $y = \cos x$ when $x = \dfrac{\pi}{4}$.

Differentiating twice we have

$$y' = -\sin x \quad \text{and} \quad y'' = -\cos x$$

substituting in the equation for R gives

$$R = \frac{(1 + \sin^2 45°)^{3/2}}{-\cos 45°} = -2.6$$

The negative sign means that the curve is concave downwards.

5.8.4   DETERMINATION OF LIMITS BY DIFFERENTIATION: L'HÔPITAL'S RULE

The determination of limits of functions by differentiation has a special significance in science and engineering. For this reason we state briefly *l'Hôpital's rule*. It gives us the values of expressions at points for which the value cannot be calculated directly because indeterminate expressions arise.

The Indeterminate Expression $\frac{0}{0}$

L'Hôpital's first rule states:

If $\lim_{x \to x_0} f(x) = 0$ and $\lim_{x \to x_0} g(x) = 0$, then

$$\lim_{x \to x_0} \frac{f(x)}{g(x)} = \lim_{x \to x_0} \frac{f'(x)}{g'(x)}$$

if the limit on the right-hand side exists.

If f' and g' at $x = x_0$ are continuous and $g'(x_0) \neq 0$, then

$$\lim_{x \to x_0} \frac{f(x)}{g(x)} = \frac{f'(x_0)}{g'(x_0)} \tag{5.21}$$

If $\lim_{x \to x_0} f'(x) = 0$ and $\lim_{x \to x_0} g'(x) = 0$, then we apply the same rule again.

EXAMPLE $L = \lim_{x \to 0} \dfrac{1 - \cos x}{x^2}$ is of the form $\dfrac{0}{0}$.

We differentiate numerator and denominator, so that

$$L = \lim_{x \to 0} \frac{\sin x}{2x} \text{ which is again } \frac{0}{0}.$$

We differentiate the top and bottom again:

$$L = \lim_{x \to 0} \frac{\cos x}{2} = \frac{1}{2}$$

The Indeterminate Expression $\frac{\infty}{\infty}$

L'Hôpital's second rule states:

If $\lim_{x \to x_0} f(x) = \infty$ and $\lim_{x \to x_0} g(x) = \infty$, then

$$\lim_{x \to x_0} \frac{f(x)}{g(x)} = \lim_{x \to x_0} \frac{f'(x)}{g'(x)}$$

if the limit on the right-hand side exists.

If $\lim_{x \to x_0} f'(x) = \infty$ and $\lim_{x \to x_0} g'(x) = \infty$, then we apply the rule once more.

EXAMPLE $\lim_{x \to \infty} \dfrac{x}{\ln x} = \dfrac{\infty}{\infty}$

By the above rule we have

$$\lim_{x \to \infty} \frac{x}{\ln x} = \lim_{x \to \infty} \frac{1}{1/x} = \infty$$

Special Forms

The expressions $0 \times \infty, \infty - \infty, 1^\infty, 0^0, \infty^0$ can be reduced to $\dfrac{0}{0}$ or $\dfrac{\infty}{\infty}$.

EXAMPLE   The expression $0 \times \infty$

$$\lim_{x \to +0} (x \ln x) = \lim_{x \to +0} \frac{\ln x}{1/x} = \lim_{x \to +0} \frac{1/x}{-1/x^2} = \lim_{x \to +0} (-x) = 0$$

EXAMPLE   The expression ∞^0

$$\lim_{x \to \infty} x^{1/x} = \lim_{x \to \infty} e^{\left(\frac{1}{x} \cdot \ln x\right)} = e^{\lim\limits_{x \to \infty} \frac{\ln x}{x}} = e^0 = 1$$

5.9   FURTHER METHODS FOR CALCULATING DIFFERENTIAL COEFFICIENTS

We now outline some methods which are useful to know when complicated functions arise.

5.9.1   IMPLICIT FUNCTIONS AND THEIR DERIVATIVES

Functions such as $y = 3x^2 + 5$, $y = \sin^{-1} x$, $y = ae^{-x}$ are referred to as *explicit functions*. Functions like $x^2 + y^2 = R^2$, $x^3 - 3xy^2 + y^3 = 10$ where the function has not been solved for y are called *implicit functions*. In this case, y is said to be an implicit function of x. Similarly, we could equally say that x is an implicit function of y.

It is occasionally possible to solve an implicit function for one of the variables. For example, the equation of a circle of radius R, $x^2 + y^2 = R^2$, can be solved for y, giving $y = \pm\sqrt{R^2 - x^2}$. Remember that in some cases it may be difficult or impossible to do this.

Differentiation of implicit functions

Differentiate all the terms of the equation as it stands and regard y as a function of x; then solve for $\dfrac{dy}{dx}$.

EXAMPLE   Obtain the derivative $\dfrac{dy}{dx}$ of $x^2 + y^2 = R^2$ (equation of a circle).

Step 1:   Differentiate all terms of the equation with respect to x. This is often expressed as applying the operator $\dfrac{d}{dx}$ to each term:

$$\frac{d}{dx}x^2 + \frac{d}{dx}y^2 = \frac{d}{dx}R^2$$

Step 2:   Carry out the differentiation

$$2x + 2y\frac{dy}{dx} = 0 \qquad (R \text{ is a constant})$$

Step 3:   Solve for $\frac{dy}{dx}$

$$\frac{dy}{dx} = -\frac{x}{y}$$

EXAMPLE   Obtain $\frac{dy}{dx}$ of $x^3 - 3xy^2 + y^3 = 10$.

Step 1:

$$\frac{d}{dx}x^3 - \frac{d}{dx}(3xy^2) + \frac{d}{dx}y^3 = \frac{d}{dx}(10)$$

Step 2:   Differentiate $3x^2 - 3\left(y^2 + x2y\frac{dy}{dx}\right) + 3y^2\frac{dy}{dx} = 0$

$$\left(Note: \text{ Treat } \frac{d}{dx}(xy^2) \text{ as a product, i.e. like } \frac{d}{dx}(uv).\right)$$

Step 3:   Solve for $\frac{dy}{dx}$ $\frac{dy}{dx}(3y^2 - 6xy) = 3(y^2 - x^2)$

Hence

$$\frac{dy}{dx} = \frac{y^2 - x^2}{y^2 - 2xy}$$

5.9.2.   LOGARITHMIC DIFFERENTIATION

Certain functions may be more easily differentiated by expressing them logarithmically first.

EXAMPLE   Differentiate $y = \sqrt{1 + x^2} \cdot \sqrt[3]{1 + x^4}$

Step 1:   Take logs to the base e on both sides

$$\ln y = \tfrac{1}{2}\ln(1 + x^2) + \tfrac{1}{3}\ln(1 + x^4)$$

Step 2:   Differentiate the new expression with respect to x

$$\frac{d}{dx}(\ln y) = \frac{1}{2}\frac{2x}{1 + x^2} + \frac{1}{3}\frac{4x^3}{1 + x^4}$$

$$\frac{1}{y}\frac{dy}{dx} = \frac{x}{1 + x^2} + \frac{4}{3}\frac{x^3}{1 + x^4}$$

Step 3:   Solve for $\frac{dy}{dx}$ $\frac{dy}{dx} = \sqrt{1 + x^2}\sqrt[3]{1 + x^4}\left(\frac{x}{1 + x^2} + \frac{4}{3}\frac{x^3}{1 + x^4}\right)$

5.10 PARAMETRIC FUNCTIONS AND THEIR DERIVATIVES

5.10.1 PARAMETRIC FORM OF AN EQUATION

A curve in a plane cartesian coordinate system has so far been represented by an equation of the form

$$y = f(x)$$

However, we frequently encounter variables x and y which are functions of a third variable, for example t or θ, which is called a parameter. We can express this in the following general form:

$$x = x(t) \qquad \text{and} \qquad y = y(t)$$

or alternatively in the form

$$x = g(t) \qquad \text{and} \qquad y = h(t)$$

For example, in order to describe the movement of a point in a plane, we can specify the components of the position vector $\mathbf{r}$ as functions of time. Then the parameter is the time t:

$$\mathbf{r} = [x(t), \quad y(t)]$$

As time passes, the components of the position vector vary and the point of the position vector moves along a curve (figure 5.30).

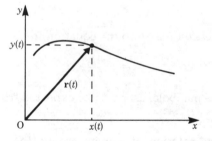

Figure 5.30

EXAMPLE   Describe the movement of a particle projected horizontally with initial velocity v_0 in a constant gravitational field (figure 5.31).

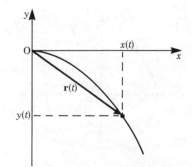

Figure 5.31

The horizontal movement of the particle is of constant velocity v_0. Thus the x-component is

$$x(t) = v_0 t$$

The vertical movement of the particle is that of a freely falling body with a gravitational acceleration g. Thus the vertical component is given by

$$y(t) = -\frac{g}{2} t^2$$

The position vector of the movement is

$$\mathbf{r}(t) = \left(v_0 t, \ -\frac{g}{2} t^2 \right)$$

Both components depend on a third variable, the time t. To each value of the parameter t there corresponds a value for x and a value for y.

Generally, when x and y are expressed as functions of a third variable, the equation is said to be in *parametric form*.

The parameter may be the time t, an angle θ or any other variable. In many cases it is possible to eliminate the parameter and to obtain the function of the curve in the familiar form. To do this in the case given above, we solve the equation $x = x(t)$ with respect to t:

$$x = v_0 t$$

$$t = \frac{x}{v_0}$$

Now we insert this expression for t into the equation for $y = y(t)$:

$$y = -\frac{g}{2v_0^2} x^2$$

This is the equation of a parabola, as was to be expected from the known properties of freely falling bodies.

EXAMPLE   Describe the rotation of a point around the circumference of a circle (figure 5.32). The equation of a circle can be expressed in parametric form:

$$x = R \cos \phi$$

$$y = R \sin \phi$$

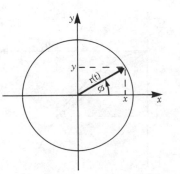

Figure 5.32

In this case R is the radius, and the parameter is the angle measured from the x-axis to the position vector. The position vector is

$$\mathbf{r}(t) = (R\cos\phi, \quad R\sin\phi)$$

In order to obtain the equation of the circle in cartesian coordinates, we take the square of both parametric equations and add them.

$$x^2 + y^2 = R^2\cos^2\phi + R^2\sin^2\phi = R^2$$

If a point rotates with uniform velocity on a circle, the angle ϕ is given by $\phi = \omega t$.

Here ω is a constant which is called angular velocity (see Chapter 1). The parameter is now t and the parametric form of these rotations is

$$x(t) = R\cos\omega t$$

$$y(t) = R\sin\omega t$$

The position vector scanning the circle is given by

$$\mathbf{r}(t) = (R\cos\omega t, \quad R\sin\omega t)$$

EXAMPLE   Describe the parametric form of a straight line in a plane (figure 5.33). **b** is a vector pointing in the direction of the line. **a** is a vector from the origin of the coordinate system to a given point on the line. Now consider the vector

$$\mathbf{r}(\lambda) = \mathbf{a} + \lambda\mathbf{b}$$

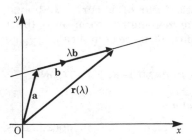

Figure 5.33

This vector $\mathbf{r}(\lambda)$ scans all points on the line if the parameter λ varies between $-\infty$ and $+\infty$. Thus, the line is given in parametric form by

$$x(\lambda) = a_x + \lambda b_x$$

$$y(\lambda) = a_y + \lambda b_y$$

If we consider straight lines in three-dimensional space, there is a third equation for the z-component:

$$z(\lambda) = a_z + \lambda b_z$$

EXAMPLE   Describe a helix in parametric form. Let us consider a point which moves on a helical curve (screw). The direction of the screw is the z-axis (figure 5.34). With one rotation, the point gains height by an amount h.

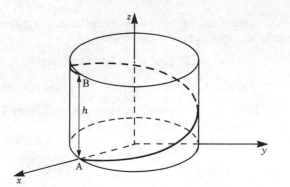

Figure 5.34

The coordinates of the point are easily given in parametric form if we use the angle θ as the parameter:

$$x = R \cos \theta$$

$$y = R \sin \theta$$

$$z = h \frac{\theta}{2\pi}$$

The position vector is

$$\mathbf{r}(\theta) = \left(R \cos \theta, \quad R \sin \theta, \quad h \frac{\theta}{2\pi} \right)$$

These examples show that it is sometimes more relevant to the nature of a problem to establish the parametric form of a curve. Although, in the plane, it is often possible to transform the paremetric form into the more familiar relationship $y = f(x)$, this may sometimes be more complicated.

5.10.2   DERIVATIVES OF PARAMETRIC FUNCTIONS

Derivative of a Position Vector

Given a position vector in plane cartesian coordinates in parametric form, the parameter being the time t,

$$\mathbf{r}(t) = [x(t), \quad y(t)] = x(t)\mathbf{i} + y(t)\mathbf{j}$$

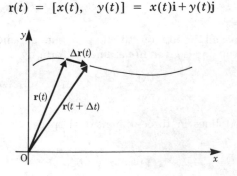

Figure 5.35

We can find the derivative of a position vector by finding its velocity. According to figure 5.35, the velocity is given by

$$v = \lim_{\Delta t \to 0} \frac{\Delta r}{\Delta t} = \lim_{\Delta t \to 0} \frac{r(t + \Delta t) - r(t)}{\Delta t}$$

The components are

$$v = \left(\lim_{\Delta t \to 0} \frac{\Delta x}{\Delta t}, \quad \lim_{\Delta t \to 0} \frac{\Delta y}{\Delta t} \right) = \left(\frac{dx}{dt}, \quad \frac{dy}{dt} \right)$$

The components of the velocity are the derivatives of the x- and y-components. Therefore if a vector is given in parametric form, its derivative can be obtained by differentiating each component with respect to the parameter.

EXAMPLE   Find the parametric form of the velocity and acceleration of a particle, acting under gravity, which is projected horizontally (figure 5.31).

We start with the known equation in parametric form:

$$r(t) = \left(v_0 t, \quad -\frac{g}{2} t^2 \right)$$

The velocity is obtained by differentiating each component with respect to the parameter t:

$$v(t) = (v_0, \quad -gt)$$

For the horizontal component of v we obtain the constant initial velocity v_0. For the vertical component we get the time-dependent velocity of a freely falling body.

If we want to know the acceleration, we have to differentiate once more with respect to t:

$$a(t) = (0, \quad -g)$$

We find there is no horizontal acceleration. But there is a vertical acceleration (due to gravity).

EXAMPLE   Find the parametric form of the velocity and acceleration of a point rotating on a circle with radius R.

We start with the known equation in parametric form with the parameter t for time:

$$r(t) = (R \cos \omega t, \quad R \sin \omega t)$$

The components of the velocity are given by the derivatives of the components with respect to the parameter t:

$$v(t) = \frac{dr(t)}{dt} = \frac{d}{dt} (R \cos \omega t, \quad R \sin \omega t)$$

$$v(t) = (-R\omega \sin \omega t, \quad R\omega \cos \omega t)$$

The magnitude of v is

$$v = \sqrt{R^2 \omega^2 \sin^2 \omega t + R^2 \omega^2 \cos^2 \omega t} = \omega R$$

The acceleration is given by the second derivative with respect to time:

$$\mathbf{a}(t) = \frac{d\mathbf{v}(t)}{dt} = (-R\omega^2 \cos \omega t, \ -R\omega^2 \sin \omega t)$$

Comparing with $\mathbf{r}$ gives

$$\mathbf{a}(t) = -\omega^2 \mathbf{r}(t)$$

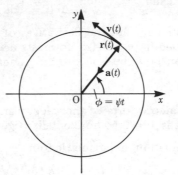

Figure 5.36

The acceleration has the opposite direction to $\mathbf{r}$ (figure 5.36). You can verify for yourself that $\mathbf{r}(t)$ is perpendicular to $\mathbf{v}(t)$ and $\mathbf{v}(t)$ is perpendicular to $\mathbf{a}(t)$.

(*Hint*: The scalar products $\mathbf{r} \cdot \mathbf{v}$ and $\mathbf{v} \cdot \mathbf{a}$ must vanish in this case.)

The magnitude of $\mathbf{a}$ is $a = \omega^2 R$.

Using the equation $v = \omega R$ we can express the acceleration in two ways:

$$a = \frac{v^2}{R}$$

$$a = v\omega = \omega^2 R$$

The Normal Vector

At each point on the curve there is a tangent vector. A vector perpendicular to the tangent vector is called a normal vector (figure 5.37). It is easy to find the formula of a normal vector: the scalar product of the tangent vector and a normal vector must vanish.

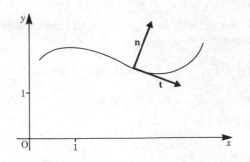

Figure 5.37

Given: a tangent vector $t = \left(\dfrac{dx}{dt}, \ \dfrac{dy}{dt} \right)$

Wanted: a normal vector $n = (n_x, n_y)$

Condition: scalar product = 0: $\dfrac{dx}{dt} n_x + \dfrac{dy}{dt} n_y = 0$

Solve for n_x: $n_x = -\dfrac{dy/dt}{dx/dt} n_y = -\dfrac{dy}{dx} n_y$

We are free to choose n_y. In figure 5.37 a normal vector is obtained by setting $n_y = 1$.

$$\mathbf{n} = \left(-\dfrac{dy}{dx}, \ 1 \right)$$

Note that in solving the equation for n_x the parameter t was eliminated. Therefore, the result is also true for any curve given in the usual form $y = f(x)$. In this case, a tangent vector is given by

$$\mathbf{t} = \left(1, \ \dfrac{dy}{dx} \right)$$

Derivative of a Curve Given in Parametric Form

Given the parametric equations

$$x = x(t)$$
$$y = y(t)$$

we wish to find the differential coefficient $\dfrac{dy}{dx}$, i.e. the slope of the curve. We proceed as follows.

Step 1:   Differentiate the equations for x and y with respect to the parameter to obtain

$$\dfrac{dx}{dt} \quad \text{and} \quad \dfrac{dy}{dt}$$

Step 2:   Rearrange to obtain the desired derivative:

$$\dfrac{dy}{dx} = \dfrac{dy/dt}{dx/dt}$$

This is the slope y' of the function $y = f(x)$ at the point (x, y). Note that we did not establish the function $y = f(x)$ to find its derivative.

EXAMPLE   Find the parametric form of the equation of a circle. The parameter is denoted by θ this time:

$$x = R \cos \theta$$
$$y = R \sin \theta$$

Step 1:   Differentiate x and y with respect to the parameter θ:

$$\frac{dx}{d\theta} = -R\sin\theta = -y, \quad \frac{dy}{d\theta} = R\cos\theta = x$$

Step 2:   Obtain the derivative of y with respect to x:

$$\frac{dy}{dx} = \frac{dy/d\theta}{dx/d\theta} = -\cot\theta = -\frac{x}{y}$$

EXAMPLE   The *cycloid* is a curve traced out by a point on the circumference of a wheel which rolls without slipping. It is conveniently expressed in parametric form. The following gives the equation for a wheel with radius a. The parameter θ is the angle of rotation as the wheel moves in the x-direction. Figure 5.38 shows the wheel at different positions. The set of parametric equation is

$$x = a(\theta - \sin\theta)$$
$$y = a(1 - \cos\theta)$$

We wish to obtain the derivative of y with respect to x.

Step 1: $$\frac{dx}{d\theta} = a(1 - \cos\theta), \quad \frac{dy}{d\theta} = a\sin\theta$$

Step 2: $$\frac{dy}{dx} = \frac{dy/d\theta}{dx/d\theta} = \frac{\sin\theta}{1 - \cos\theta}$$

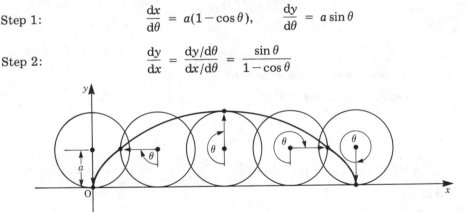

Figure 5.38

APPENDIX: DIFFERENTIATION RULES

General rules	Function $y = f(x)$	Derivative $y' = f'(x)$
1. Constant factor	$y = cf(x)$	$y' = cf'(x)$
2. Sum (algebraic) rule	$y = u(x) + v(x)$	$y' = u'(x) + v'(x)$
3. Product rule	$y = u(x)\,v(x)$	$y' = u'(x)v(x) + u(x)v'(x)$
4. Quotient rule	$y = \dfrac{u(x)}{v(x)}$	$y' = \dfrac{u'(x)v(x) - u(x)v'(x)}{v(x)^2}$
5. Chain rule	$y = f[g(x)]$	$y' = \dfrac{df}{dg}\,g'(x)$
6. Inverse functions	$y = f^{-1}(x)$	$y' = \dfrac{1}{dx/dy} = \dfrac{1}{f'(y)}$

i.e. $x = f(y)$

Derivatives of fundamental functions	Function $y = f(x)$	Derivative $y' = f'(x)$
1. Constant factor	$y = \text{constant}$	$y' = 0$
2. Power function	$y = x^n$	$y' = nx^{n-1}$
3. Trigonometric functions	$y = \sin x$	$y' = \cos x$
	$y = \cos x$	$y' = -\sin x$
	$y = \tan x$	$y' = \dfrac{1}{\cos^2 x} = 1 + \tan^2 x$
	$y = \cot x$	$y' = \dfrac{-1}{\sin^2 x} = -1 - \cot^2 x$
4. Inverse trigonometric functions	$y = \sin^{-1} x$	$y' = \dfrac{1}{\sqrt{1-x^2}}$
	$y = \cos^{-1} x$	$y' = -\dfrac{1}{\sqrt{1-x^2}}$
	$y = \tan^{-1} x$	$y' = \dfrac{1}{1+x^2}$
	$y = \cot^{-1} x$	$y' = -\dfrac{1}{1+x^2}$
5. Exponential function	$y = e^x$	$y' = e^x$
Logarithmic function	$y = \ln x$	$y' = \dfrac{1}{x}$
6. Hyperbolic trigonometric functions	$y = \sinh x$	$y' = \cosh x$
	$y = \cosh x$	$y' = \sinh x$
	$y = \tanh x$	$y' = \dfrac{1}{\cosh^2 x} = 1 - \tanh^2 x$
	$y = \coth x$	$y' = -\dfrac{1}{\sinh^2 x} = 1 - \coth^2 x$
7. Inverse hyperbolic trigonometric functions	$y = \sinh^{-1} x$	$y' = \dfrac{1}{\sqrt{1+x^2}}$
	$y = \cosh^{-1} x$	$y' = \dfrac{1}{\sqrt{x^2-1}} \quad (x > 1)$
	$y = \tanh^{-1} x$	$y' = \dfrac{1}{1-x^2} \quad (|x| < 1)$
	$y = \coth^{-1} x$	$y' = -\dfrac{1}{x^2-1} \quad (|x| > 1)$

EXERCISES

5.1   Sequences and Limits

1.   Calculate the limiting value of the following sequences for $n \to \infty$:

(a) $a_n = \dfrac{\sqrt{n}}{n}$

(b) $a_n = \dfrac{5+n}{2n}$

(c) $a_n = (-\tfrac{1}{4})^n - 1$

(d) $a_n = \dfrac{2}{n} + 1$

(e) $a_n = \dfrac{n^3 + 1}{2n^3 + n^2 + n}$

(f) $a_n = 2 + 2^{-n}$

(g) $a_n = \dfrac{n^2 - 1}{(n+1)^2} + 5$

2.   Calculate the following limits:

(a) $\lim\limits_{x \to 0} \dfrac{x^2 + 1}{x - 1}$

(b) $\lim\limits_{x \to 2} \dfrac{1}{x}$

(c) $\lim\limits_{x \to 0} \dfrac{x^2 + 10x}{2x}$

(d) $\lim\limits_{x \to \infty} e^{-x}$

(e) $\lim\limits_{x \to 0} \dfrac{\sqrt{1+x} - \sqrt{1-x}}{x}$

(f) $\lim\limits_{x \to 0} \dfrac{1 - \cos x}{x}$

5.2   Continuity

3.   (a) Is the function $y = 1 + |x|$ continuous at the point $x = 0$?

(b) Determine the points for which the following function is discontinuous:

$$f(x) = \begin{cases} 1 & \text{for} & 2k \leqslant x \leqslant 2k + 1 \\ -1 & \text{for} & 2k + 1 < x < 2(k + 1) \end{cases}, \qquad k = 0, 1, 2, 3 \ldots$$

(c) At which points is the function $f(x)$ shown in figure 5.39 discontinuous?

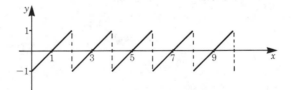

Figure 5.39

5.3   Series

4.   Obtain the values of the following sums:

(a) $S_5 = \sum\limits_{v=1}^{5} \left(1 + \dfrac{1}{v}\right)$

(b) $S_{10} = \sum\limits_{n=0}^{9} 3(\tfrac{1}{2})^n$

(c) What is the value of the sum $S = \sum\limits_{n=0}^{\infty} (\tfrac{1}{2})^n$?

5.4   Differentiation of a Function;   5.5   Calculating Differential Coefficients

5. (a) Given the curve $y = x^3 - 2x$, calculate the slope of the secant to the curve between the points $x_1 = 1$ and $x_2 = \frac{3}{2}$. Compare the slope of the secant with that of the tangent at the point $x_1 = 1$.

  (b) The distance–time law for a particular motion is given by $s(t) = 3t^2 - 8t$ m. Evaluate the velocity at $t = 3$ seconds.

  (c) Determine the differential dy of the following functions:

   (i) $f(x) = x^2 + 7x$ (ii) $f(x) = x^5 - 2x^4 + 3$ (iii) $f(x) = 2(x^2 + 3)$

6. Differentiate with respect to x the following expressions:

  (a) $3x^5$ (b) $8x - 3$ (c) $x^{7/3}$

  (d) $7x^3 - 4x^{3/2}$ (e) $\dfrac{x^3 - 2x}{5x^2}$

7. Obtain the derivatives of the following:

  (a) $y = 2x^3$ (b) $y = \sqrt[3]{x}$ (c) $y = \dfrac{1}{x^2}$

  (d) $y = \dfrac{2x}{4+x}$ (e) $y = (x^2 + 2)^3$ (f) $y = x^4 + \dfrac{1}{x}$

  (g) $y = \sqrt{1 + x^2}$ (h) $y = \left(a - \dfrac{b}{x}\right)^3$

8. Differentiate

  (a) $y = 3\cos(6x)$ (b) $y = 4\sin(2\pi x)$ (c) $y = Ae^{-x}\sin(2\pi x)$
  (d) $y = \ln(x+1)$ (e) $y = \sin x \cos x$ (f) $y = \sin x^2$
  (g) $y = (3x^2 + 2)^2$ (h) $y = a\sin(bx + c)$ (i) $y = e^{2x^3 - 4}$

9. Differentiate (inverse trigonometric functions)

  (a) $y = \cos^{-1}(cx)$ (b) $y = A\tan^{-1}(x + 2)$
  (c) $y = \sin^{-1}(x^2)$ (d) $y = \coth^{-1}(\sqrt{x})$

10. Differentiate (hyperbolic trigonometric functions)

  (a) $y = C\sinh(0.1x)$ (b) $u = \pi\tanh(v + 1)$
  (c) $\eta = \ln(\cosh\xi)$ (d) $s = \ln(\cosh t)$
  (e) $y = \sinh^2 x - \cosh^2 x$ (f) $y = 2x\coth x - x^2$

11. Differentiate (inverse hyperbolic trigonometric functions)

  (a) $y = A\sinh^{-1}(10x)$ (b) $u = C\coth^{-1}(v + 1)$
  (c) $\eta = \tanh^{-1}(\sin\xi)$ (d) $y = \sinh^{-1}\left(\dfrac{x-1}{x}\right)$

5.6   Higher Derivatives;   5.7   Extreme Values and Point of Inflexion

12.   Obtain the following derivatives:
  (a) $g(\phi) = a \sin \phi + \tan \phi,$ required $g'(\phi)$, 1st derivative
  (b) $v(u) = u\, e^u,$ required $v''(u)$, 2nd derivative
  (c) $f(x) = \ln x,$ required $f''(x)$, 2nd derivative
  (d) $h(x) = x^5 + 2x^2,$ required $h^{(iv)}(x)$, 4th derivative

13.   Find the zeros and the extreme values for the following functions:
  (a) $y = 2x^4 - 8x^2$ (b) $y = 3 \sin \phi$
  (c) $y = \sin(0.5x)$ (d) $y = 2 + \tfrac{1}{2}x^3$
  (e) $y = 2 \cos(\phi + 2)$ (f) $y = \tfrac{2}{3}x^3 - 2x^2 - 6x$

14.   *Points of inflexion.* Show that the following functions have a point of inflexion. Calculate the value of the function at such a point.
  (a) $y = x^3 - 9x^2 + 24x - 7$ (b) $y = x^4 - 8x^2$

15.   *Curve sketching.* Sketch the following functions:
  (a) $y = x^2 + \dfrac{1}{x^3}$ (b) $y = \dfrac{4x + 1}{2x + 3}$ (c) $y = \dfrac{x^2 - 6x + 8}{x^2 - 6x + 5}$

5.8   Applications of Differential Calculus

16.   *Errors.* (a) A tray in the form of a cube is to be manufactured out of sheet metal. It is to have a cubic capacity of $0.05\,\text{m}^3$. If the tolerance on the linear dimensions is not to exceed 3 mm, calculate the change in the volume and in the area of metal as a percentage.

  (b) The height of a tower is calculated from its angles of elevation of $35°$ and $28°$, observed at two points 100 m apart in a horizontal straight line through its base. If the measurement of the larger angle is found to have an error of $0.5°$, what will be the error in the calculated height?

17.   *Curvature.* Calculate the radius of curvature for the following functions:
  (a) $y = x^3$ at $x = 1$ (b) $y^2 = 10x$ at $x = 2.5$

18.   *L'Hôpital's rule*

  (a) $\displaystyle \lim_{x \to 0} \frac{\sin x}{x}$ (b) $\displaystyle \lim_{x \to +0} \frac{\ln(1 + 1/x)}{1/x}$

  (c) $\displaystyle \lim_{x \to +0} x^{\sin x}$ (d) $\displaystyle \lim_{x \to 0} \frac{1}{x}\left(\frac{1}{\sinh x} - \frac{1}{\tanh x} \right)$

  (e) $\displaystyle \lim_{x \to 1} x^{\tan\left(\frac{\pi}{2}x\right)}$ (f) $\displaystyle \lim_{x \to \infty} \frac{\ln x^3}{\sqrt[3]{x}}$

  (g) $\displaystyle \lim_{\theta \to 0} \frac{\cos a\theta - \cos b\theta}{\theta^2}$

5.9   Further Methods

19.   *Implicit functions.* Obtain the derivative y' for the following expressions:
    (a) $2x^2 + 3y^2 = 5$ (b) $3x^3y^2 + x \cos y = 0$
    (c) $(x + y)^2 + 2x + y = 1$ at $x = 1,\ y = -1$

20.   *Logarithmic differentiation.* Obtain y' for the following expressions:
    (a) $y = (5x + 2)(3x - 7)$ (b) $y = x^{\sin x}$
    (c) $y = x^{x}$

5.10   Parametric Functions and their Derivatives

21.   *Parametric equations.* Obtain $\dfrac{dy}{dx}$ for the following expressions:

    (a) $x = ut$ and $y = vt - \frac{1}{2}gt^2$
        u, v and g are constants
    (b) $x = a(\cos t + t \sin t)$
        $y = a(\sin t - t \cos t)$

22.   A point rotates in the x–y plane with a radius R around the origin of the coordinate system with constant angular velocity. In 2 s it completes 3 revolutions. Give the parametric form of the movement.

23.   (a) The parametric form of a curve is
$$x(t) = t$$
$$y(t) = t$$
$$z(t) = t$$
    What curve is it?

    (b) What curve is described by the following equations?
$$x(t) = a \cos t$$
$$y(t) = b \sin t$$

24.   (a) Calculate the acceleration vector
$$v_x(t) = -v_0 \sin \omega t$$
$$v_y(t) = v_0 \cos \omega t$$

    (b) The position of a point in three-dimensional space is given by
$$r(t) = (R \cos \omega t, \quad R \sin \omega t, \quad t)$$
    Calculate the velocity for $t = \dfrac{2\pi}{\omega}$

    (c) The acceleration of a freely falling body is
$$\mathbf{a} = (0, 0, -g)$$
    Calculate the velocity $v(t)$ if
$$\mathbf{v}(0) = (v_0, 0, 0)$$

CHAPTER 6

Integral calculus

6.1 THE PRIMITIVE FUNCTION

6.1.1 FUNDAMENTAL PROBLEM OF INTEGRAL CALCULUS

In Chapter 5 we started from a graph of a function which could be differentiated and obtained its slope or gradient.

The problem was to find the derivative $f'(x) = \dfrac{dy}{dx}$ of a given function $y = f(x)$

This problem can also be reversed. Let us assume that the derivative of a function is known. Can we find the function?

EXAMPLE A function is known to have the same slope throughout its range of definition (see figure 6.1) i.e.

$$y' = m$$

Can we find the function? To do so we review all functions known to us to find out whether there is among them one which has a constant slope.

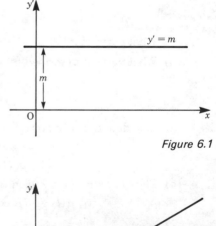

Figure 6.1

In this case, we know such a function: it is a straight line. Hence, one possible solution of our problem is a straight line with a slope m through the origin, i.e.

$$y(x) = mx$$

Figure 6.2 shows such a function. The process of finding the function from its derivative is called *integration* and the result an *indefinite integral* or *primitive function*.

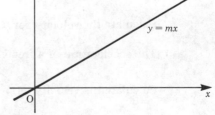

Figure 6.2

124

A general statement of the problem is as follows:

Let a given function $f(x)$ be the derivative of a function $F(x)$ which we wish to find. Then $F(x)$ has to satisfy the condition that

$$F'(x) = f(x)$$

DEFINITION $F(x)$ is a *primitive function* of $f(x)$ if the following holds true:
$$F'(x) = f(x) \tag{6.1}$$

EXAMPLE   Let $f(x) = m$, a constant. We have already found a solution to be

$$F(x) = mx$$

We can easily verify the result by differentiating:

$$f(x) = \frac{d}{dx}(mx) = m$$

Remember that the derivative of a constant term is zero. Hence, if a constant term is added to the function $F(x)$ just found, the function obtained will also have the same derivative. Therefore we can put

$$F(x) = mx + C$$

where C is any constant.

It obviously follows that there is not just one solution but many others which differ only by constants. Thus the solution of the equation $F'(x) = f(x)$ gives rise to the family of curves given by

$$y = F(x) + C$$

In our simple example, all straight lines with the slope m are primitive functions of $f(x) = m$, as shown in figure 6.3.

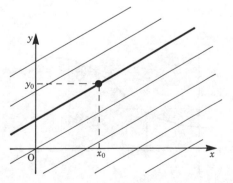

Figure 6.3

In order to obtain a particular primitive function from the whole family of primitive functions, we need to specify certain conditions.

We may, for example, specify that the function must pass through a particular point given by a set of coordinates. Such conditions are known as *boundary conditions*.

In our example, suppose we specify that when $x = x_0$, $y = y_0$. Substituting in the equation $y = mx + C$, we find

$$y_0 = mx_0 + C$$

Hence
$$C = y_0 - mx_0$$

The final solution is

$$y = mx + y_0 - mx_0$$

This is shown by the solid line in figure 6.3.

All primitive functions differ from each other by a constant which can be determined from specified boundary conditions.

6.2 THE AREA PROBLEM: THE DEFINITE INTEGRAL

Consider the problem of calculating the area under a curve. The area F (shown shaded in figure 6.4) is bounded by the graph of the function $f(x)$, the x-axis and the lines parallel to the y-axis at $x = a$ and $x = b$. If $f(x)$ is a straight line, then the area F is easily calculated. We now develop a method for the evaluation of F which is applicable to any function, provided that it is continuous in the interval $a \leqslant x \leqslant b$. For the time being, we will assume that $f(x)$ is positive in the interval considered.

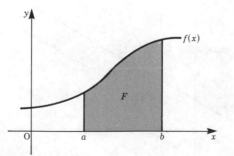

Figure 6.4

We divide the interval into n sub-intervals of lengths Δx_1, Δx_2, ..., Δx_n and select from each sub-interval a value for the variable x_i, as shown in figure 6.5. The value of the function, or height, is $f(x_i)$.

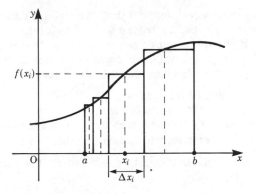

Figure 6.5

The area F is approximately given by the sum of the rectangles

$$F \approx f(x_1)\Delta x_1 + f(x_2)\Delta x_2 + \ldots + f(x_i)\Delta x_i + \ldots + f(x_n)\Delta x_n$$

This sum is more compactly written

$$F \approx \sum_{i=1}^{n} f(x_i)\,\Delta x_i$$

We now wish to find the limit of this sum for n increasing indefinitely. Each Δx will diminish indefinitely at the same time (figure 6.6).

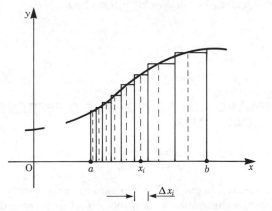

Figure 6.6

We know intuitively that by this process we shall obtain the exact value for the area F.

Hence

$$F = \lim_{\substack{n \to \infty \\ \Delta x_i \to 0}} \sum_{i=1}^{n} f(x_i)\,\Delta x_i$$

A new symbol is now introduced to denote this limiting value and write

$$F = \int_a^b f(x)\,dx$$

The symbol $\int$ is called an *integral sign* and the expression is called a *definite integral*. It is read as 'the integral of $f(x)\,dx$ from a to b'. The integral sign is an elongated S and stands for 'sum'. It should be remembered, however, that the integral is the limit of a sum. The limiting process is valid for continuous functions. For discontinuous functions we have to prove in each case that a limit exists.

If we take the greatest value of the function in each sub-interval as the height of the rectangles, then the sum is called an *upper sum*; if we take the smallest value of the function in each sub-interval, then the sum is called a *lower sum*. For continuous functions the upper and lower sums will coincide in the limiting process.

The dx after the integral sign should not be left out as it is part of the process we have just examined.

DEFINITION $$F = \lim_{\substack{n \to \infty \\ \Delta x_i \to 0}} \sum_{i=1}^{n} f(x_i)\,\Delta x_i = \int_a^b f(x)\,dx \qquad (6.2)$$

The symbol $\int_a^b f(x)\,dx$ is called the *definite integral* of $f(x)$ between the values $x = a$ and $x = b$.

a is called the *lower limit of integration*,
b is called the *upper limit of integration*,
$f(x)$ is called the *integrand*,
x is called the *variable of integration*.

6.3   FUNDAMENTAL THEOREM OF THE DIFFERENTIAL AND INTEGRAL CALCULUS

The fundamental theorem states: The area function is a primitive of the function $f(x)$.

How is it arrived at? We begin with a continuous and positive function $f(x)$ and consider the area below the graph of the function (shown shaded in figure 6.7).

In contrast to the previous area problem whose limits were fixed, here we consider the upper limit as a variable. It follows, therefore, that the area is no longer constant but is a function of the upper limit x.

x now has two meanings:

(1) x is the upper limit of integration;
(2) x is also the variable in the function $f(x)$.

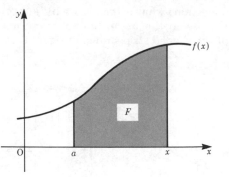

Figure 6.7

To avoid any difficulties that might arise because of this double meaning, we will change the notation and use t as the variable in $y = f(t)$ and consider the area below the graph of this function between the fixed lower limit $t = a$ and the variable upper limit $t = x$, as shown in figure 6.8.

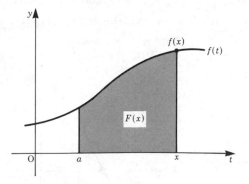

Figure 6.8

The area under the curve is

$$F(x) = \int_a^x f(t)\,dt$$

The function $F(x)$ defines the area below the curve of $f(t)$ bounded by $t = a$ and $t = x$. Figure 6.9 shows the function $F(x)$ for the curve $f(t)$ depicted in figure 6.8. We call the function $F(x)$ the *area function*.

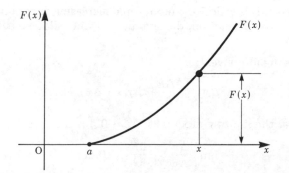

Figure 6.9

We now ask ourselves the question: what is the nature of $F(x)$? To answer this, first of all consider a small increase in the upper limit x by an amount Δx. The area increases by the amount of the shaded strip shown in figure 6.10.

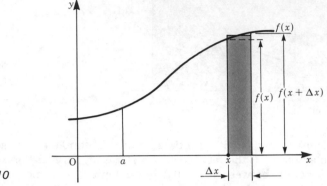

Figure 6.10

The area function increases by ΔF as shown in Figure 6.11.

This increase in the area, ΔF, lies between the values $f(x)\,\Delta x$ and $f(x+\Delta x)\,\Delta x$, i.e.

$$f(x)\,\Delta x \leqslant \Delta F \leqslant f(x+\Delta x)\,\Delta x$$

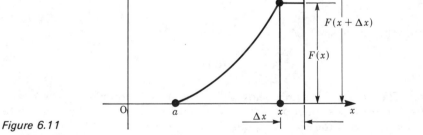

Figure 6.11

We should note that this is valid for a monotonic increasing function. If the function is a decreasing monotonic function, the argument is still valid, except for a change in the inequalities.

Dividing by Δx throughout gives

$$f(x) \leqslant \frac{\Delta F}{\Delta x} \leqslant f(x+\Delta x)$$

Let us now consider the limiting process as $\Delta x \to 0$ for the area function

$$\lim_{\Delta x \to 0} \frac{\Delta F}{\Delta x} = \frac{\mathrm{d}F}{\mathrm{d}x} = F'(x)$$

We also have

$$\lim_{\Delta x \to 0} f(x + \Delta x) = f(x)$$

Hence we find

$$f(x) \leqslant F'(x) \leqslant f(x)$$

This means that

$$F'(x) = f(x)$$

The derivative of the area function $F(x)$ is equal to $f(x)$. In other words, *the area function is a primitive of the function* $f(x)$.

This is the fundamental theorem of the differential and integral calculus. It embodies the relationship between the two.

The area function is given by

$$F(x) = \int_a^x f(t)\,dt$$

Differentiating gives

$$F'(x) = \frac{d}{dx}\left(\int_a^x f(t)\,dt\right) = f(x)$$

If we carry out an integration, followed by a differentiation, the operations cancel one another out. Thus, loosely speaking, differentiation and integration are inverse processes.

The fundamental theorem of the differential and integral calculus

If
$$F(x) = \int_a^x f(t)\,dt$$

then
$$F'(x) = f(x)$$

(6.3)

So far we have not paid much attention to the choice of the lower limit a. We will now investigate how the area function changes if we replace the lower limit a by a new one, a', as shown in figure 6.12.

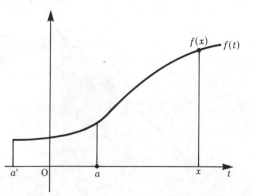

Figure 6.12

Let $F_1(x)$ be the new area function. $F_1(a')$ is zero; $F_1(a)$ corresponds to the area between a' and a. If we now consider the course followed by the original area function $F(x)$, we see that the new area function $F_1(x)$ is made up of two parts:

$$F_1(a) = \text{area between } a' \text{ and } a \text{ (which is a constant)}$$

and $\qquad F(x) = \text{area between } a \text{ and } x$

Hence $\qquad F_1(x) = F(x) + F_1(a)$

Thus a change in the lower limit leads to a new area function which differs from the original function by a constant. This is illustrated in figure 6.13. It is in accordance with the fact we found earlier that primitive functions differ from each other by a constant.

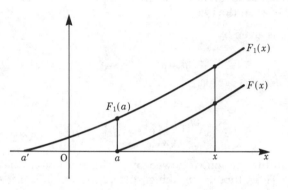

Figure 6.13

6.4 THE DEFINITE INTEGRAL

6.4.1 CALCULATION OF DEFINITE INTEGRALS FROM INDEFINITE INTEGRALS

We will now proceed to calculate the value of the definite integral from the geometrical meaning of the primitive function as an area function.

We require the area shown shaded below the function $y = x$ and between the limits a and b in figure 6.14.

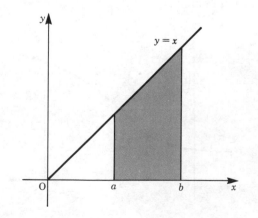

Figure 6.14

A primitive function which satisfies the condition $F'(x) = f(x)$ is

$$F(x) = \tfrac{1}{2}x^2$$

This can easily be verified by differentiation.

This primitive function is the area function for the lower limit $x = 0$ and represents the area below the graph and between the limits $x = 0$ and x. We wish to calculate the area for $x = a$ to $x = b$. We know from section 6.3 that this area is the difference between two areas, namely the area $F(b)$ bounded by the graph and the limits $x = 0$, $x = b$, and the area $F(a)$ bounded by the graph and the limits $x = 0$, $x = a$ (see figure 6.15). Hence the area required is

$$A = F(b) - F(a)$$

In our example, this area is $A = \dfrac{b^2}{2} - \dfrac{a^2}{2}$

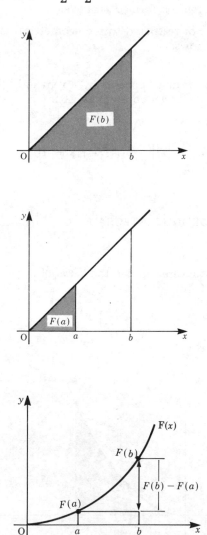

Figure 6.15

Having obtained the value of a definite integral by means of a particular example, we can now generalise the procedure.

We obtain a primitive function $F(x)$ for a given $f(x)$ and then form the difference of the values of the primitive function $F(x)$ at the positions of the upper and lower limit. As a shorthand notation, a square bracket with the limits as shown is often used.

Calculation of a definite integral:

$$\int_a^b f(x)\,dx = \Big[F(x)\Big]_a^b = F(b) - F(a) \tag{6.4}$$

Primitive functions for a given $f(x)$ differ only by an additive constant which cancels out when a difference is formed as above.

The definite integral is not restricted to the geometrical case of calculating area, as the following example shows.

EXAMPLE   What is the distance covered by a vehicle in the time interval $t = 0$ to $t = 12$ seconds if the velocity v is constant at $10\,\text{m/s}$?

The distance is given by

$$s = \int_{t_1}^{t_2} v\,dt = \Big[vt\Big]_{t_1}^{t_2} = v(t_2 - t_1) = 10 \times 12 = 120\,\text{m}$$

6.4.2   EXAMPLES OF DEFINITE INTEGRALS

Integration of x^2

We wish to calculate the area under the parabola $y = x^2$ between $x_1 = 1$ and $x_2 = 2$ shown in figure 6.16.

The area required is
$$A = \int_1^2 x^2\,dx$$

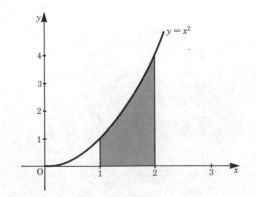

Figure 6.16

First we have to find a primitive function whose derivative is $f(x) = x^2$; such a function is

$$F(x) = \frac{x^3}{3}$$

We can easily verify this statement by differentiating $F(x)$.

The area required is then given by

$$A = \int_1^2 x^2 \, dx = \left[\frac{x^3}{3}\right]_1^2 = \frac{8}{3} - \frac{1}{3} = \frac{7}{3} \quad \text{units of area}$$

Integration of the Cosine Function

We want the area under the cosine function in the interval $0 \leqslant x \leqslant \pi/2$ (figure 6.17).

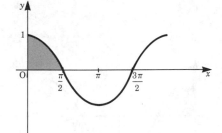

Figure 6.17

The required area is
$$A = \int_0^{\pi/2} \cos x \, dx$$

From our knowledge, we are able to identify a primitive function of $f(x) = \cos x$. It is

$$F(x) = \sin x$$

Hence we have

$$A = \int_0^{\pi/2} \cos x \, dx = \left[\sin x\right]_0^{\pi/2} = \sin\frac{\pi}{2} - \sin 0 = 1$$

If the area lies below the x-axis, the definite integral is negative.

Consider the area under the x-axis in the interval $\pi/2 \leqslant x \leqslant 3\pi/2$ (figure 6.18). It is

$$A = \int_{\pi/2}^{3\pi/2} \cos x \, dx = \left[\sin x\right]_{\pi/2}^{3\pi/2} = \sin\frac{3\pi}{2} - \sin\frac{\pi}{2} = -1 - 1 = -2 \quad \text{units of area}$$

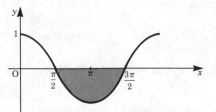

Figure 6.18

If we want to find the absolute value of an area, we must pay attention to the value of the function between the limits of integration, i.e. whether the function lies entirely above the x-axis, below it, or partly above and partly below.

If the function is partly positive and partly negative, it is necessary to split it into parts, as shown in the following example and illustrated in figure 6.19.

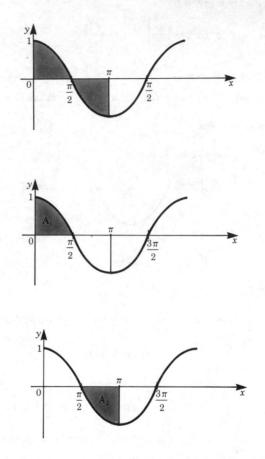

Figure 6.19

Suppose we require the absolute value of the area bounded by the function $\cos x$ for $0 \leqslant x \leqslant \pi$. We proceed as follows:

First: Since the function is positive for $0 \leqslant x \leqslant \pi/2$, the area $A_1 = 1$ (see above).

Second: Since the function is negative for $\pi/2 \leqslant x \leqslant \pi$, the area is given by

$$\int_{\pi/2}^{\pi} \cos x \; dx = \left[\sin x\right]_{\pi/2}^{\pi} = \sin \pi - \sin \frac{\pi}{2} = -1$$

If this area is to be taken positive, we must take its absolute value, i.e.

$$A_2 = |-1| = 1$$

The total absolute area is $A = A_1 + A_2 = 2$ units of area.

Uniformly Accelerated Motion

EXAMPLE   This is an example which differs from the calculation of an area and shows an application of integral calculus to science.

We first calculate the velocity of a rocket moving with a constant acceleration of 15 m/s^2, 40 s after starting from rest.

The relationship between acceleration a and velocity v is

$$a = \frac{dv}{dt}, \quad \text{at any instant } t$$

The velocity is given by

$$v = \int_{t_1}^{t_2} a \, dt = \left[at \right]_{t_1}^{t_2} = a(t_2 - t_1) = 15(40 - 0) = 600 \text{ m/s}$$

In addition we wish to find the distance covered by the rocket during that time. The relationship between velocity and distance is $v = \dfrac{ds}{dt}$ at any instant t. Substituting in the expression

$$v = \int_0^t a \, dt = at$$

gives

$$at = \frac{ds}{dt}$$

Hence $$s = \int_{t_1}^{t_2} at \, dt = \left[\frac{at^2}{2} \right]_{t_1}^{t_2} = \frac{a}{2}(t_2^2 - t_1^2) = \frac{15}{2}(40^2 - 0) = 12\,000 \text{ m}$$

6.5 METHODS OF INTEGRATION

We now consider general methods for determining primitive functions. The primitive function is an *indefinite integral*, indicated by the symbol

$$\int f(x) \, dx$$

6.5.1 PRINCIPLE OF VERIFICATION

The problem of obtaining the derivatives of functions can always be solved. But for the inverse problem, integration, a solution cannot always be found. Integration is a more difficult problem, and this makes the principle of verification important. We try to guess a solution and check whether it is a solution by differentiating, since

$$F'(x) = \frac{d}{dx} \left(\int f(x) \, dx \right) = f(x)$$

The function $f(x)$ is given. We assume that $F(x)$ is a primitive function.

Next we test the assumption by differentiating $F(x)$ and comparing $F'(x)$ with $f(x)$. Our assumption is valid if $F'(x) = f(x)$. In this case, $F(x)$ is a primitive function of $f(x)$. If our assumption is incorrect, i.e. $F'(x) \neq f(x)$, we have to make a new assumption and repeat the procedure until a solution is found. For this task we shall find tables of integrals a great help. These tables, which are universally available, cover a great many cases.

6.5.2 STANDARD INTEGRALS

The integration of basic functions can be found very easily by applying the fundamental theorem of differential and integral calculus, integration being the inverse of differentiation. Table 6.1 gives some elementary functions and their integrals; it is easy to verify the results by differentiation. A more comprehensive table will be found in the appendix at the end of this chapter.

Function	Integral
$x^n \ (n \neq -1)$	$\dfrac{x^{n+1}}{n+1} + C$
$\sin x$	$-\cos x + C$
$\cos x$	$\sin x + C$
e^x	$e^x + C$
$\dfrac{1}{x} \ (x \neq 0)$	$\ln|x| + C$

$$(6.5)$$

6.5.3 CONSTANT FACTOR AND THE SUM OF FUNCTIONS

Many integrals can be simplified before carrying out the integration. It is very wise to carry out the initial step of simplifying integrals as it reduces the amount of work involved and time is therefore saved.

Constant Factor

If k is a constant then

$$\int k f(x) \, dx = k \int f(x) \, dx \qquad\qquad (6.6)$$

PROOF   Let $F(x) = \int f(x) \, dx$. Then it is true to write

$$k F(x) = k \int f(x) \, dx$$

Differentiating both sides gives

$$k F' = k f(x)$$

Sum and Difference of Functions

The integral of the sum of two or more functions is equal to the sum of the integrals of the individual functions, i.e.

$$\int \{f(x) + g(x)\}\, dx \;=\; \int f(x)\, dx + \int g(x)\, dx \tag{6.7}$$

PROOF   Let $F(x)$ be the primitive function of $f(x)$ and $G(x)$ that of $g(x)$. Then

$$F(x) + G(x) \;=\; \int f(x)\, dx + \int g(x)\, dx$$

Differentiating both sides gives

$$F'(x) + G'(x) \;=\; f(x) + g(x)$$

For the difference of two functions it follows that

$$\int \{f(x) - g(x)\}\, dx \;=\; \int f(x)\, dx - \int g(x)\, dx \tag{6.8}$$

6.5.4 INTEGRATION BY PARTS: PRODUCT OF TWO FUNCTIONS

This follows directly from the product rule for differentiation. Let the functions be $u(x)$ and $v(x)$. If we differentiate the product uv we have

$$\frac{d}{dx}\{u(x)\ v(x)\} \;=\; \frac{du}{dx}\, v(x) + \frac{dv}{dx}\, u(x)$$

or, written in a more concise way,

$$(uv)' \;=\; u'v + uv'$$

By transposing we get

$$uv' \;=\; (uv)' - u'v$$

Now we integrate this equation and can hence write down the primitive function straight away as follows:

$$\int uv'\, dx \;=\; uv - \int vu'\, dx \tag{6.9}$$

The integral on the right-hand side is frequently much easier to evaluate than the one on the left-hand side. This method is particularly useful when the expression to be integrated contains functions such as $\log x$ and inverse functions. The right choice of u and v' is decisive.

EXAMPLE

$$\int x\, e^x\, dx$$

Let $u = x, \qquad v' = e^x.$

Then $u' = 1$ and we know that $v = e^x$.

Substituting in the equation gives

$$\int uv' \, dx \; = \; uv - \int vu' \, dx$$

$$\int x \, e^x \, dx \; = \; x \, e^x - \int e^x \, dx \; = \; e^x(x-1) + C$$

We must not forget the constant of integration C.

Suppose we had chosen $u = e^x$ and $v' = x$.

Then $u' = e^x$ and $v = \frac{1}{2}x^2$.

Hence

$$\int x \, e^x \, dx \; = \; \frac{x^2 \, e^x}{2} - \frac{1}{2}\int x^2 \, e^x \, dx$$

We note that the right-hand integral will be more difficult to solve than in the previous case.

EXAMPLE $\int x^2 \, e^x \, dx$

Let us try $u = x^2$ and $v' = e^x$.

Before going any further we must consider whether our choice is going to be favourable. We need to consider the vu' product in $\int vu' \, dx$ and do a rough calculation.

From $u = x^2$ it follows by differentiation that

$$u' \; = \; 2x$$

And from $v' = e^x$ it follows by integration that

$$v \; = \; e^x$$

Aside: If you are meeting this for the first time, you should get into the habit of making quick and rough calculations, and not leave the choice of u and v' to chance. We know that powers of x are reduced by one when differentiated. The term e^x remains unchanged when differentiated. If we had set $u = e^x$ and $v' = x^2$, then we would have found an increase of one in the power of x, i.e. an x^3, leading to a more difficult integral on the right-hand side.

Going back to our example, we have

$$\int x^2 \, e^x \, dx \; = \; x^2 \, e^x - 2\int x \, e^x \, dx$$

We have already computed the integral on the right-hand side in the previous example. It is

$$e^x(x-1)+C'$$

Thus $$\int x^2 e^x \, dx = x^2 e^x - 2e^x(x-1)+C = e^x(x^2-2x+2)+C$$

We can easily verify the result by differentiating — you are advised to do so.

EXAMPLE $$\int \sin^2 x \, dx = \int \sin x \sin x \, dx$$

Let $u = \sin x$ and $v' = \sin x$

Then $u' = \cos x$ and $v = -\cos x$

$$\int \sin^2 dx = -\sin x \cos x + \int \cos^2 x \, dx$$

Since $\cos^2 x = 1 - \sin^2 x$, then, by substituting, we find that

$$\int \sin^2 x \, dx = -\sin x \cos x + \int (1-\sin^2 x) \, dx$$

$$= -\sin x \cos x + x - \int \sin^2 x \, dx$$

Transposing gives

$$2\int \sin^2 x \, dx = x - \sin x \cos x + C' = x - \tfrac{1}{2}\sin 2x + C'$$

Hence we obtain the final solution:

$$\int \sin^2 dx = \frac{x}{2} - \frac{1}{2}\sin x \cos x + C = \frac{x}{2} - \frac{1}{4}\sin 2x + C$$

With the help of integration by parts, it is often possible to simplify integrals whose integrands are of the nth power, since the exponent can be reduced by one each time. Applying the method successively may lead either to a standard integral or to another integrable expression. This was demonstrated in the last two examples.

For the sake of completeness a further point should be noted. If the exponent of the sine function is some arbitrary number ($\neq 0$), then the integral can be solved step by step. This process leads to what is known as a *reduction formula*.

EXAMPLE Reduction formula for $\int \sin^n x \, dx$.

Let $u = \sin^{n-1} x$ and $v' = \sin x$

Then $u' = (n-1)\sin^{n-2} x \cos x$ and $v = -\cos x$

The integral now becomes

$$\int \sin^n x \, dx = -\cos x \sin^{n-1} x + (n-1)\int \sin^{n-2} x \cos^2 x \, dx$$

Recalling the identity $\cos^2 x = 1 - \sin^2 x$, the integral on the right-hand side can be split into the sum of two integrals:

$$\int \sin^n x \, dx = -\cos x \sin^{n-1} x + (n-1) \int \sin^{n-2} x \, dx - (n-1) \int \sin^n x \, dx$$

Observe that the integral to be solved now appears on both sides of the identity. Rearranging leads to the desired formula:

$$\int \sin^n x \, dx = -\frac{1}{n} \cos x \sin^{n-1} x + \frac{n-1}{n} \int \sin^{n-2} x \, dx \qquad (6.10)$$

Remember that this formula is valid for any exponent $n \neq 0$. If n is negative then it is to be read from the right to the left. In the exercises, the reader will be invited to derive the reduction formula for $\int \cos^n x \, dx$.

6.5.5   INTEGRATION BY SUBSTITUTION

Suppose we want to evaluate the following integral:

$$\int \sin(ax + b) \, dx$$

How should we proceed?

A minute's thought leads us to try to reduce it to a standard form, such as $\int \sin u \, du$.

We can *substitute* $u = ax + b$

But we still have to find a substitution for dx. To do this we differentiate $u = ax + b$ with respect to x and find that

$$\frac{du}{dx} = a$$

From which $$dx = \frac{1}{a} du$$

The integral now becomes

$$\int \sin u \, \frac{1}{a} \, du = \frac{1}{a} \int \sin u \, du = -\frac{1}{a} \cos u + C$$

To express the result in terms of the original variable x, we substitute back for $u = ax + b$ and obtain

$$\int \sin(ax + b) \, dx = -\frac{1}{a} \cos(ax + b) + C$$

Substitution then enables us to reduce the function into a standard form. Further-more, it also often transforms a difficult or an apparently unsolvable problem into a solvable one.

The method of substitution is carried out in four steps.

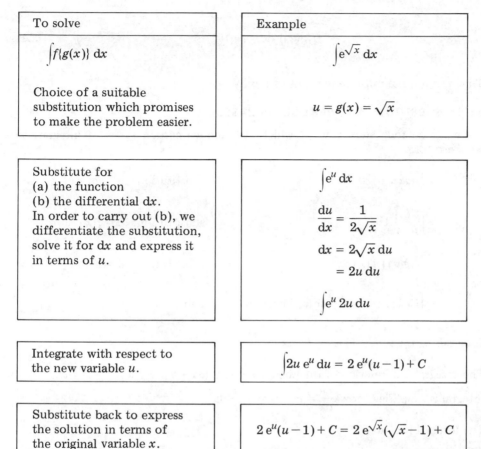

To solve	Example
$\int f\{g(x)\}\,dx$ Choice of a suitable substitution which promises to make the problem easier.	$\int e^{\sqrt{x}}\,dx$ $u = g(x) = \sqrt{x}$
Substitute for (a) the function (b) the differential dx. In order to carry out (b), we differentiate the substitution, solve it for dx and express it in terms of u.	$\int e^{u}\,dx$ $\dfrac{du}{dx} = \dfrac{1}{2\sqrt{x}}$ $dx = 2\sqrt{x}\,du$ $\quad = 2u\,du$ $\int e^{u}\,2u\,du$
Integrate with respect to the new variable u.	$\int 2u\,e^{u}\,du = 2\,e^{u}(u-1)+C$
Substitute back to express the solution in terms of the original variable x.	$2\,e^{u}(u-1)+C = 2\,e^{\sqrt{x}}(\sqrt{x}-1)+C$

There are no general rules for finding suitable substitutions.

To complete the discussion on this method, let us assume that the integrand can be put in the form $f\{g(x)\}$ as a function of a function. Then the integral we have to solve is

$$\int f\{g(x)\}\,dx$$

We now introduce a new variable by a *substitution*.

Let $u = g(x)$ for the inner function.

Differentiating u with respect to x gives

$$\frac{du}{dx} = \frac{dg}{dx} = g'(x)$$

We solve for dx to give

$$dx = \frac{1}{g'(x)}\,du$$

The integral becomes

$$\int f\{g(x)\}\, dx = \int f(u)\frac{du}{g'} \qquad (6.11)$$

This new integral must be in terms of u only.

* ## 6.5.6 SUBSTITUTION IN PARTICULAR CASES

We examine four substitutions which will reduce certain types of integrals to a standard form.

$$\int f(ax+b)\, dx \qquad\qquad\qquad \text{substitution: } u = ax+b$$

$$\int \frac{f'(x)}{f(x)}\, dx \qquad\qquad\qquad \text{substitution: } u = f(x)$$

$$\int f[g(x)]\, g'(x)\, dx \qquad\qquad \text{substitution: } u = g(x)$$

$$\int R(\sin x,\ \cos x,\ \tan x,\ \cot x)\, dx \qquad \text{substitution } u = \tan\frac{x}{2}$$

Integrals of the type $\displaystyle\int f(ax+b)\, dx$

The integral is a function of the linear function $(ax+b)$. By letting $u = ax+b$, the integral is simplified; now $\dfrac{du}{dx} = a$ or $dx = \dfrac{1}{a}\, du$.

The integral becomes

$$\int f(ax+b)\, dx = \frac{1}{a}\int f(u)\, du \qquad (6.12)$$

It is now much simpler than the original integral. If $f(u)$ is a simple function its solution is known. We found this to be the case when we introduced the substitution method and considered the integral

$$\int \sin(ax+b)\, dx$$

To illustrate this point let us look at some further examples:

EXAMPLE $\displaystyle\int \frac{2}{\cos^2(4x-12)}\, dx$

Let $u = 4x-12$, then $du = 4\, dx$; hence $dx = \dfrac{1}{4}\, du$. The integral becomes

$$\frac{2}{4}\int \frac{du}{\cos^2 u}$$

This is a standard integral given in the appendix at the end of this chapter:

$$\frac{1}{2}\int\frac{du}{\cos^2 u} = \frac{1}{2}\tan u + C$$

Substituting back in terms of x gives the final solution:

$$\int\frac{2}{\cos^2(4x-12)}\,dx = \frac{1}{2}\tan(4x-12) + C$$

EXAMPLE $\quad\displaystyle\int\frac{5dx}{1+(ax+b)^2}$

Let $u = ax + b$, then $dx = \dfrac{1}{a}du$

Hence we get one of the standard integrals given in the appendix at the end of this chapter:

$$\int\frac{5dx}{1+(ax+b)^2} = \frac{5}{a}\int\frac{du}{1+u^2} = \frac{5}{a}\tan^{-1}u + C$$

$$= \frac{5}{a}\tan^{-1}(ax+b) + C$$

Integrals of the type $\displaystyle\int\frac{f'(x)}{f(x)}\,dx$

The integrand is a fraction whose numerator is the differential coefficient of the denominator.

Let $u = f(x)$, then $f'(x)\,dx = du$

Hence we have

$$\int\frac{f'(x)}{f(x)}\,dx = \int\frac{du}{u} = \ln|u| + C = \ln|f(x)| + C \qquad (6.13)$$

In many cases we will have first of all to put the function to be integrated in the above form.

EXAMPLE $\quad\displaystyle\int\frac{x+a/2}{x^2+ax+b}\,dx = \frac{1}{2}\int\frac{2x+a}{x^2+ax+b}\,dx$

Let $u = x^2 + ax + b$, then $du = (2x+a)\,dx$

Hence

$$\frac{1}{2}\int\frac{2x+a}{x^2+ax+b}\,dx = \frac{1}{2}\int\frac{du}{u} = \frac{1}{2}\ln|u| + C$$

$$= \frac{1}{2}\ln|x^2+ax+b| + C$$

EXAMPLE $\displaystyle\int\frac{\cos x}{a + b\sin x}\,dx$

Let $u = a + b\sin x$, then $du = b\cos x\,dx$ or $\cos x\,dx = \dfrac{1}{b}\,du$

Hence

$$\int\frac{\cos x}{a + b\sin x}\,dx = \frac{1}{b}\int\frac{du}{u} = \frac{1}{b}\ln|u| + C$$

$$= \frac{1}{b}\ln|a + b\sin x| + C$$

Integrals of the type $\displaystyle\int f(g(x))g'(x)\,dx$

The integrand is a product, but what is important is the fact that the second function is the differential coefficient of the inner function.

To solve the integral, let

$$u = g(x), \qquad g'(x)\,dx = du$$

Hence we have

$$\int f(g(x))g'(x)\,dx = \int f(u)\,du \tag{6.14}$$

EXAMPLE $\displaystyle\int\sin^2 x\cos x\,dx$

Let $u = \sin x$, then $du = \cos x\,dx$

Hence

$$\int\sin^2 x\cos x\,dx = \int u^2\,du = \frac{1}{3}u^3 + C$$

$$= \frac{1}{3}\sin^3 x + C$$

In many cases we first have to generate the form of the integrand which corresponds to equation 6.14, as the following example shows.

EXAMPLE $\displaystyle\int(\tan^4 x + \tan^2 x + 1)\,dx$

The integrand does not contain the factor $\dfrac{1}{\cos^2 x}$ which we need in order to apply the method explained. We therefore expand with $\cos^2 x$ by using the relation

$$\frac{1}{\cos^2 x} = 1 + \tan^2 x \quad\text{or}\quad 1 = (1 + \tan^2 x)\cos^2 x$$

We find

$$\int (\tan^4 x + \tan^2 x + 1)\, dx \;=\; \int \frac{(\tan^4 x + \tan^2 x + 1)}{(1 + \tan^2 x)\cos^2 x}\, dx$$

Now let $u = \tan x$, then $du = \dfrac{1}{\cos^2 x}\, dx$.

We obtain a new integral, namely

$$\int \frac{u^4 + u^2 + 1}{1 + u^2}\, du \;=\; \int \left(u^2 + \frac{1}{1 + u^2} \right) du \;=\; \int u^2\, du + \int \frac{du}{1 + u^2}$$

i.e. two standard forms.

The final solution is

$$\frac{1}{3} u^3 + \tan^{-1} u + C \;=\; \frac{1}{3}\tan^3 x + \tan^{-1}(\tan x) + C$$

$$=\; \frac{1}{3}\tan^3 x + x + C$$

Integrals of the type $\displaystyle \int R(\sin x, \cos x, \tan x, \cot x)\, dx$

The integrand is a rational expression, denoted by R, of the trigonometric functions. It can be transformed into a more accessible form by substitution $u = \tan \dfrac{x}{2}$, i.e. $x = 2\tan^{-1} u$.

By differentiating we get
$$dx = \frac{2du}{1 + u^2} \tag{6.15a}$$

The trigonometric functions can all be expressed in terms of u. Thus

$$\sin x = 2\sin\frac{x}{2}\cos\frac{x}{2} = \frac{2\tan(x/2)}{1 + \tan^2(x/2)} = \frac{2u}{1 + u^2} \tag{6.15b}$$

$$\cos x = \cos^2(x/2) - \sin^2(x/2) = \frac{1 - \tan^2(x/2)}{1 + \tan^2(x/2)} = \frac{1 - u^2}{1 + u^2} \tag{6.15c}$$

$$\tan x = \frac{\sin x}{\cos x} = \frac{2u}{1 - u^2} \tag{6.15d}$$

$$\cot x = \frac{\cos x}{\sin x} = \frac{1 - u^2}{2u} \tag{6.15e}$$

Thus the integral is transformed into one whose integrand is a function of u.

EXAMPLE
$$\int \frac{dx}{\sin x} = \int \frac{1 + u^2}{2u} \frac{2\, du}{1 + u^2} = \int \frac{du}{u} = \ln|u| + C = \ln\left| \tan\frac{x}{2} \right| + C$$

EXAMPLE $\displaystyle\int \frac{dx}{1+\sin x} = 2\int \frac{1}{1+2u/(1+u^2)}\ \frac{du}{1+u^2} = 2\int \frac{du}{1+2u+u^2} = 2\int \frac{du}{(1+u)^2}$

$$= \frac{-2}{1+\tan(x/2)} + C$$

In the table of fundamental standard integrals (at the end of this chapter) we find

that $\displaystyle\int \frac{dx}{1+\sin x} = \tan\left(\frac{x}{2} - \frac{\pi}{4}\right) + C$. The reader should verify for him- or herself that

these results differ by a constant only $(= 1)$.

6.5.7   INTEGRATION BY PARTIAL FRACTIONS

We now consider the integration of functions where the numerator and denominator are polynomials. Such functions are called *fractional rational functions* and have the following form:

$$R(x) = \frac{P(x)}{Q(x)} = \frac{a_n x^n + a_{n-1} x^{n-1} + \ldots + a_1 x + a_0}{b_m x^m + b_{m-1} x^{m-1} + \ldots + b_1 x + b_0}$$

where m and n are integers and a_n and $b_m \neq 0$. The coefficients a_i and b_i are real.

If $n < m$, $R(x)$ is a proper fractional rational function; in short, a proper fraction.

If $n > m$, $R(x)$ is an improper fraction, e.g. $\dfrac{x^4}{x^3+1}$

Any improper fraction can be transformed into a sum of a polynomial and a proper fraction by simple division. For example

$$\frac{x^4}{x^3+1} = x - \frac{x}{x^3+1}$$

Our discussion is restricted to proper fractional rational functions, such as the second expression in the above example.

In order to understand the expansion of such a function, we have to remember the fundamental theorem of algebra. This theorem states that a rational function of degree n, such as

$$P(x) = a_n x^n + a_{n-1} x^{n-1} + \ldots + a_1 x + a_0$$

can be resolved into a product of factors, each of which is linear:

$$P(x) = a_n (x - x_1)(x - x_2) \ldots (x - x_n)$$

a_i are constant real coefficients $(a_n \neq 0)$.

x_i are the real or complex roots of the equation $P(x) = 0$.

Complex roots occur in pairs and are conjugate, e.g. $\alpha + j\beta$ and $\alpha - j\beta$. (Complex numbers are treated in detail in Chapter 9.)

EXAMPLE   The cubic equation

$$x^3 - x^2 - 4x + 4 = 0$$

has the following roots: $x_1 = 1$, $x_2 = 2$, $x_3 = -2$.

Hence, $$x^3 - x^2 - 4x + 4 = (x-1)(x-2)(x+2)$$

Such an expansion into linear factors will help us to integrate fractional rational functions. Assume that the integral to be solved is of the form:

$$\int \frac{a_n x^n + a_{n-1} x^{n-1} + \ldots + a_0}{x^m + b_{m-1} x^{m-1} + \ldots + b_0} \, dx$$

(Note that $b_m = 1$ is easily obtained by division, and that $n < m$.)

The integral is then resolved into a sum of proper partial fractions. The form of the partial fractions is dictated by the roots of the denominator, which we will call $D(x)$. There are three cases to consider:

Case 1: $D(x)$ has real and unequal roots

Case 2: $D(x)$ has real and repeated roots

Case 3: $D(x)$ has complex roots

Real and Unequal Roots

The denominator takes on the form

$$D(x) = x^m + b_{m-1} x^{m-1} + \ldots + b_0 = (x - x_1)(x - x_2) \ldots (x - x_m)$$

EXAMPLE $\int \dfrac{3x - 5}{x^2 - 2x - 8} \, dx$

The denominator has two real and unequal roots, $x_1 = -2$ and $x_2 = 4$.

Hence $x^2 - 2x - 8 = (x + 2)(x - 4)$ and the integral becomes

$$\int \frac{3x - 5}{x^2 - 2x - 8} \, dx = \int \frac{3x - 5}{(x + 2)(x - 4)} \, dx$$

In this form, the integral is not easily solved. Now we will show that integrals of this type can be solved if we expand the integrand into partial fractions, i.e.

$$\frac{3x - 5}{(x + 2)(x - 4)} = \frac{A}{x + 2} + \frac{B}{x - 4}$$

A and B are constants to be determined. Multiplying both sides of the identity by $(x + 2)(x - 4)$ also referred to as *'clearing the fractions'*, yields

$$3x - 5 = A(x - 4) + B(x + 2)$$

As this is an identity, it must hold for all values of x.

To calculate the values of A and B it is sufficient to insert any value for x we care to choose. If, in this example, we insert $x = x_1 = -2$ and then $x = x_2 = 4$, we find

$$x = -2, \qquad 3 \times (-2) - 5 = A(-2-4); \qquad \text{hence } A = \frac{11}{6}$$

$$x = 4, \qquad 3 \times 4 - 5 = B(4+2); \qquad \text{hence } B = \frac{7}{6}$$

The integral becomes

$$\int \frac{3x-5}{(x-2)(x+4)} \, dx = \int \left(\frac{11/6}{x+2} + \frac{7/6}{x-4} \right) dx$$

$$= \frac{11}{6} \ln |x+2| + \frac{7}{6} \ln |x-4| + C$$

The expansion of a function into partial fractions is carried out in three steps.

(1) Find the roots of the denominator and express it as the product of factors of the lowest possible degree.

(2) Rewrite the original integrand as the sum of partial fractions.

(3) Multiply both sides of the identity by the denominator and then calculate the values of the constants of the partial fractions $A, B, C, \ldots, M$ by inserting successively the roots of the denominator, $x_1, x_2, x_3, \ldots, x_n$.

Rule   If the roots of the denominator $D(x)$, $\quad x_1, x_2, x_3, \ldots, x_n$, are real and unequal, then we set up

$$\frac{N(x)}{D(x)} = \frac{A}{(x-x_1)} + \frac{B}{(x-x_2)} + \ldots + \frac{M}{(x-x_n)} \qquad (6.16)$$

Real and Repeated Roots

Let us consider the integral

$$\int \frac{dx}{x^3 - 3x^2 + 4} = \int \frac{dx}{(x+1)(x-2)^2}$$

The denominator has roots $x_1 = x_2 = 2$ and $x_3 = -1$. The roots x_1 and x_2 are equal; they are called repeated roots. To every r-fold linear factor $(x-x_i)^r$ of $D(x)$ there correspond r partial fractions of the form

$$\frac{A_1}{(x-x_i)} + \frac{A_2}{(x-x_i)^2} + \frac{A_3}{(x-x_i)^3} + \ldots + \frac{A_r}{(x-x_i)^r}$$

Let us return to our example. The integrand is now

$$\frac{1}{(x+1)(x-2)^2} = \frac{A}{x+1} + \frac{B_1}{(x-2)} + \frac{B_2}{(x-2)^2}$$

To calculate A_1, B_1 and B_2 we insert three particular values of x. We have, by clearing the fractions,

$$1 = A(x-2)^2 + B_1(x+1)(x-2) + B_2(x+1)$$

With $x = x_1 = 2$ $\qquad 1 = 3B_2$ $\qquad\qquad$ hence $\quad B_2 = \dfrac{1}{3}$

with $x = x_3 = -1$ $\qquad 1 = 9A$ $\qquad\qquad$ hence $\quad A = \dfrac{1}{9}$

with $x = 0$ (say) $\qquad 1 = 4A + (-2B_1) + B_2$ $\qquad$ hence $\quad B_1 = -\dfrac{1}{9}$

Thus we obtain

$$\int \frac{dx}{x^3 - 3x + 4} = \frac{1}{9} \int \frac{dx}{(x+1)} - \frac{1}{9} \int \frac{dx}{(x-2)} + \frac{1}{3} \int \frac{dx}{(x-2)^2}$$

Apart from integrals of the type $\int \dfrac{du}{u}$, we should recognise the standard integral $\int u^n \, du$.

According to section 6.5.5, we will have with the substitution $u = x - 2$:

$$\frac{1}{3} \int \frac{dx}{(x-2)^2} = \frac{1}{3} \int \frac{du}{u^2} = \frac{1}{3} \int u^{-2} \, du = -\frac{1}{3} \frac{1}{u} + C = -\frac{1}{3(x-2)} + C$$

The solution to our example is

$$\int \frac{dx}{x^3 - 3x^2 + 4} = \frac{1}{9}(\ln|x+1| - \ln|x-2|) - \frac{1}{3(x-2)} + C$$

> **Rule** If the denominator has a real root x_0 repeated r times (and some other distinct real roots $x_1, \ldots, x_n$), then the integrand takes on the form
>
> $$\frac{N(x)}{D(x)} = \frac{A_1}{x - x_0} + \frac{A_2}{(x - x_0)^2} + \ldots + \frac{A_r}{(x - x_0)^r} + \frac{B_1}{x - x_1} + \frac{B_2}{x - x_2} + \ldots + \frac{B_n}{x - x_n}$$
>
> $$(6.17)$$

Complex Roots

If $D(x)$ has complex roots, the method outlined above must be altered.

For each quadratic expression we will set $\dfrac{Px + Q}{x^2 + ax + b}$; P and Q are constants to be determined.

The procedure is best illustrated by an example.

EXAMPLE $\int \dfrac{2x^2 - 13x + 20}{x(x^2 - 4x + 5)}\,dx$

Again we set $D(x) = x(x^2 - 4x + 5) = 0$

The roots of the denominator are $x_1 = 0$, $x_2 = 2 - j$, $x_3 = 2 + j$, i.e. there is only one real root. Thus we can write

$$\frac{2x^2 - 13x + 20}{x(x^2 - 4x + 5)} = \frac{A}{x} + \frac{Px + Q}{x^2 - 4x + 5}$$

In the case of complex roots, the denominator is not split up into linear factors.

Clearing the fractions we have

$$2x^2 - 13x + 20 = A(x^2 - 4x + 5) + Px^2 + Qx$$

The constants A, P and Q are calculated by inserting particular values for x.

Hence

$$\text{putting} \quad x = x_1 = 0, \qquad 20 = 5A$$
$$\text{putting} \quad x = 1, \qquad\qquad 9 = 2A + P + Q$$
$$\text{putting} \quad x = -1, \qquad 35 = 10A + P - Q$$

Solving for A, P and Q gives

$$A = 4, \quad P = -2, \quad Q = 3$$

The integral becomes

$$\int \frac{2x^2 - 13x + 20}{x(x^2 - 4x + 5)}\,dx = 4\int \frac{dx}{x} - \int \frac{2x - 3}{x^2 - 4x + 5}\,dx$$

The second integral can be solved by the methods already seen.

The reader is invited to verify that

$$\int \frac{2x - 4}{x^2 - 4x + 5}\,dx = \ln|x^2 - 4x + 5| + \tan^{-1}(x - 2) + C'$$

The solution of our integral is

$$\int \frac{2x^2 - 13x + 20}{x(x^2 - 4x + 5)}\,dx = 4\ln|x| + \ln|x^2 - 4x + 5| + \tan^{-1}(x - 2) + C$$

This last example illustrates the fact that integration of partial fractions needs careful consideration and that, in the end, we may have to use the whole range of integration techniques.

Rule If the denominator $D(x)$ of the integrand has, e.g. the two conjugates, complex roots x_1 and x_2, then the expansion into partial fractions takes on the following form:

$$\frac{N(x)}{D(x)} = \ldots + \frac{Px + Q}{(x - x_1)(x - x_2)} + \ldots = \ldots + \frac{Px + Q}{x^2 + ax + b} + \ldots \qquad (6.18)$$

6.6 RULES FOR SOLVING DEFINITE INTEGRALS

The rules for solving indefinite integrals apply equally to definite integrals, e.g. a constant factor in the integrand can be placed in front of the integral. The integral of a sum or difference of functions is equal to the sum or difference of the integrals of the individual functions, and so forth.

We have interpreted the definite integral geometrically as the area under a curve. From this interpretation we can easily derive certain characteristics of the definite integral which are geometrically evident.

Expansion of an Integral Into a Sum by Subdividing the Range of Integration

The value of the integral $\int_a^b f(x)\,dx$ remains unchanged if we insert another limit c between the limits a and b and calculate its value not from a to b but from a to c and then from c to b. Thus

$$
\begin{aligned}
\int_a^b f(x)\,dx &= \int_a^c f(x)\,dx + \int_c^b f(x)\,dx \\
&= F(c) - F(a) + F(b) - F(c) \\
&= F(b) - F(a)
\end{aligned}
\tag{6.19}
$$

The rule becomes evident if we consider the problem geometrically (see figure 6.20).

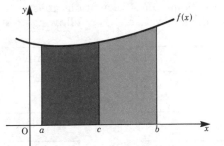

Figure 6.20

Interchanging the Limits of Integration

If we interchange the limits of integration the integral changes sign:

$$
\int_a^b f(x)\,dx = -\int_b^a f(x)\,dx
\tag{6.20}
$$

PROOF
$$
\int_a^b f(x)\,dx = F(b) - F(a) = -[F(a) - F(b)]
$$

$$
= -\int_b^a f(x)\,dx
$$

Upper and Lower Limits of Integration are Equal

If the upper and lower limits of integration are equal, the integral vanishes:

$$\int_a^b f(x)\,dx = 0 \qquad \text{if} \qquad a = b \tag{6.21}$$

Designation

The value of a definite integral is independent of the designation of the variable:

$$\int_a^b f(x)\,dx = \int_a^b f(z)\,dz = \int_a^b f(u)\,du \tag{6.22}$$

The value of a definite integral depends only on its limits and not on the designation of the variable of integration. We can use whatever designation we like for convenience; this is often done in engineering and science.

Substitution of Limits of Integration

By means of a suitable substitution, it is often possible to transform a given integral into a standard one. The method of substitution discussed in section 6.5.5 is applicable to definite integrals. Sometimes calculations are shortened by working out new limits corresponding to the new variable. The following illustrates the procedure.

To solve	Example
$\displaystyle\int_a^b f(g(x))\,dx$	$\displaystyle\int_1^5 \sqrt{2x-1}\,dx$

Select a substitution.	$u = g(x) = \sqrt{2x-1}$

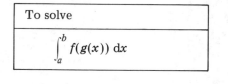

Substitute and change the limits: lower limit $u_1 = u(a)$, upper limit $u_2 = u(b)$.	$u_1 = \sqrt{2 \times 1 - 1} = 1$ $u_2 = \sqrt{2 \times 5 - 1} = 3$ $\displaystyle\int_1^5 \sqrt{2x-1}\,dx = \int_1^3 u^2\,du$

Integrate $\displaystyle\int_{u_1}^{u_2} f(u)\,du$	$\displaystyle\int_1^3 u^2\,du = \left[\frac{1}{3}u^3\right]_1^3 = \frac{26}{3}$

6.7 MEAN VALUE THEOREM

If $f(x)$ is a continuous function throughout the range $x = a$ to $x = b$, then

$$\int_a^b f(x)\,\mathrm{d}x = f(x_0)(b-a)$$

Within the interval a to b, there exists at least one value x_0 of x for which the area of the rectangle of width $(b-a)$ and height $f(x_0)$ is equal to the area under the curve within that same interval (figure 6.21).

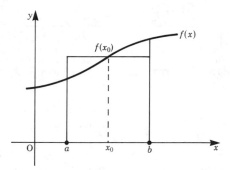

Figure 6.21

It follows, therefore, that the value $f(x_0)$ is the *mean value of the function* in the interval considered. Hence

$$f(x_0) = y_m = \frac{1}{b-a}\int_a^b f(x)\,\mathrm{d}x \tag{6.23}$$

In engineering and science, we frequently find it necessary to calculate the mean value of a varying quantity. For example, in the case of a variable force acting against some resistance, the work done will depend on the mean value of that force; the power in an electrical network is the mean value of the product of the alternating current and voltage.

EXAMPLE   A force applied to a body from $s_1 = 1\,\mathrm{m}$ to $s_2 = 8\,\mathrm{m}$ is given by $F = \dfrac{s^2}{2}$ N. Calculate its mean value.

If F_m = mean force in the interval $s_2 - s_1$, then

$$F_m = \frac{1}{s_2 - s_1}\int_{s_1}^{s_2} F(s)\,\mathrm{d}s = \frac{1}{(8-1)}\int_1^8 \frac{s^2}{2}\,\mathrm{d}s = \frac{1}{7}\frac{\left[s^3\right]_1^8}{6} = \frac{1}{42}(8^3 - 1) = 12.2\,\mathrm{N}$$

This force, F_m, represents the constant force applied to the body which produces the same amount of work in the interval as the actual force.

6.8 IMPROPER INTEGRALS

Let the function $y = f(x) = \dfrac{1}{x^2}$, with $x \neq 0$ (figure 6.22).

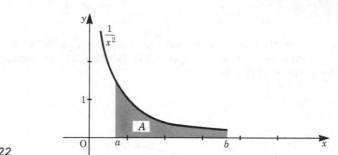

Figure 6.22

It is required to calculate the area shown shaded in the figure within the interval $x = a$ to $x = b$.

Let A be this area. Then

$$A = \int_a^b \frac{dx}{x^2} = \left[-\frac{1}{x} \right]_a^b = \frac{1}{a} - \frac{1}{b}$$

There is no particular difficulty in this instance. Suppose we now extend the upper limit to the right (figure 6.23): the value of the area will increase, and if we allow b to grow indefinitely we find that

$$A = \lim_{b \to \infty} \int_a^b \frac{dx}{x^2} = \lim_{b \to \infty} \left(\frac{1}{a} - \frac{1}{b} \right) = \frac{1}{a}$$

or, more simply,

$$F = \int_a^\infty \frac{dx}{x^2} = \frac{1}{a}$$

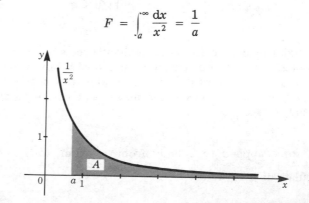

Figure 6.23

Such an integral is called an *improper integral*, but its value can be finite.

DEFINITION   Integrals with infinite limits of integration are called *improper integrals*. An improper integral is said to be *convergent* if its value is finite, and *divergent* if its value is infinite.

Integrals whose integrands tend to infinity for some value of the variable are also called improper,

e.g. $\int \dfrac{dx}{\sqrt{x-1}}$ is improper because the function $f(x) = \dfrac{1}{\sqrt{x-1}}$ at $x = 1$ is infinite.

However, the area under the curve in the interval $x = 1$ to $x = 2$ is finite since

$$\int_1^2 \frac{dx}{\sqrt{x-1}} \;=\; \left[2\sqrt{x-1}\right]_1^2 \;=\; 2 \qquad \text{(see figure 6.24)}$$

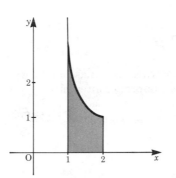

Figure 6.24

By the above definition we have extended to infinite limits the concept of the definite integral, which was originally established for finite limits.

It should be borne in mind that not all integrals with infinite limits are convergent.

EXAMPLE $\displaystyle\int_a^\infty \frac{dx}{x},\qquad a > 0$

First consider $\displaystyle\int_a^b \frac{dx}{x}$ where b is finite. Its value is $\ln b - \ln a$.

If we now allow b to grow beyond all bounds, the term $\ln b$ tends to infinity. Hence the integral has no finite value: it is an improper divergent integral, i.e.

$$\int_a^\infty \frac{dx}{x} \;\to\; \infty$$

EXAMPLE   Work done in the gravitational field.

If U is the work required to move a body of mass m through a given distance against the gravitational field produced by a body of mass M (figure 6.25), then, by Newton's law of gravitation, the force F between the two bodies is

$$F \;=\; \gamma \frac{mM}{r^2}$$

where γ = universal gravitational constant, r = distance between the centres of the bodies.

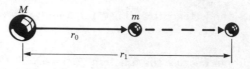

Figure 6.25

For a small displacement dr, the work done, dU, is

$$dU = F\,dr = \gamma \frac{mM}{r^2}\,dr$$

If the body of mass m is moved from a distance r_0 to a distance r_1, the total work done during the displacement is

$$U = \int_{r_0}^{r_1} \gamma \frac{mM}{r^2}\,dr = \gamma\,mM \int_{r_0}^{r_1} \frac{dr}{r^2} = \gamma mM \left(\frac{1}{r_0} - \frac{1}{r_1} \right)$$

An interesting case occurs when the mass m 'leaves' the gravitational field, i.e. $r_1 \to \infty$. We find a convergent improper integral:

$$v = \gamma mM \int_{r_0}^{\infty} \frac{dr}{r^2} = \frac{\gamma mM}{r_0}$$

6.9   LINE INTEGRALS

As an example, we will consider the force **F** exerted on a body which depends on the position **r** in space (see figure 6.26). This could be, e.g. a gravitational force on a mass point or an electric force on a charged particle. We want to determine the work U corresponding to the body's movement along some curve from a point P_1 to a point P_2. This movement can be described in parametric form

$$\mathbf{r}(t) = (x(t),\ y(t),\ z(t))$$

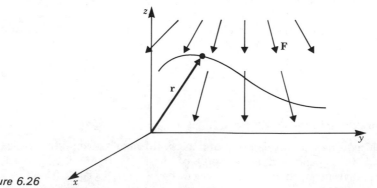

Figure 6.26

The components of the force are

$$\mathbf{F}(\mathbf{r}) = (F_x(\mathbf{r}),\ F_y(\mathbf{r}),\ F_z(\mathbf{r}))$$

The curve may be thought of as being split up into n small segments. As an approximation for the work, we take the sum of all fractional amounts of work, assuming that the force is approximately constant along each tiny segment (see figure 6.27).

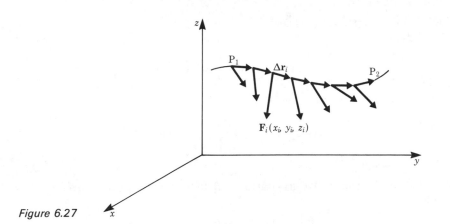

Figure 6.27

The work ΔU_i corresponding to the ith segment is determined by the scalar product of the force $\mathbf{F}_i$ and the vector pointing along the segment $\Delta \mathbf{r}_i$.

$$\Delta U_i = \mathbf{F}_i \cdot \Delta \mathbf{r}_i$$

Thus the whole work is the sum

$$U = \sum_i \mathbf{F}_i \cdot \Delta \mathbf{r}_i$$

If we make the elements smaller and smaller we get, in the limiting case, the integral

$$U = \int_{P_1}^{P_2} \mathbf{F}(\mathbf{r}) \cdot d\mathbf{r}$$

This type of integral is called a *line integral*. The name is based on the fact that the path of integration is a curve or a line in space. Let us look at the integral in more detail. The force is given by

$$\mathbf{F} = (F_x(\mathbf{r}), \quad F_y(\mathbf{r}), \quad F_z(\mathbf{r}))$$

Now our problem is to determine an expression for the path element $d\mathbf{r}$. We start with the expression

$$\mathbf{r} = (x(t), \quad y(t), \quad z(t))$$

As t varies from t_1 to t_2, the position vector $\mathbf{r}$ moves from P_1 to P_2. The path element is given by

$$d\mathbf{r} = \left(\frac{dx(t)}{dt} dt, \quad \frac{dy(t)}{dt} dt, \quad \frac{dz(t)}{dt} dt \right)$$

$$d\mathbf{r} = (dx, \quad dy, \quad dz)$$

Now we calculate the line integral:

$$U = \int_{P_1}^{P_2} \mathbf{F}((\mathbf{r}(t))) \cdot d\mathbf{r}$$

$$= \int_{t_1}^{t_2} \mathbf{F}(x(t),\, y(t),\, z(t)) \cdot d\mathbf{r}$$

$$= \int_{t_1}^{t_2} \left(F_x(\mathbf{r}) \cdot \mathbf{i} + F_y(\mathbf{r}) \cdot \mathbf{j} + F_z(\mathbf{r}) \cdot \mathbf{k} \right) \left(\frac{dx}{dt}\, dt \cdot \mathbf{i} + \frac{dy}{dt}\, dt \cdot \mathbf{j} + \frac{dz}{dt}\, dt \cdot \mathbf{k} \right)$$

Hence the formula for the work U reads

$$U = \int_{t_1}^{t_2} \left(F_x(\mathbf{r})\, \frac{dx}{dt}\, dt + F_y(\mathbf{r})\, \frac{dy}{dt}\, dt + F_z(\mathbf{r})\, \frac{dz}{dt}\, dt \right)$$

EXAMPLE   Let us consider the gravitational field near the Earth's surface. It is expressed by

$$\mathbf{F} = (0,\quad 0,\quad -mg)$$

Consider the fairground Ferris wheel in figure 6.28. We want to find the work done during the ascent of the Ferris wheel (mass m). The path is the semicircle from P_1 to P_2. Its parametric form with parameter ϕ is

$$\mathbf{r} = (0,\quad R \sin \phi,\quad -R \cos \phi)$$

$$d\mathbf{r} = (0,\quad R \cos \phi,\quad R \sin \phi)\, d\phi$$

$$U = \int_{\phi=0}^{\pi} (0,\quad 0,\quad -mg)\,(0,\quad R \cos \phi,\quad R \sin \phi)\, d\phi$$

$$= \int_{\phi=0}^{\pi} (-mgR \sin \phi)\, d\phi = \left[mgR \cos \phi \right]_0^{\pi} = 2mgR$$

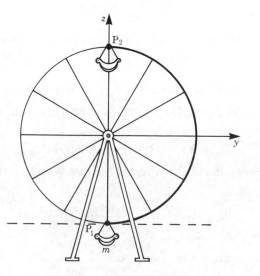

Figure 6.28

Forces like gravitational or electrostatic forces are due to *conservative fields*. This means that the work done on a body depends not on the path but only on the points P_1 and P_2, and that it is independent of time.

Examples of non-conservative fields are electrical fields caused by induction processes.

In the case of conservative fields, the line integral can be calculated easily if the path is chosen in such a way that force and path are either perpendicular or parallel to each other. This means that the path is divided into segments which are easy to work with.

APPENDIX

TABLE OF FUNDAMENTAL STANDARD INTEGRALS

The constant of integration has been omitted

$f(x)$	$\int f(x)\,dx$		$f(x)$	$\int f(x)\,dx$	
c	cx		$\dfrac{1}{x^2+a^2}$	$\dfrac{1}{a}\tan^{-1}\dfrac{x}{a}$ or $\dfrac{-1}{a}\cot^{-1}\dfrac{x}{a}$	
x^n	$\dfrac{x^{n+1}}{n+1}$	$(n\neq -1)$	$\dfrac{1}{x^2+2ax+b}$	$\dfrac{1}{\sqrt{b-a^2}}\tan^{-1}\left(\dfrac{x+a}{\sqrt{b-a^2}}\right)(b>a^2)$	
$\dfrac{1}{x}$	$\ln|x|$	$(x\neq 0)$	$\dfrac{2x+a}{x^2+ax+b}$	$\ln|x^2+ax+b|$	
e^x	e^x				
a^x	$\dfrac{a^x}{\ln a}$	$\left(\begin{matrix}a>0\\a\neq 1\end{matrix}\right)$	$\sqrt{ax+b}$	$\dfrac{2}{3a}\sqrt{(ax+b)^3}$	
			$\dfrac{1}{\sqrt{ax+b}}$	$\dfrac{2}{a}\sqrt{ax+b}$	
$\ln x$	$x\ln x-x$	$(x>0)$			
$\dfrac{1}{x-a}$	$\ln|x-a|$		$\dfrac{1}{\sqrt{a^2-x^2}}$	$\sin^{-1}\dfrac{x}{a}$	
$\dfrac{1}{(x-a)^2}$	$-\dfrac{1}{x-a}$		$\sqrt{a^2-x^2}$	$\dfrac{x}{2}\sqrt{a^2-x^2}+\dfrac{a^2}{2}\sin^{-1}\dfrac{x}{a}$	
$\dfrac{1}{x^2-a^2}$	$\dfrac{1}{2a}\ln\left|\dfrac{x-a}{x+a}\right|=\begin{cases}\dfrac{-1}{a}\tanh^{-1}\dfrac{x}{a},\\ |x|<|a|\\ \dfrac{-1}{a}\coth^{-1}\dfrac{x}{a},\\ |x|>|a|\end{cases}$				

TABLE OF STANDARD INTEGRALS *CONTINUED*

$f(x)$	$\int f(x)\,dx$	$f(x)$	$\int f(x)\,dx$
$\dfrac{1}{\sqrt{x^2+a^2}}$	$\ln\left(\dfrac{x+\sqrt{x^2+a^2}}{\lvert a\rvert}\right)=\sinh^{-1}\dfrac{x}{a}$	$\dfrac{1}{1-\sin x}$	$-\cot\left(\dfrac{x}{2}-\dfrac{\pi}{4}\right)=\tan\left(\dfrac{x}{2}+\dfrac{\pi}{4}\right)$
$\sqrt{x^2+a^2}$	$\dfrac{x}{2}\sqrt{x^2+a^2}+\dfrac{a^2}{2}\ln(x+\sqrt{a^2+x^2})$	$\dfrac{1}{1+\cos x}$	$\tan\dfrac{x}{2}$
$\dfrac{1}{\sqrt{x^2-a^2}}$	$\ln\left\lvert\dfrac{x+\sqrt{x^2-a^2}}{a}\right\rvert=\cosh^{-1}\dfrac{x}{a}$	$\dfrac{1}{1-\cos x}$	$-\cot\dfrac{x}{2}$
		$\tan x$	$-\ln\lvert\cos x\rvert$
$\sin x$	$-\cos x$	$\tan^2 x$	$\tan x-x$
$\sin^2 x$	$\dfrac{1}{2}(x-\sin x\cos x)$	$\cot x$	$\ln\lvert\sin x\rvert$
		$\cot^2 x$	$-\cot x-x$
$\dfrac{1}{\sin x}$	$\ln\left\lvert\tan\dfrac{x}{2}\right\rvert$		
$\dfrac{1}{\sin^2 x}$	$-\cot x$	$\sin^{-1}x$	$x\sin^{-1}x+\sqrt{1-x^2}$
		$\cos^{-1}x$	$x\cos^{-1}x-\sqrt{1-x^2}$
$\cos x$	$\sin x$	$\tan^{-1}x$	$x\tan^{-1}x-\ln\sqrt{1+x^2}$
$\cos^2 x$	$\dfrac{1}{2}(x+\sin x\cos x)$	$\cot^{-1}x$	$x\cot^{-1}x+\ln\sqrt{1+x^2}$
$\dfrac{1}{\cos x}$	$\ln\left\lvert\tan\left(\dfrac{x}{2}+\dfrac{\pi}{4}\right)\right\rvert$	$\sinh x$	$\cosh x$
$\dfrac{1}{\cos^2 x}$	$\tan x$	$\cosh x$	$\sinh x$
		$\tanh x$	$\ln\lvert\cosh x\rvert$
$\dfrac{1}{1+\sin x}$	$\tan\left(\dfrac{x}{2}-\dfrac{\pi}{4}\right)$	$\coth x$	$\ln\lvert\sinh x\rvert$
		$\sinh^{-1}x$	$x\sinh^{-1}x-\sqrt{x^2+1}$
		$\cosh^{-1}x$	$x\cosh^{-1}x-\sqrt{x^2-1}$
		$\tanh^{-1}x$	$x\tanh^{-1}x+\ln\sqrt{1-x^2}$
		$\coth^{-1}x$	$x\coth^{-1}x+\ln\sqrt{x^2-1}$

RULES AND TECHNIQUES OF INTEGRATION

1.1
$$\int_a^b f(x)\,dx \;=\; \int_a^c f(x)\,dx + \int_c^b f(x)\,dx$$

1.2
$$\int_a^b k\,f(x)\,dx \;=\; k\int_a^b f(x)\,dx \qquad (k = \text{constant})$$

1.3
$$\int_a^b f(x)\,dx \;=\; -\int_b^a f(x)\,dx$$

1.4
$$\int_a^a f(x)\,dx \;=\; 0$$

1.5
$$\int_a^b [f(x)+g(x)]\,dx \;=\; \int_a^b f(x)\,dx + \int_a^b g(x)\,dx$$

2.1   *Integration by parts*

$$\int_a^b u(x)\,v'(x)\,dx \;=\; \Big[u(x)\,v(x)\Big]_a^b - \int_a^b u'(x)\,v(x)\,dx$$

2.2   *Integration by substitution*

The integrand is a function of a function; the inner function is taken as the new variable.

$$\int_a^b f(g(x))\,dx \;=\; \int_{g(a)}^{g(b)} f(u)\,\frac{du}{g'}$$

By substitution

$$u \;=\; g(x)$$

2.3   *Integration by partial fractions*

Proper fractional, rational functions are expanded into the sum of partial fractions.

$$\int \frac{P(x)}{Q(x)} dx \;=\; \int \frac{P(x)}{(x-a)(x-b)^2(x^2+cx+d)}\,dx$$

$$=\; \int \left[\frac{A}{x-a} + \frac{B_1}{x-b} + \frac{B_2}{(x-b)^2} + \frac{Cx+D}{x^2+cx+d} \right] dx$$

($P(x)$ must be of lower order than $Q(x)$)

EXERCISES

6.1   The Primitive Function

1.   Find the primitives of the following functions and the value of the constant:

(a) $f(x) = 3x$ given $F(1) = 2$

(b) $f(x) = 2x + 3$ given $F(1) = 0$

6.4   The Definite Integral

2.   Evaluate the following definite integrals:

(a) $\displaystyle\int_0^{\pi/2} 3 \cos x \, dx$

(b) $\displaystyle\int_{-\pi/2}^{\pi/2} 3 \cos x \, dx$

(c) $\displaystyle\int_0^{\pi} 3 \cos x \, dx$

3.   Obtain the absolute values of the areas corresponding to the following integrals:

(a) $\displaystyle\int_{-2}^{0} (x - 2) \, dx$

(b) $\displaystyle\int_0^2 (x - 2) \, dx$

(c) $\displaystyle\int_0^4 (x - 2) \, dx$

6.5   Methods of Integration

4.   Integrate and verify the result by differentiating

(a) $\displaystyle\int \frac{2 \, dx}{(x + 1)^2} = \frac{x - 1}{x + 1} + C$

(b) $\displaystyle 2\int \sin^2 (4x - 1) \, dx = x - \frac{1}{8} \sin (8x - 2) + C$

(c) $\displaystyle\int \frac{1 - x^2}{(1 + x^2)^2} \, dx = \frac{x}{1 + x^2} + C$

5.   Evaluate the following integrals by using the table of standard integrals given at the end of chapter 6.

(a) $\displaystyle\int \frac{dx}{x - a}$

(b) $\displaystyle\int \frac{1}{\cos^2 x} \, dx$

(c) $\displaystyle\int \frac{a}{\sqrt{x^2 + a^2}} \, dx$

(d) $\displaystyle\int \sin^2 \alpha \, d\alpha$

(e) $\displaystyle\int a^t \, dt$

(f) $\displaystyle\int \sqrt[3]{x^7} \, dx$

(g) $\displaystyle\int 5(x^2 + x^3) \, dx$

(h) $\displaystyle\int \left(\frac{3}{2} t^3 + 4t \right) dt$

6.   Integrate by parts the following integrals:

(a) $\displaystyle\int x \ln x \, dx$

(b) $\displaystyle\int x^2 \cos x \, dx$

(c) $\displaystyle\int x^2 \ln x \, dx$

(d) $\displaystyle\int x^2 \cosh \frac{x}{a} \, dx$

(e) Find the reduction formula for $\displaystyle\int \cos^n x \, dx$ $(n \neq 0)$

(f) Find the general formula for $\displaystyle\int x^n \ln x \, dx$

Use a suitable substitution to evaluate the following integrals:

7. (a) $\int \sin(\pi x)\, dx$

    (c) $\int \dfrac{dx}{2x+a}$

    (b) $\int 3e^{3x-6}\, dx$

    (d) $\int (ax+b)^5\, dx$

8. (a) $\int \cot 2x\, dx$

    (c) $\int \dfrac{x^{39}}{x^{40}+21}\, dx$

    (b) $\int \dfrac{2x}{a+x^2}\, dx$

    (d) $\int \dfrac{\sinh u}{\cosh^2 u}\, du$

9. (a) $\int (\sin^4 x + 8\sin^3 x + \sin x)\cos x\, dx$ (b) $\int x^4 \sqrt{3x^5-1}\, dx$

    (c) $\int \dfrac{-x}{\sqrt{a-x^2}}\, dx$

    (d) $\int x \cos x^2\, dx$

10. Mixed questions

    (a) $\int \dfrac{e^x}{e^x+1}\, dx$

    (b) $\int \cos\left(x-\dfrac{\pi}{2}\right) dx$

    (c) $\int \cos^3 x\, dx$

    (d) $\int \dfrac{1}{x \ln x}\, dx$

    (e) $\int \dfrac{3x^2-1}{x^3-x}\, dx$

    (f) $\int \dfrac{1}{(1+x^2)\tan^{-1}x}\, dx$

11. Using partial fractions, integrate the following functions:

    (a) $\dfrac{1}{2-x-x^2}$

    (b) $\dfrac{2x+3}{x(x-1)(x+2)}$

    (c) $\dfrac{x^2}{(x-1)(x-2)(x-3)}$

    (d) $\dfrac{x}{x^4-x^2-2}$

    (e) $\dfrac{1}{x^3+3x^2-4}$

    (f) $\dfrac{x^2-1}{x^4+x^2+1}$

    (g) $\dfrac{x^2+15}{(x-1)(x^2+2x+5)}$

6.6 Rules for Solving Definite Integrals

12. Evaluate the following definite integrals:

    (a) $\displaystyle\int_{-2}^{2} (x^5 - 8x^3 + x + 7)\, dx$

    (b) $\displaystyle\int_{0}^{1} \dfrac{1}{1+x}\, dx$

    (c) $\displaystyle\int_{0}^{2} \sin t\, dt$

    (d) $3\displaystyle\int_{100}^{125} dt$

13. Find the value of the absolute area between the following boundary lines:

    (a) $y = x^3$; x-axis; $a = \dfrac{1}{2}$; $b = 2$

    (b) $y = \cos x$; x-axis; $a = -\dfrac{3\pi}{2}$; $b = \dfrac{5}{6}\pi$

    (c) What is the value of the area between the curves $y = 4x^3$ and $y = 6x^2 - 2$?
        (*Hint*: Sketch the graphs of both functions first. Note that for $x = 1$ both curves have a point in common, but do not intersect.)

6.8   Improper Integrals;   6.9   Line Integrals

14.   Integrate the following:

(a) $\displaystyle\int_4^\infty \frac{d\rho}{\rho^2}$ (b) $\displaystyle\int_{10}^\infty \frac{dx}{x}$ (c) $\displaystyle\gamma\int_{r_0}^\infty \frac{dr}{r^2}$ (d) $\displaystyle\int_1^\infty \frac{d\lambda}{\lambda}$

(e) $\displaystyle\int_1^\infty \frac{dr}{r^3}$ (f) $\displaystyle\int_1^\infty \left(1+\frac{1}{x^2}\right) dx$ (g) $\displaystyle\int_{-\infty}^{-1} \frac{dx}{x^2}$

(h) $\displaystyle\int_1^\infty \frac{1}{\sqrt{x}}\, dx$

15.   A force in a conservative field is given by
$$\mathbf{F} = (2, 6, 1)\,\mathrm{N}$$
A body is moved along the line given by
$$\mathbf{r}(t) = \mathbf{r}_0 + t\mathbf{i}$$
from point $\mathbf{r}(0) = \mathbf{r}_0$ to point $\mathbf{r}(2) = \mathbf{r}_0 + 2\mathbf{i}$. Calculate the work done.

16.   A force in a conservative field is given by
$$\mathbf{F} = (x, y, z)\,\mathrm{N}$$
A body moves from the origin of the coordinate system to the point
$$P = (5, 0, 0)$$
Calculate the work done.

17.   Given the force
$$\mathbf{F} = \left(\frac{x}{\sqrt{x^2+y^2}}, \frac{y}{\sqrt{x^2+y^2}}\right)$$
Evaluate the line integral along a semicircle around the origin of the coordinate system with radius R. Can you give the answer without computing?

18.   Given a force $\mathbf{F} = (0, -z, y)$. calculate the line integral along the curve
$$\mathbf{r}(t) = \left(\sqrt{2}\cos t, \quad \cos 2t, \quad \frac{2t}{\pi}\right)$$
from $t = 0$ to $t = \dfrac{\pi}{2}$.

CHAPTER 7

Applications of integration

The purpose of this chapter is to consider some of the important applications of integration as applied to problems in engineering and science. Its objective is two-fold. Firstly, it demonstrates the practical use of the integral calculus to readers who are particularly interested in applications. Secondly, other readers may use this chapter as a reference when practical problems are encountered.

You will remember the calculation of areas discussed in Chapter 6 as one typical example. We will consider this problem again and move on to the calculation of volumes, lengths of curves, centroids, centres of mass, moments of inertia, and centres of pressure, all of which are frequently encountered in practice. In line with the notation used in most technical books, we will now use the symbol δ instead of Δ. Both refer to the same concept, that of a very small but finite increment.

7.1 AREAS

By definition, the area of a plane figure is the product of two linear dimensions, e.g. the area A of a rectangle of width W and length L is $A = WL$ square units. We will calculate areas bounded by curves. Consider the curve CC_1 shown in figure 7.1. We wish to calculate the area bounded by a portion P_1P_2 of the curve and the x-axis.

P_1 has coordinates (x_1, y_1) and P_2 coordinates (x_2, y_2). At x and $x + \delta x$ we erect two perpendiculars to the x-axis to meet the curve at B and B' respectively; the

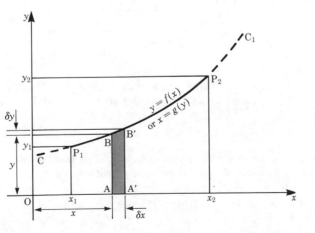

Figure 7.1

strip thus formed is a rectangle (or nearly so) whose area is approximately given by $y\,\delta x$ where y is the mean height of the rectangle. As we saw in the previous chapter, the area lies between $\overrightarrow{AB}\cdot\delta x$ and $\overrightarrow{A'B'}\cdot\delta x$; however, δx can be as small as we like. The total area A under the curve from x_1 to x_2 is the sum of all such rectangles, and as we take δx smaller and smaller the area is given by the following definite integral:

$$A = \int_{x_1}^{x_2} y\,dx \qquad (7.1)$$

To evaluate it, we must know y as a function of x, i.e. $y = f(x)$:

$$A = \int_{x_1}^{x_2} f(x)\,dx = F(x_2) - F(x_1)$$

EXAMPLE   Calculate the area A bounded by the parabola $y = 2 + 0.5x^2$ and the x-axis between $x = 1.5$ and $x = 3.5$ to 3 decimal places.

$$A = \int_{1.5}^{3.5} (2 + 0.5x^2)\,dx = \left[2x + \frac{0.5}{3}x^3\right]_{1.5}^{3.5}$$

$$= 2(3.5 - 1.5) + \frac{0.5}{3}(3.5^3 - 1.5^3) = 10.583 \text{ square units}$$

EXAMPLE   This example is taken from a problem in thermodynamics. Figure 7.2 shows the path corresponding to a gas as if it expands in a cylinder against a piston; p is the pressure and V the volume.

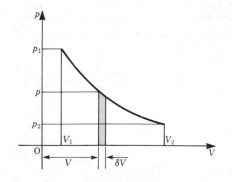

Figure 7.2

The work done by the gas during expansion is given by the area under the p-V curve. Thus the work done, δW, in expanding from volume V to a volume $V + \delta V$ is given by

$$\delta W = p\,\delta V$$

where p is the mean pressure in the interval δV. The total work done in expanding from a pressure p_1, volume V_1, to a pressure p_2, volume V_2, is

$$W = \int_{V_1}^{V_2} p\,dV \qquad \text{(units of work, i.e. Joules in SI units)}$$

To evaluate this work we need to know the expansion law relating pressure and volume. There are two important cases to consider: (a) the isothermal case in which the temperature is constant throughout the whole process and pV is constant; (b) the adiabatic case in which there is no flow of energy through the walls and pV^n is constant.

Case (a): $pV = \text{constant} = C$;   hence $p = \dfrac{C}{V}$.

The work done is

$$W = \int_{V_1}^{V_2} \frac{C}{V}\, dV = C \int_{V_1}^{V_2} \frac{dV}{V}$$

$$= [C \ln V]_{V_1}^{V_2} = C \ln \frac{V_2}{V_1} = pV \ln \frac{V_2}{V_1}$$

Case (b) $pV^n = C$, $n > 1$, e.g. $n = 1.4$ for air

Hence $p = \dfrac{C}{V^n}$, and the work done is

$$W = C \int_{V_1}^{V_2} \frac{dV}{V^n} = C \int_{V_1}^{V_2} V^{-n}\, dV = \frac{C}{1-n}(V_2^{1-n} - V_1^{1-n})$$

$$= \frac{1}{1-n}(CV_2^{1-n} - CV_1^{1-n})$$

Since C is a constant, we can write $C = p_2 V_2^n = p_1 V_1^n$.

Substitution gives $W = \dfrac{p_2 V_2 - p_1 V_1}{1-n}$ for the work done.

COMPLEMENTARY AREA Referring once more to figure 7.1, we may in particular cases wish to calculate the area bounded by the curve and the y-axis between $y = y_1$ and $y = y_2$ as shown. We proceed as before and consider a small strip of mean length x and width δy whose area δA_2 is $x\, \delta y$. The total area is

$$A_2 = \int_{y_1}^{y_2} x\, dy \tag{7.2}$$

To evaluate it, we must know the functional relationship, i.e. $x = g(y)$.

The area A_2 is often referred to as the *complementary area*.

7.1.1 AREAS FOR PARAMETRIC FUNCTIONS

Occasionally a curve is defined by parametric equations of the form

$$x = f(t) \qquad \text{and} \qquad y = g(t) \qquad \text{(cf. Chapter 5, section 5.10)}$$

In this case, the areas are given by the following integrals: (7.3)

$$A = \int_{x_1}^{x_2} y \, dx = \int_{t_1}^{t_2} y \frac{dx}{dt} \, dt = \int_{t_1}^{t_2} g(t) \frac{dx}{dt} \, dt$$

The limits t_1 and t_2 are those values of t which correspond to x_1 and x_2.

Similarly, the complementary area:

$$A_1 = \int_{y_1}^{y_2} x \, dy = \int_{t_1}^{t_2} x \frac{dy}{dt} \, dt = \int_{t_1}^{t_2} f(t) \frac{dy}{dt} \, dt \qquad (7.4)$$

EXAMPLE   The cycloid (figure 7.3) is given by the equations $x = a(\theta - \sin\theta)$, $y = a(1 - \cos\theta)$. Calculate the area between the x-axis and one arc of the curve. The angle turned through is 2π.

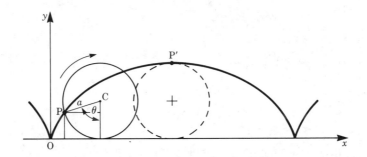

Figure 7.3

Remember that the cycloid is a curve traced out by a point P on the circumference of a circle which rolls without slipping along the x-axis. It has already been introduced in Chapter 5 (cf. figure 5.38).

The area required is

$$A = \int_0^{2\pi a} y \, dx = \int_0^{2\pi} a(1 - \cos\theta) \frac{dx}{d\theta} \, d\theta$$

But $\dfrac{dx}{d\theta} = a(1 - \cos\theta)$, so that

$$A = a^2 \int_0^{2\pi} (1 - \cos\theta)^2 \, d\theta = a^2 \int_0^{2\pi} (1 - 2\cos\theta + \cos^2\theta) \, d\theta$$

Using the table of standard integrals, appendix chapter 6:

$$= a^2 \left[\theta - 2\sin\theta + \frac{\theta}{2} + \frac{\sin 2\theta}{4} \right]_0^{2\pi} = 3\pi a^2$$

7.1.2 AREAS IN POLAR COORDINATES

The equation of a curve is in some cases expressed in polar coordinates (r, θ), being the length of the radius vector measured from the origin O and θ, the angle it makes with a known direction, as shown in figure 7.4.

Suppose that we require the area bounded by the radii OC and OD and the curve CD. Consider a small sector OC'D':

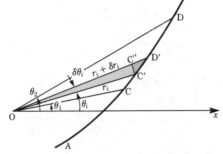

Figure 7.4

$$OC' = r_i, \qquad OD' = r_i + \delta r_i$$

The angle between OC' and OD' is $\delta\theta$.

Now we will work out an approximation for the area of OC'D'.

Let C'' cut the line OD' so that $OC'' = r_i$ and $C'C'' = r_i \delta\theta_i$

The area δA, OC'C'', is given by

$$\delta A = \tfrac{1}{2}\,\text{base} \times \text{height}$$

$$= \tfrac{1}{2}C'C'' \times OC' = \tfrac{1}{2}r_i^2\,\delta\theta_i$$

An approximation for the total area A will be the sum of all such small areas, i.e.

$$A = \sum_{i=1}^{n} \tfrac{1}{2}r_i^2\,\delta\theta_i$$

Now let us take $\delta\theta_i$ smaller and smaller so that n becomes very large. Then, in the limit, the area is given by an integral:

$$A = \lim_{n \to \infty} \sum_{i=1}^{n} \tfrac{1}{2}r_i^2\,\delta\theta_i = \tfrac{1}{2}\int_{\theta_1}^{\theta_2} r^2\,d\theta \tag{7.5}$$

EXAMPLE   The area A of the circle may be considered as being generated by a line of constant length, its radius, rotating through 2π radians about its centre. The area is then given by

$$A = \tfrac{1}{2}\int_{0}^{2\pi} r^2\,d\theta = \tfrac{1}{2}r^2\int_{0}^{2\pi} d\theta = \tfrac{1}{2}r^2\,2\pi = \pi r^2$$

EXAMPLE   The curve represented by the equation $r = a \sin 3\theta$ consists of three loops, as shown in figure 7.5, lying within a circle of radius a. As θ varies from 0 to $\pi/3$, the radius r traces the loop OABC. Calculate its area if $a = 250 \,\text{mm}$.

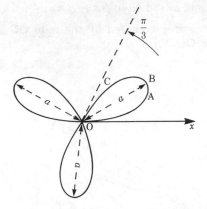

Figure 7.5

The area of one loop is $A = \frac{1}{2} \int_0^{\pi/3} r^2 \, d\theta = \frac{1}{2} a^2 \int_0^{\pi/3} \sin^2 3\theta \, d\theta$

Using the table of integrals we find

$$\int \sin^2 kx \, dx = \frac{x}{2} - \frac{\sin 2kx}{4k} + C$$

Hence

$$A = \frac{1}{2} a^2 \left[\frac{\theta}{2} - \frac{\sin 6\theta}{12} \right]_0^{\pi/3} = \frac{1}{2} a^2 \left(\frac{\pi}{6} - 0 \right) = \frac{\pi a^2}{12} = \frac{\pi}{12} \times 0.25^2 \approx 0.016 \,\text{m}^2$$

7.1.3   AREAS OF CLOSED CURVES

Let ABCD be a closed curve (see figure 7.6) such that it cannot be cut by any line parallel to the y-axis at more than two points, and all ordinates are positive. AA$'$ and CC$'$ are tangents parallel to the y-axis and OA$' = a$, OC$' = b$.

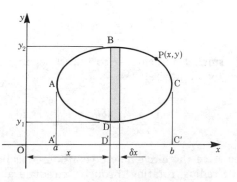

Figure 7.6

The area A enclosed by the curve is

$$A = \int_a^b \text{BD}' \, dx - \int_a^b \text{DD}' \, dx$$

where the points B and D move along ABC and ADC respectively. Let us denote BD' by $f_2(x)$ and DD' by $f_1(x)$.

The area is

$$A = \int_a^b f_2(x)\,dx - \int_a^b f_1(x)\,dx = \int_a^b (f_2 - f_1)\,dx$$

EXAMPLE   Calculate the area enclosed between the straight line $y = 4x$ and the parabola $y = 2 + x^2$.

It is wise to sketch a graph of the two functions, as shown in figure 7.7.

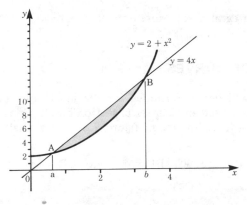

Figure 7.7

The required area is shown shaded. The curves cross each other at A and B, corresponding to $x = a$ and $x = b$, respectively. We need to calculate the values of a and b, our limits of integration. These are given by solving the equation

$$4x = 2 + x^2$$

This is a quadratic equation whose roots are $x_1 = a = 0.59$, $x_2 = b = 3.41$ to 2 d.p.

Since the straight line between A and B is above the parabola, we have

$$f_2(x) = 4x, \qquad f_1(x) = 2 + x^2$$

Hence the area is given by

$$A = \int_{0.59}^{3.41} (4x - 2 - x^2)\,dx = \left[2x^2 - 2x - \frac{1}{3}x^3\right]_{0.59}^{3.41} = 3.77 \text{ square units}$$

Let us suppose that the coordinates of a point $P(x, y)$ on a closed curve (see figure 7.6) are given in terms of a parameter t, such that t increases from t_1 to t_2 as we travel round the curve once. The point travels from A to C, via B, and from C back to A, via D. The equation for the area A of the closed curve becomes

$$A = \int_{t_1}^{t_2} y\,\frac{dx}{dt}\,dt$$

EXAMPLE   Suppose that the closed curve ABCD (figure 7.6) is an ellipse whose equation is

$$\frac{(x-h)^2}{a^2} + \frac{(y-k)^2}{b^2} = 1 \qquad (h, k, a \text{ and } b \text{ are constants})$$

What is the area of the ellipse?

Let $x = h - a\cos t$ and $y = k + b\sin t$.

Then, as t varies from 0 to 2π, a point P(x, y) goes round the curve in the direction ABCDA.

The area is

$$\int_0^{2\pi} (k + b\sin t)\, a\sin t\, dt = ka\int_0^{2\pi} \sin t\, dt + ab\int_0^{2\pi} \sin^2 t\, dt = \pi ab$$

Note that the first integral $= ka\int_0^{2\pi} \sin t\, dt = 0$.

7.2 LENGTHS OF CURVES

In this section we will derive formulae for the length of a curve. (In fact, one of these has already been used in Chapter 5, section 5.8.3.) Consider the curve defined by the equation $y = f(x)$, shown in figure 7.8, and a small portion BC, B and C being close to each other.

Let δs = length of the arc BC, BD = δx and CD = δy, as shown by the small triangle BCD.

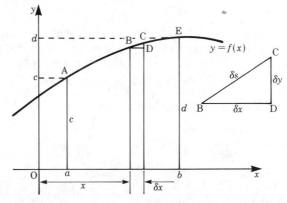

Figure 7.8

Then the arc BC is nearly equal to the chord BC, so we may write

$$(\delta s)^2 \approx (\text{chord BC})^2 = (\delta x)^2 + (\delta y)^2$$

Therefore

$$\left(\frac{\delta s}{\delta x}\right)^2 \approx 1 + \left(\frac{\delta y}{\delta x}\right)^2 \qquad \text{or} \qquad \left(\frac{\delta s}{\delta y}\right)^2 \approx \left(\frac{\delta x}{\delta y}\right)^2 + 1$$

$$\frac{\delta s}{\delta x} \approx \sqrt{1 + \left(\frac{\delta y}{\delta x}\right)^2} \qquad \text{or} \qquad \frac{\delta s}{\delta y} \approx \sqrt{1 + \left(\frac{\delta x}{\delta y}\right)^2}$$

Hence, as $\delta x \to 0$, $\dfrac{\delta s}{\delta x} \to \dfrac{ds}{dx}$ and $\dfrac{\delta y}{\delta x} \to \dfrac{dy}{dx}$

and $\dfrac{ds}{dx} = \sqrt{1 + \left(\dfrac{dy}{dx}\right)^2}$ and $\dfrac{ds}{dy} = \sqrt{1 + \left(\dfrac{dx}{dy}\right)^2}$

The total length s of the curve from A to E, corresponding to $x = a$ and $x = b$, respectively, is

$$s = \int_a^b \sqrt{1 + \left(\frac{dy}{dx}\right)^2}\, dx = \int_a^b (1 + y'^2)^{1/2}\, dx \qquad (7.6)$$

The length is also given by

$$s = \int_c^d (1 + x'^2)^{1/2}\, dy$$

EXAMPLE   Let us find the length of the circumference of a circle of radius R, which, of course, is well known to us.

The equation of a circle is

$$x^2 + y^2 = R^2$$

Differentiating implicitly with respect to x gives

$$2x + 2y \frac{dy}{dx} = 0 \qquad \text{or} \qquad yy' = -x$$

Hence $y' = -\dfrac{x}{y} = -\dfrac{x}{\sqrt{R^2 - x^2}}$ and $(1 + y'^2) = \dfrac{R^2}{R^2 - x^2}$

The length of the circumference

$$L = 4 \times \text{length of } \tfrac{1}{4} \text{ circumference} = 4R \int_0^R \frac{dx}{\sqrt{R^2 - x^2}}$$

Note that to evaluate the integral we can substitute $x = R \sin \theta$.

Then $dx = R \cos \theta\, d\theta$, so that

$$L = 4R \int_0^{\pi/2} \frac{R \cos \theta\, d\theta}{R \cos \theta} = 4R \int_0^{\pi/2} d\theta = 2\pi R$$

EXAMPLE   Evaluate the length of a parabola from the origin to $x = 2$. The equation of the parabola is $y = \tfrac{1}{4} x^2$.

$$y = \tfrac{1}{4} x^2, \qquad y' = \tfrac{1}{2} x$$

The required length of curve is

$$\int_0^2 (1 + y'^2)^{1/2}\, dx = \int_0^2 \left(1 + \frac{x^2}{4}\right)^{1/2} dx$$

$$= \frac{1}{2} \int_0^2 (4 + x^2)^{1/2}\, dx \approx 2.3 \text{ units of length}$$

Note that the integral is of the form $\int \sqrt{a^2 + x^2}\, dx$ which is included in the table of standard integrals on p. 162.

$$\int \sqrt{a^2 + x^2}\, dx \;=\; \frac{1}{2} x \sqrt{a^2 + x^2} + \frac{a^2}{2} \ln\left(x + \sqrt{a^2 + x^2}\right) + C$$

You will soon discover that evaluating lengths of curves can be very laborious; in fact, there are few curves whose length can be expressed by means of simple functions. This is due to the presence of the square root. In most cases the lengths are calculated by approximate means.

7.2.1   LENGTHS OF CURVES IN POLAR COORDINATES

Referring to figure 7.9, which shows a detail from figure 7.4, consider the small triangle $C'D'C''$. We have

$$C'C'' = r\,\delta\theta, \qquad C''D' = \delta r \qquad \text{and} \qquad C'D' = \delta s$$

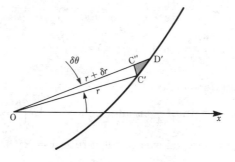

Figure 7.9

Using Pythagoras' theorem, $(C'D')^2 = (C'C'')^2 + (C''D')^2$, we can write

$$(\delta s)^2 \;=\; r^2(\delta\theta)^2 + (\delta r)^2$$

We obtain one expression with respect to θ and one with respect to r:

$$\delta s \;=\; \sqrt{r^2 + \left(\frac{\delta r}{\delta\theta}\right)^2}\;\delta\theta \qquad \text{or} \qquad \delta s \;=\; \sqrt{1 + r^2\left(\frac{\delta\theta}{\delta r}\right)^2}\;\delta r$$

As $\delta r \to 0$, $\delta\theta \to 0$. The length of the curve is then given by an integral:

$$s = \int_{\theta_1}^{\theta_2} \left(r^2 + \left(\frac{dr}{d\theta}\right)^2\right)^{1/2} d\theta \qquad \text{or} \qquad s = \int_{r_1}^{r_2} \left(1 + r^2\left(\frac{d\theta}{dr}\right)^2\right)^{1/2} dr \qquad (7.7)$$

EXAMPLE   Calculate the length of the cardioid whose equation is $r = a(1 + \cos\theta)$. θ varies from 0 to 2π. The curve is symmetrical about the x-axis (figure 7.10).

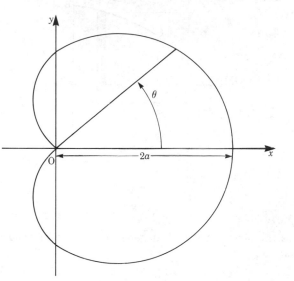

Figure 7.10

Because of symmetry, the length will be twice that given by letting θ vary from 0 to π.

Since $r = a(1 + \cos\theta)$, $\dfrac{dr}{d\theta} = -a\sin\theta$

The length is given by

$$L = 2\int_0^\pi [a^2(1 + \cos\theta)^2 + a^2\sin^2\theta]^{1/2}\,d\theta$$

$$= 2a\int_0^\pi (2 + 2\cos\theta)^{1/2}\,d\theta = 4a\int_0^\pi \cos\frac{\theta}{2}\,d\theta = 8a$$

7.3 SURFACE AREA AND VOLUME OF A SOLID OF REVOLUTION

When a solid of revolution is generated, the boundary of the revolving figure sweeps out the surface of the solid. The volume of the solid depends on the area of the revolving figure, and the surface generated depends on the perimeter of the revolving figure. Consider the curve AB, defined by $y = f(x)$, and shown in figure 7.11 between $x = a$ and $x = b$.

Let us revolve the curve AB about the x-axis. Two figures are generated: (a) a surface and (b) a solid. If we consider a small strip of width δx and height y, then the small surface generated is given by $\delta A = 2\pi y\,\delta s$, where δs is the length of the curve corresponding to δx. The total surface will be the sum of all such elements, i.e. surface $\approx \Sigma 2\pi y\,\delta s$. If δx becomes smaller and smaller we have, in the limit,

$$A = \int_a^b 2\pi y\,ds = 2\pi\int_a^b y\left(1 + \left(\frac{dy}{dx}\right)^2\right)^{1/2}dx \tag{7.8}$$

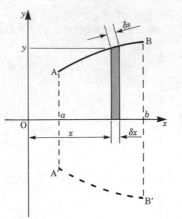

Figure 7.11

Furthermore, as the strip is rotated, it generates a thin circular slice whose volume δV is approximately

$$\delta V = \pi y^2 \, \delta x$$

For the whole curve, as $\delta x \to 0$, the volume of the solid generated is

$$V = \pi \int_a^b y^2 \, \mathrm{d}x \qquad (7.9)$$

EXAMPLE   The straight line $y = mx$ is rotated about the x-axis, thus generating a right circular cone, as shown in figure 7.12. Calculate (a) its surface area and (b) its volume. (Of course, the results are well known. They are usually obtained without using integral calculus.)

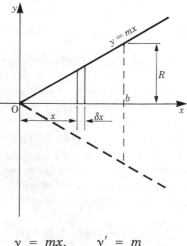

Figure 7.12 $y = mx, \qquad y' = m$

(a) The surface area is

$$A \;=\; 2\pi \int_0^b mx\sqrt{1+m^2}\,dx \;=\; \left[2\pi m\sqrt{1+m^2}\,\frac{x^2}{2}\right]_0^b$$

$$=\; \pi m b^2\sqrt{1+m^2} \;=\; \pi R\sqrt{b^2+R^2}$$

where R is the radius of the base of the cone and $m = \dfrac{R}{b}$.

We can express the surface area of the cone as

$$A \;=\; \pi R L, \quad \text{where} \quad L \;=\; \text{slant height} \;=\; \sqrt{b^2+R^2}$$

(b) The volume is

$$V \;=\; \int_0^b \pi y^2\,dx \;=\; \pi m^2 \int_0^b x^2\,dx$$

$$=\; \pi m^2\left[\frac{1}{3}x^3\right]_0^b \;=\; \frac{1}{3}\pi m^2 b^3 \;=\; \frac{1}{3}\pi \frac{R^2}{b^2}b^3 \;=\; \frac{1}{3}\pi R^2 b$$

Hence

Surface area of a cone $= \frac{1}{2} \times$ circumference of base $\times$ slant height

Volume of a cone $= \frac{1}{3} \times$ area of base $\times$ height ($\frac{1}{3}$ of the volume of
a cylinder having same base and height)

EXAMPLE   Calculate (a) the surface area and (b) the volume of a lune of a sphere of radius R and thickness h (see figure 7.13).

The surface will be generated by rotating the arc **AB** and the volume by rotating the area **ABCD** about the x-axis.

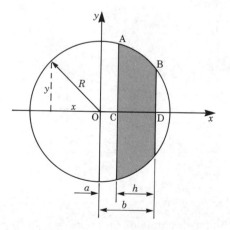

Figure 7.13

From the figure, we have

$$y^2 \;=\; R^2 - x^2$$

Differentiating implicitly gives

$$yy' \;=\; -x$$

Thus

$$y'^2 = \frac{x^2}{y^2} \quad \text{and} \quad 1+y'^2 = \frac{y^2+x^2}{y^2} = \frac{R^2}{R^2-x^2}$$

(a) The surface area is

$$A = 2\pi \int_a^b (R^2-x^2)^{1/2} \frac{R}{(R^2-x^2)^{1/2}} \, dx = 2\pi R \int_a^b dx = 2\pi R(b-a)$$

Hence $A = 2\pi Rh$

(b) The volume V is

$$V = \pi \int_a^b y^2 \, dx = \pi \int_a^b (R^2-x^2) \, dx = \pi \left[R^2x - \frac{x^3}{3} \right]_a^b$$

For the special case where $b = R$ and $a = 0$, we have

$$V = \tfrac{2}{3}\pi R^3$$

This is the volume of a half sphere or hemisphere. Hence the volume of a sphere is $V = \tfrac{4}{3}\pi R^3$.

EXAMPLE   Small aluminium alloy pillars having a parabolic profile are manufactured by turning down cylinders 125 mm in length and 50 mm in diameter. The diameter of the pillars at the thinner end is to be 30 mm. Calculate the amount of metal removed. (The density of aluminium is 2720 kg/m³.)

Figure 7.14 shows the required profile of each pillar and the amount of material to be removed is indicated by the shading.

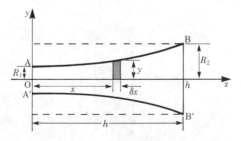

Figure 7.14

The volume of material removed is

Volume of the cylinder − volume of the pillar $= \pi R_2^2 h -$ volume ABB′A′

With axes as shown, the equation of the parabola AB is

$$y = a + bx^2$$

To find the values of a and b, we note that when $x = 0$, $y = R_1$ (= 15 mm) and when $x = h$, $y = R_2$ (= 25 mm); also $h = 125$ mm

Hence $a = R_1$ and $b = \dfrac{R_2 - R_1}{h^2}$, i.e. $a = 15$, $b = 10/125^2 = 0.00064$

Rotating the element of width δx about the x-axis gives the volume of the slice as $\pi y^2 \delta x$; the volume of the pillar is

$$V = \pi \int_0^h y^2 \, dx = \pi \int_0^h (a + bx^2)^2 \, dx = \pi \int_0^h (a^2 + 2abx^2 + b^2 x^4) \, dx$$

$$= \pi h \left(a^2 + \frac{2abh^2}{3} + \frac{b^2}{5} h^4 \right)$$

Substituting numerical values gives

$$V = 125\pi \left(15^2 + \frac{2}{3} \times 15 \times 0.00064 \times 125^2 + \frac{0.00064^2}{5} \times 125^4 \right)$$

$$\approx 0.135 \times 10^6 \, \text{mm}^3 = 0.135 \times 10^{-3} \, \text{m}^3$$

$$\text{Volume of cylinder} = \pi R_2^2 h = \pi \times 25^2 \times 125 \approx 0.245 \times 10^6 \, \text{mm}^3$$

$$= 0.245 \times 10^{-3} \, \text{m}^3$$

$$\text{Material removed} \approx (0.245 - 0.135) \times 10^{-3} \times 2720 \approx 0.3 \, \text{kg}$$

If some arbitrary closed curve is rotated about the x-axis it generates a solid, i.e. a ring of irregular cross section (see figure 7.6).

Referring to figure 7.6, the volume generated by the thin slice BD is a hollow circular plate having radii $D'B = y_2$ and $DD' = y_1$ and thickness δx.

Hence its volume V is

$$\delta V = \pi (y_2{}^2 - y_1{}^2) \, \delta x$$

The volume V of the hollow solid of revolution is

$$V = \pi \int_a^b (y_2{}^2 - y_1{}^2) \, dx \qquad (7.10)$$

where $y_1 = f_1(x)$ and $y_2 = f_2(x)$ are the equations of the curves ADC and ABC respectively.

EXAMPLE   Calculate the volume of a solid ring (*torus* or *anchor ring*) obtained by rotating a circle of radius R about an axis distant h from its centre ($h > R$).

The circle is conveniently positioned, as shown in figure 7.15, relative to the x- and y-axes. O' is the centre of the circle.

Consider any point P with coordinates (x, y). Consider the triangle $O'A'P$; we have $(y - h)^2 + x^2 = R^2$. Solving for y gives $y = h \pm \sqrt{R^2 - x^2}$.

The two functions $f_1(x)$ and $f_2(x)$ are $f_2(x) = h + \sqrt{R^2 - x^2}$ for portion ABC of the circle, and $f_1(x) = h - \sqrt{R^2 - x^2}$ for portion ADC of the circle.

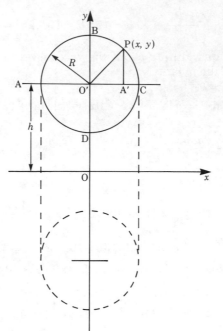

Figure 7.15

Therefore

$$\delta V = \pi(f_2^2 - f_1^2)\,\delta x = 4\pi h\sqrt{R^2 - x^2}\,\delta x$$

and the volume of the ring formed is

$$V = 4\pi h\int_{-R}^{R}\sqrt{R^2 - x^2}\,\mathrm{d}x = 2\pi^2 R^2 h$$

Note that the integral can be solved by letting $x = R\sin\theta$ (cf. the table of integrals).

7.4 APPLICATIONS TO MECHANICS

7.4.1 BASIC CONCEPTS OF MECHANICS

When a rigid body moves in two or three dimensions under the action of forces it is the same as if the whole mass of the body was concentrated in one point with all the forces acting through that point, giving the body a translation in the direction of the resultant force. Furthermore, the body rotates about an axis through that point under the action of the resultant moment of the forces about that axis. The point referred to is called the *centre of mass* of the body.

If M is the total mass of the body and $\mathbf{F}$ the resultant force, then Newton's second law of motion states $\mathbf{F} = M\ddot{\mathbf{r}}$ (vector equation), where $\ddot{\mathbf{r}}$ is the acceleration of translation.

Also, if **H** is the angular momentum of the body, then

$$\text{Moment of the external forces } = \frac{d}{dt}(\mathbf{H})$$

It can be shown that $\mathbf{H} = I\omega$, where I is the *moment of inertia* of the body about an axis through the centre of mass and ω is the angular velocity of the body. Remember that the moment of inertia is (the sum of) the product of a mass by the square of its distance from the axis of rotation.

When studying the motion of a rigid body, e.g. a car, an aircraft, a link in a mechanism, etc., we need to know the position of the centre of mass and the moment of inertia.

Another important point is met when studying the forces acting on a body which is immersed in a liquid. This point, called the *centre of pressure*, is the point where the total pressure on the body is supposed to act.

We will consider these three concepts in some detail.

7.4.2 CENTRE OF MASS AND CENTROID

Consider a system of n particles P_i whose masses are m_i $(i = 1, 2, \ldots, n)$; let the coordinates of these particles, referred to a cartesian set of axes x, y, z, as shown in figure 7.16, be x_i, y_i, z_i. If M is the total mass of the particles, the position of the centre of mass G is given by the following equations:

$$\bar{x} = \frac{1}{M}\sum_{i=1}^{n} m_i x_i, \qquad \bar{y} = \frac{1}{M}\sum_{i=1}^{n} m_i y_i, \qquad \bar{z} = \frac{1}{M}\sum_{i=1}^{n} m_i z_i$$

where
$$M = \sum_{i=1}^{n} m_i$$

The product mass × distance is often referred to as the *first moment*.

When the particles form a solid body, the above summations become integrals. If δm is the mass of a typical particle in the body at distances x, y and z from the planes, then the centre of mass of the body is given by

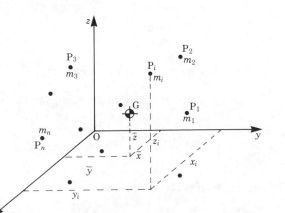

Figure 7.16

$$\bar{x} = \frac{\int x \, dm}{\int dm}, \qquad \bar{y} = \frac{\int y \, dm}{\int dm}, \qquad \bar{z} = \frac{\int z \, dm}{\int dm}$$

between appropriate limits.

$$\int dm = M = \text{total mass of the body}$$

A plane figure of area ABCD may be considered as a thin lamina. Its centre of mass is found by taking moments about the x- and y-axes (figure 7.17).

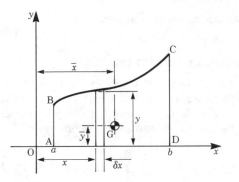

Figure 7.17

Let it be a requirement to find the position of the centre of mass, G, of the thin lamina ABCD of mass m per unit area shown in the figure. The small strip has a mass $my \, \delta x$. Hence, by the above equations, if $\bar{x}$ is the x-coordinate of the centre of mass, G

$$\bar{x} = \frac{\int_a^b xmy \, dx}{\int_a^b my \, dx} = \frac{1}{A}\int_a^b yx \, dx \qquad (7.11a)$$

A is the total area $= \int_a^b y \, dx$. We note that m cancels out.

If we now take moments about the x-axis, we have, for the y-coordinate $\bar{y}$ of the centre of mass, G

$$\bar{y} = \frac{1}{2A}\int_a^b y^2 \, dx \qquad \text{(independent of } m\text{)} \qquad (7.11b)$$

$\left(\text{The moment of the small strip about the } x\text{-axis is } \dfrac{y}{2}(my \, \delta x) = \dfrac{1}{2} my^2 \, \delta x.\right)$

G in the case of an area or a volume is usually referred to as the *centroid*.

EXAMPLE   Find the centre of mass G of a thin strip AB bent into a circular arc, as shown in figure 7.18. The mass per unit length is m and the radius r. The arc subtends an angle 2θ at the centre O.

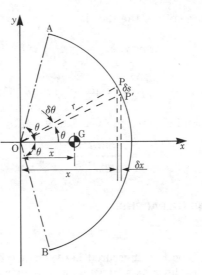

Figure 7.18

Taking axes as shown, it follows that the centre of mass G lies along the x-axis.

Consider a small element PP' of length δs; its mass is $m\,\delta s = mr\,\delta\theta$, using polar coordinates.

The moment of PP' about the y-axis is $xmr\,\delta\theta = mr^2\cos\theta\,\delta\theta$.

Hence the position of G is

$$\bar{x} = \frac{\int_{-\theta}^{\theta} mr^2\cos\theta\,\mathrm{d}\theta}{\int_{-\theta}^{\theta} mr\,\mathrm{d}\theta} = \frac{2mr^2\sin\theta}{2mr\theta} = \frac{r\sin\theta}{\theta}$$

If the strip is bent into a semicircle, $\theta = \dfrac{\pi}{2}$ and

$$\bar{x} = \frac{2r}{\pi}$$

EXAMPLE   Determine the centre of mass G of the solid cone shown in figure 7.12.

The equation of the straight line is $y = \dfrac{R}{b}x$.

The mass of the thin slice obtained by rotating the element δx about the x-axis is $m\pi y^2\,\delta x$, where m is the mass per unit volume. The total mass of the cone is

$$M = m\pi \int_0^b y^2 \, dx = m\pi \frac{R^2}{b^2} \int_0^b x^2 \, dx$$

$$= \frac{1}{3} m\pi R^2 b$$

The moment about the y-axis of the slice is

$$x m\pi y^2 \, \delta x = m\pi \frac{R^2}{b^2} x^3 \, \delta x$$

Hence the total moment $= \dfrac{m\pi R^2}{b^2} \displaystyle\int_0^b x^3 \, dx = \dfrac{1}{4} m\pi R^2 b^2$

The position of G, which lies along the x-axis, is given by

$$\bar{x} = \frac{\frac{1}{4} m\pi R^2 b^2}{\frac{1}{3} m\pi R^2 b} = \frac{3}{4} b$$

7.4.3   THE THEOREMS OF PAPPUS

Pappus' First Theorem

Let AB be an arc of length L measured between $x = a$ and $x = b$ (figure 7.19). When it revolves about the x-axis, it generates a surface of revolution whose area is $S = 2\pi \displaystyle\int_a^b y \, ds$. If $\bar{y}$ is the ordinate of the centroid (of the arc) G then

$$\bar{y}L = \int_a^b y \, ds$$

Multiplying both sides by 2π gives

$$S = 2\pi \bar{y} L = 2\pi \int_a^b y \, ds \tag{7.12}$$

This is known as *Pappus' first theorem*. It states that the area of a surface of revolution is equal to the product of the path travelled by the centroid (of the arc) and the length of the generating arc.

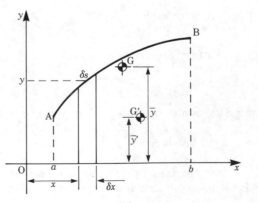

Figure 7.19

Pappus' Second Theorem

The volume of the solid of revolution generated by the area bounded by the arc, the ordinates at $x = a$ and $x = b$ and the x-axis is given by

$$V = \pi \int_a^b y^2 \, dx$$

If we denote the area by A, the centroid (of area) by G' and its ordinate by $\overline{y}'$, then

$$\overline{y}'A = \frac{1}{2} \int_a^b y^2 \, dx$$

Multiplying both sides by 2π gives

$$V = 2\pi\overline{y}'A = \pi \int_a^b y^2 \, dx \qquad (7.13)$$

This is known as *Pappus' second theorem*. It states that the volume of a solid of revolution is equal to the product of the path travelled by the centroid and the generating area.

These two theorems apply also to a closed curve, provided it does not cut the axis about which it is rotated.

EXAMPLE   Calculate (a) the surface area and (b) the volume of a torus (cf. figure 7.15). O' is the centroid in this case.

(a) By the first of Pappus' theorems for the surface area we have

$$S = 2\pi\overline{y}L$$

But $L = 2\pi R$, $\overline{y} = h$; hence $S = 2\pi h \times 2\pi R = 4\pi^2 Rh$

(b) By the second of Pappus' theorems for the volume we have

$$V = 2\pi\overline{y}A$$

But $A = \pi R^2$, $\overline{y} = h$; hence $V = 2\pi h\pi R^2 = 2\pi^2 R^2 h$

Pappus' theorems are also useful in obtaining the centroid of a curve or an area when we know the surface or volume generated. This is illustrated by the following example.

EXAMPLE   Find the position of the centroid of one quarter of a circular area of radius R.

Rotating the quarter circle about the x-axis gives a hemisphere. If $\overline{y}$ is the position of the centroid, we find

$$2\pi\overline{y} \frac{\pi R^2}{4} = \frac{2}{3}\pi R^3$$

Hence

$$\overline{y} = \frac{4R}{3\pi} \qquad \text{(note that } \overline{x} = \overline{y} \text{ by symmetry)}$$

7.4.4   MOMENTS OF INERTIA; SECOND MOMENT OF AREA

Moments of inertia play an important role in the study of the motion of rigid bodies. Equally important is the concept of the *second moment* of area which arises, for example, in the study of beam bending, torsion of bars and in problems involving surfaces immersed in fluids.

Moments of Inertia

> **DEFINITION**   The moment of inertia I of a mass M at a distance l from a fixed axis is given by
>
> $$I = Ml^2$$

If we have a system of n masses M_i at distance l_i from the fixed axis, then the total moment of inertia is

$$I = \sum_{i=1}^{n} M_i l_i^2$$

When the number of masses is infinite, i.e. when they merge into one mass forming a rigid body, the summation becomes an integral.

Figure 7.20 shows a rigid body with P a typical particle of mass δm at distances x and y from a cartesian set of axes.

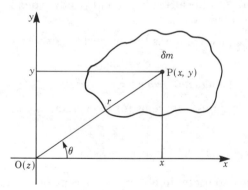

Figure 7.20

By definition, the moment of inertia of that particle about the x-axis is $y^2 \delta m$ and the moment of inertia of the whole body about that axis is

$$I_x = \int_A y^2 \, dm \qquad \text{limits are defined by area } A. \tag{7.14}$$

Similarly, the moment of inertia about the y-axis is

$$I_y = \int_A x^2 \, dm \tag{7.15}$$

It is important to specify the axis by a subscript unless it is obvious from the nature of the problem.

EXAMPLE   Obtain the moment of inertia of a thin disc of radius R, thickness h and density ρ about a diameter (figure 7.21).

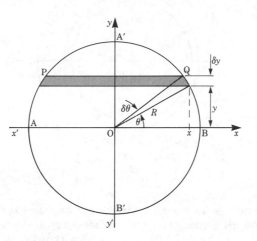

Figure 7.21

Consider a strip PQ of length $2x$ and of width δy, parallel to the x-axis and at a distance y from it. The moment of inertia of the strip about the x-axis is $2xy^2h\rho \, \delta y$.

Hence, for the whole disc

$$I_x = 2\rho h \int_{-R}^{R} xy^2 \mathrm{d}y$$

The integral may be solved by substituting $x = R\cos\theta$, $y = R\sin\theta$ and consequently $\mathrm{d}y = R\cos\theta \, \mathrm{d}\theta$:

$$I_x = 4\rho R^4 h \int_{0}^{\pi/2} \cos^2\theta \sin^2\theta \, \mathrm{d}\theta = \frac{\pi\rho R^4 h}{4}$$

If we denote the mass of the disc, $\pi R^2 h\rho$, by M, then $I_x = \frac{1}{4}MR^2$. This result is obviously true for any diameter.

EXAMPLE   For an axis through O perpendicular to the disc (the z-axis) let us take a thin ring at a distance r from the axis and of width δr (figure 7.22). Its moment of inertia is, by definition, $r^2\delta m = r^2 2\pi r h \, \delta r\rho = 2\pi\rho h r^3 \, \delta r$.

The total moment of inertia of the disc about the z-axis is

$$I_z = 2\pi\rho h \int_{0}^{R} r^3 \mathrm{d}r = \frac{1}{2}\pi\rho h R^4 = \frac{1}{2}MR^2$$

This result is also valid for a long cylinder. I_z is often referred to as the *polar moment of inertia*.

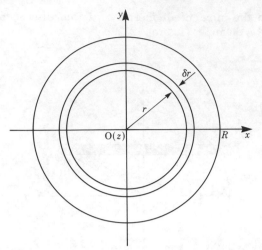

Figure 7.22

EXAMPLE A flywheel is an element with many practical applications. It consists basically of a hollow thin disc. The hole is necessary so that the flywheel can be supported by a shaft. In figure 7.23 the cylinder (known as a boss) FHJG is to ensure a good support on the shaft.

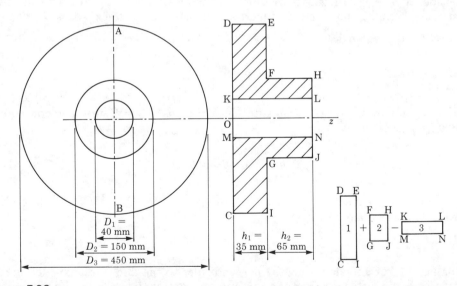

Figure 7.23

Calculate the moment of inertia of the flywheel about its central z-axis. It is made of steel whose density is $7800\ \text{kg/m}^3$. The sketch on the right of the figure shows a cross section through the diameter AB.

The moment of inertia I_z is the sum of the moments of inertia of the discs that make up the flywheel:

$$I_z = I(\text{CDEI}) + I(\text{FHJG}) - I(\text{KLMN}) = I_1 + I_2 - I_3$$

The moment of inertia of the hole has to be subtracted. Each element is a thin disc whose moment of inertia about the z-axis is $\frac{1}{2}MR^2$ (previous example) or $\frac{1}{8}MD^2$, where D is the diameter of the element.

Let us first calculate the masses:

$$M_1 = \frac{\pi}{4}D_1^2 h_1 \rho = \frac{\pi}{4} \times 0.45^2 \times 0.035 \times 7800 = 43.42 \text{ kg to 2 d.p.}$$

Similarly, we find $M_2 = 8.96 \text{ kg}$, $M_3 = 0.98 \text{ kg}$.

The moment of inertia about the z-axis is

$$I_z = \frac{1}{8}(43.42 \times 0.45^2 + 8.96 \times 0.15^2 - 0.98 \times 0.04^2) = 1.124 \text{ kg/m}^2$$

Perpendicular and Parallel Axis Theorems

Consider the particle of mass δm at P shown in figure 7.20. Its coordinates are x and y. The body is a thin plate in the x–y plane.

The moments of inertia of the small element are $y^2 \delta m$ about Ox, $x^2 \delta m$ about Oy and $r^2 \delta m$ about Oz.

Since $r^2 = x^2 + y^2$, multiplying through by δm gives

$$r^2 \delta m = x^2 \delta m + y^2 \delta m$$

By integration, we find

$$\int r^2 \, dm = \int x^2 \, dm + \int y^2 \, dm$$

Hence
$$I_z = I_y + I_x \tag{7.16}$$

This is known as the *perpendicular axis theorem*.

For example, we saw that the moment of inertia of the thin disc in figure 7.21 was $\frac{1}{4}MR^2$ about a diameter. It follows from the perpendicular axis theorem that

$$I_z = 2I_x = 2 \times \frac{1}{4}MR^2 = \frac{1}{2}MR^2$$

Now let us consider the body shown in figure 7.24. Its centre of mass is at G, and there are two parallel axes, one through G and another AB at a fixed distance d. The moment of inertia about AB of a small strip of mass δm is

$$\delta m (x + d)^2$$

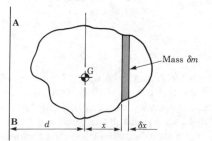

Figure 7.24

The moment of inertia of the whole body about AB is

$$I_{AB} = \int (x+d)^2 \, dm = \int x^2 \, dm + \int d^2 \, dm + \int 2xd \, dm$$

$$= \int x^2 \, dm + d^2 \int dm + 2d \int x \, dm$$

Since $\int x \, dm = 0$ by definition of the centre of mass

$$I_{AB} = I_G + Md^2$$

This is known as the *parallel axis theorem* (*Steiner's theorem*).

Radius of Gyration

If M is the mass of the body and k a distance such that the moment of inertia of the body I is expressed by

$$I = Mk^2$$

then k is known as the *radius of gyration* about the axis. Physically it means that we regard the whole mass to be concentrated at a radius k.

EXAMPLE   We saw earlier that the moment of inertia of a cylinder about a central axis was given by $I = \frac{1}{2}MR^2$. If we write $I = Mk^2$, it follows that the radius of gyration of the cylinder is $K = \dfrac{R}{\sqrt{2}}$.

EXAMPLE   Calculate the radius of gyration of the flywheel analysed in a previous example on page 190.

Total mass $M = 43.42 + 8.96 - 0.98 = 51.40 \, \text{kg}$.

Moment of inertia $I = 1.124 \, \text{kg/m}^2$.

Since $I = Mk^2$, solving for k gives

$$k = \sqrt{\frac{I}{M}} = \sqrt{\frac{1.124}{51.40}} = 0.148 \, \text{m}$$

EXAMPLE   In practice we often need to calculate the moment of inertia of a body. Its value can be estimated from drawings, but an experimental verification of the value might be required. One way of achieving this is to suspend the body on an axis and allow it to oscillate like a pendulum. The time for a number of complete oscillations is observed and, by a simple calculation, the radius of gyration is obtained.

Figure 7.25 shows a body, such as a connecting rod in an internal combustion engine, pivoted at O. The total mass is M and its centre of mass is at G, which can both be obtained experimentally.

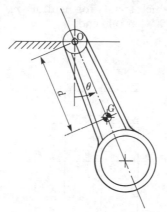

Figure 7.25

The connecting rod is allowed to oscillate about the axis O through a small angle θ. The distance between O and G is d. In this way we have a compound pendulum. It can be shown that the period of one oscillation is given by $t = 2\pi \sqrt{\dfrac{k_0{}^2}{gd}}$ (g is the acceleration due to gravity and k_0 is the radius of gyration about an axis through O). If I_G is the moment of inertia about an axis through G parallel to the axis through O, then by the parallel axis theorem

$$I_0 = I_G + Md^2 \qquad \text{or} \qquad Mk_0^2 = Mk_G^2 + Md^2$$

Hence $k_0{}^2 = k_G{}^2 + d^2$, and the required radius of gyration is

$$k_G = \sqrt{\frac{t^2 gd}{4\pi^2} - d^2}$$

Hence the moment of inertia about G is Mk_G^2.

Second Moment of Area

When studying the deflection of loaded beams or the twist in a shaft subjected to a torque, we encounter the following expression

$$r^2 \, \delta A$$

where δA is an element of area in the cross section of the beam or the shaft and r is a distance from some axis.

This product is known as the *second moment of area* and is similar to the moment of inertia of a body.

The second moment of area of a plane figure of finite size is

$$I = \int r^2 \, dA \quad \text{between appropriate limits} \tag{7.17}$$

The perpendicular axis and parallel axis theorems are valid for second moments of area as can readily be verified.

EXAMPLE   The rectangle plays an important role in beams. Let us calculate its second moment of area about various axes.

Figure 7.26 shows a rectangle, of width B and depth D, and two axes: one through the centroid denoted NA (which stands for neutral axis where the stress in a beam would be zero) and another, XX, at one end.

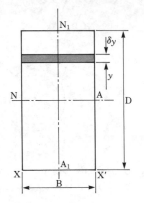

Figure 7.26

(a) To find I_{NA}, consider the small strip of thickness δy. Its second moment of area about NA is

$$B\delta y y^2 \qquad \text{or} \qquad By^2\delta y$$

Hence, for the whole rectangle

$$I_{NA} = B\int_{-D/2}^{D/2} y^2\, dy = \frac{BD^3}{12} \qquad \text{or} \qquad I_{NA} = Ak_{NA}^2$$

where A = cross sectional area = BD, and k_{NA} = radius of gyration = $\dfrac{D}{\sqrt{12}}$

By symmetry, it is easily verified that

$$I_{N_1A_1} = \frac{DB^3}{12}$$

(b) To find I_{XX}: the distance between the axis NA and XX is $\dfrac{D}{2}$; hence, by the parallel axis theorem, we have

$$I_{XX} = I_{NA} + A\left(\frac{D}{2}\right)^2$$

$$= \frac{1}{12}BD^3 + BD\frac{D^2}{4} = \frac{1}{3}BD^3$$

and $\qquad\qquad k_{XX} = \dfrac{D}{\sqrt{3}}$

EXAMPLE   Figure 7.27 shows the cross section of a type of beam known as an *I*-section. Using the dimensions given on the sketch, calculate the second moment of area about an axis through its centroid.

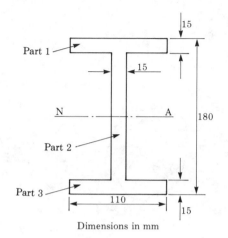

Figure 7.27

Dimensions in mm

The centroid lies half way up the section which has been indicated as NA. To calculate the second moment of area, I_{NA}, we can divide the area into three parts.

Parts 1 and 3 are of dimensions $B_1 = 110$ mm and $D_1 = 15$ mm. By the parallel axis theorem and where D = total depth of the beam:

$$I_{\mathrm{NA}_1} = \frac{1}{12}B_1 D_1^3 + B_1 D_1 \left(\frac{D}{2} - \frac{D_1}{2}\right)^2$$

Substituting numerical values gives

$$I_{\mathrm{NA}_1} = \frac{1}{12} \times 110 \times 15^3 + 110 \times 15 \times 82.5^2 = 11.26 \times 10^6 \, \mathrm{mm}^4$$

Part 2 is of dimensions $B_2 = 15$ mm and $D_2 = 150$ mm. Thus

$$I_{\mathrm{NA}_2} = \frac{1}{12}B_2 D_2^3 = \frac{1}{12} \times 15 \times 150^3 = 4.22 \times 10^6 \, \mathrm{mm}^4$$

Hence, for the whole section

$$I_{\mathrm{NA}} = 2I_{\mathrm{NA}_1} + I_{\mathrm{NA}_2} = (2 \times 11.26 + 4.22) \times 10^6 = 26.74 \times 10^6 \, \mathrm{mm}^4$$

Centre of Pressure

If a body is immersed in a fluid, e.g. water, then the pressure per unit area of surface is not uniform over the body because the pressure is proportional to the depth. The point at which the total pressure may be assumed to act is known as the centre of pressure.

Consider the plane surface S immersed in a fluid of density ρ and making an angle θ with the free surface QP, as shown in figure 7.28.

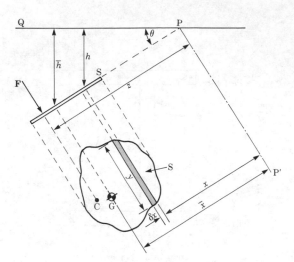

Figure 7.28

For a small area of length y and width δx, the pressure on it is $g\rho h$ where h is the vertical distance from the surface QP to the element. The force δF on this element is

$$\delta F = g\rho h y\,\delta x \qquad \text{and} \qquad h = x\sin\theta$$

The total force on the surface is given by

$$F = g\int \rho\sin\theta\, xy\,dx \text{ between appropriate limits}$$

$$= g\rho\sin\theta \int xy\,dx$$

But $\int xy\,dx$ = first moment of area about PP' $= A\bar{x}$, where A is the area of the surface.

Hence $F = g\rho A\bar{x}\sin\theta = g\rho A\bar{h}$

To find the point where the resultant force F acts, we take moments about PP'. For the element we have

$$g\rho h y\,\delta x\, x = g\rho\sin\theta\, yx^2\,\delta x$$

Hence, if z is the position of the point C where the resultant force acts, then

$$Fz = g\rho\sin\theta \int yx^2\,dx \qquad \text{or} \qquad g\rho A\bar{x}\sin\theta z = g\rho\sin\theta \int yx^2\,dx$$

Solving for z gives

$$z = \frac{\int yx^2\,dx}{A\bar{x}} \qquad\qquad (7.18)$$

Hence, the point C where the resultant force due to the pressure of the fluid acts is given by

$$z = \frac{\text{Second moment of area about PP}'}{\text{First moment of area about PP}'}$$

C is known as the *centre of pressure*.

EXERCISES

7.1   Areas

1. Calculate the area bounded by the positive branch of the parabola $y^2 = 25x$, the x-axis and the ordinates where $x = 0$ and $x = 36$.

2. Calculate the area bounded by the positive branch of the curve $y^2 = (7-x)(5+x)$, the x-axis and the ordinates where $x = -5$ and $x = 1$.

3. Calculate the area bounded by the parabola $20y = 3(2x^2 - 3x - 5)$ and the x-axis between the points where the curve cuts the x-axis.

4. Calculate the area bounded by the curve $y^2(x^2 + 6x - 55) = 1$, the x-axis and the ordinates where $x = 7$ and $x = 14$.

5. Sketch the curve $y = 2x^3 - 15x^2 + 24x + 25$ between $x = 0$ and $x = 4$ and then calculate the area enclosed by the ordinates at these points, the x-axis and the portion of the curve.

6. Calculate the area bounded by the hyperbola $r^2 \cos 2\theta = 9$ and the radial lines $\theta = 0$ and $\theta = 30°$.

7. Calculate the entire area of the curve $r = 3.5 \sin 2\theta$.

8. Calculate the area bounded by the following curves:
    (a) $y^2 = 4x$ and $x^2 = 6y$
    (b) $y = 4 - x^2$, $y = 4 - 4x$
    (c) $y = 6 + 4x - x^2$ and the line joining the points $(-2, -6)$ and $(4, 6)$.

7.2   Length of Curves

9. Calculate the lengths of the curves given in exercises 1, 2 and 3.

7.3   Surface Area and Volume of a Solid of Revolution

10. Calculate the area of the surface generated by the revolution of the curve $y = x^3$ about the x-axis between the ordinates $x = 0.5$ and $x = 0$.

11. The curve $y = x(6-x) - 7.56$ is rotated about the x-axis between the points where it crosses the x-axis. Calculate (a) the surface area and (b) the volume of the solid thus generated.

12. Calculate (a) the surface area and (b) the volume generated by rotating the cycloid $x = \theta - \sin \theta$, $y = 1 - \cos \theta$ about the x-axis.

13. Obtain an expression for the volume generated by revolving the ellipse $\frac{x^2}{9} + \frac{y^2}{25} = 1$ about the x-axis.

7.4   Applications to Mechanics

14. Find the position of the centroid of the area of one quarter of an ellipse. The equation of the ellipse is

$$\frac{x^2}{a^2} + \frac{y^2}{b^2} = 1$$

15. A plate is cut into a circular sector of 375 mm radius and 65° included angle. Find the position of the centroid along the axis of symmetry.

16. The density of the material of which a right circular cone is made varies as the square of the distance from the vertex. Find the position of the centre of mass.

17. A hemisphere has a radius of 125 mm. Calculate the position of its centroid.

18. A cylindrical shell has a mass M, a radius R and a length L. Calculate its moment of inertia about

    (a) a central axis

    (b) an axis about a diameter at one end

    (c) an axis through its centroid and along a diameter.

19. A steel rod is 3.75 m long and of circular cross section of 35 mm diameter. The density of steel is 7800 kg/m³. Calculate the moment of inertia about (a) the centroid and (b) one end.

20. A solid right circular cone has a mass of 165 kg, a base radius of 175 mm and a height of 650 mm. Calculate its moment of inertia about a central axis.

21. A beam has the cross section shown in figure 7.29. Calculate its second moment of area about an axis through its centroid (NA) and the corresponding radius of gyration. The dimensions are in millimetres.

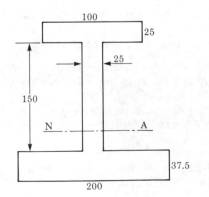

Figure 7.29

22. Calculate the total pressure on the gate in the dam shown in figure 7.30 at a depth of 5 m. The gate is 2.5 m high and 1.5 m wide. Calculate also the position of the centre of pressure. Density of water = 1000 kg/m^3

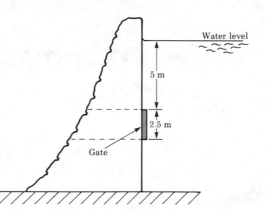

Water level

5 m

2.5 m

Gate

Figure 7.30

23. A triangular plate of base 5 m and height 8 m is immersed in a lake with its base along the water level. Calculate the total pressure on the plate and the depth of the centre of pressure if the plate is vertical. Density of water = 1000 kg/m^3

CHAPTER 8

Taylor series and power series

8.1 INTRODUCTION

In chapter 5 we showed that the sum of a geometric series is given by

$$1 + x + x^2 + x^3 + \ldots = \frac{1}{1-x}$$

This formula holds true for $-1 < x < 1$. We will now consider this result from a different point of view. The left-hand side of the equation is an infinite series in powers of x, while the right-hand side is a simple function of x. Figure 8.1 shows the graph of this function.

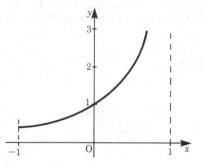

Figure 8.1

The series and the function are identical for a certain interval. It is also possible to represent other functions by means of series in ascending powers of x.

In this chapter we investigate functions which can be expressed as infinite power series. An *infinite power series* is an expression of the form

$$a_0 + a_1 x + a_2 x^2 + a_3 x^3 + \ldots = \sum_{n=0}^{\infty} a_n x^n$$

Power series have an infinite number of terms, each term being a simple power of an independent variable such as x. We will try to express functions already discussed, like the trigonometric functions, as power series. Power series are easy to handle numerically. For small values of x, the higher terms of the series decrease rapidly. In this case an approximate value of a function can be obtained by taking the first terms of the series.

The expansion of a function as a power series is useful for the following reasons:

(1) *Evaluation*. The functional values of exponential, trigonometric and logarithmic functions, for instance, can be computed numerically with the help of power series to a high degree of accuracy.

200

(2) *Approximation*. The first terms of a power series can be used to obtain an approximate value for a given function.

(3) *Term-by-term integration*. It is not always possible to integrate a function as it stands. If, however, the function can be represented by an absolutely convergent power series it can then be integrated term by term to give the value of the integral to a high degree of accuracy.

8.2   EXPANSION OF A FUNCTION IN A POWER SERIES

The following relationship holds true for many functions:

$$f(x) = \sum_{n=0}^{\infty} a_n x^n = a_0 + a_1 x + a_2 x^2 + a_3 x^3 + \ldots + a_n x^n + \ldots \tag{8.1}$$

The coefficients have to be evaluated for each function separately. One important property of such a series is that it is differentiable. It is a necessary condition but not a sufficient one. The function $f(x) = e^{-1/x^2}$ for $x \neq 0$, for example, can be differentiated, but it cannot be expanded in a series. In section 8.3 we will investigate the range of values of x for which the expansion is valid. For the time being we will assume that such an expansion is possible.

Consider equation 8.1. The fundamental assumption is that the value of the function and the power series coincide at $x = 0$ and that the values of their derivatives coincide as well. This gives rise to an algorithm for evaluating the coefficients a_i of the power series.

Step 0: The function and the series must coincide at $x = 0$. This gives

$$f(0) = a_0 + a_1 \times 0 + a_2 \times 0 + \ldots + a_n \times 0 + \ldots$$

Thus a_0 is known to be $a_0 = f(0)$.

Step 1: The first derivatives of the function and the series must coincide at $x = 0$. Obtain the first derivative:

$$f'(x) = a_1 + 2a_2 x + 3a_3 x^2 + \ldots + n a_n x^{n-1} + \ldots$$

For $x = 0$ we get

$$f'(0) = a_1 \quad \text{(all other terms are zero)}$$

Thus a_1 is known to be $a_1 = f'(0)$.

Step 2: The second derivatives of the function and the series must coincide at $x = 0$. Obtain the second derivative:

$$f''(x) = 2a_2 + 3 \times 2a_3 x + \ldots + n(n-1)a_n x^{n-2} + \ldots$$

For $x = 0$ we get

$$f''(0) = 2a_2 \quad \text{(all other terms are zero)}$$

Thus a_2 is known to be $a_2 = \frac{1}{2} f''(0)$.

Step n: The nth derivatives of the function and the series must coincide at $x = 0$. Obtain the nth derivative:

$$f^{(n)}(x) = n(n-1)(n-2) \times \ldots \times 3 \times 2 \times 1 \times a_n + (n+1) n(n-1) \times \ldots \times a_{n+1} x + \ldots$$

For $x = 0$ we get

$$f^{(n)}(0) = n(n-1)(n-2) \times \ldots \times 3 \times 2 \times 1 \times a_n \qquad \text{(all other terms are zero)}$$

Hence

$$a_n = \frac{f^{(n)}(0)}{n(n-1)(n-2)\ldots \times 3 \times 2 \times 1} = \frac{1}{n!} f^{(n)}(0)$$

Note that $n! = n(n-1)(n-2) \times \ldots \times 3 \times 2 \times 1$ ($n!$ is read 'n factorial')

$$1! = 1 \qquad\qquad = 1$$
$$2! = 1 \times 2 \qquad\quad = 2$$
$$3! = 1 \times 2 \times 3 \quad\; = 6$$
$$4! = 1 \times 2 \times 3 \times 4 = 24 \qquad \text{and so on}$$

By definition, $0! = 1$

DEFINITION   The expansion of a function $f(x)$ expressed in a power series is given by

$$f(x) = f(0) + \frac{1}{1!} f'(0)x + \frac{1}{2!} f''(0)x^2 + \frac{1}{3!} f'''(0)x^3 + \ldots$$

or, more simply

$$f(x) = \sum_{n=0}^{\infty} \frac{f^n(0)}{n!} x^n \qquad\qquad (8.2)$$

This is known as *Maclaurin's series*.

Expansion of the Exponential Function $f(x) = e^x$

The derivatives are

$$f'(x) = e^x, \quad f''(x) = e^x, \quad \ldots, \quad f^n(x) = e^x$$

Substituting in equation 8.2 gives

$$e^x = 1 + \frac{x}{1!} + \frac{x^2}{2!} + \frac{x^3}{3!} + \ldots + \frac{x^n}{n!} + \ldots$$

$$= \sum_{n=0}^{\infty} \frac{x^n}{n!} \qquad\qquad (8.3)$$

For any given value of x, the factorial $n!$ increases more rapidly than the power function x^n; hence the terms get smaller as n increases.

For $x = 1$, for example, we have

$$e^1 = 1 + 1 + \frac{1}{2} + \frac{1}{6} + \frac{1}{24} + \frac{1}{120} + \frac{1}{720} + \frac{1}{5040} + \ldots$$

so that

$$e \approx 2.7182 \ldots$$

Similarly, the expansion for e^{-x} is obtained by replacing x with $-x$:

$$e^{-x} = 1 - \frac{x}{1!} + \frac{x^2}{2!} - \frac{x^3}{3!} + \frac{x^4}{4!} - + \ldots$$

Expansion of the Sine Function $f(x) = \sin x$

$$
\begin{aligned}
f(x) &= \sin x & f(0) &= 0 \\
f'(x) &= \cos x & f'(0) &= 1 \\
f''(x) &= -\sin x & f''(0) &= 0 \\
f'''(x) &= -\cos x & f'''(0) &= -1
\end{aligned}
$$

Substituting in (8.2) gives

$$\sin x = x - \frac{x^3}{3!} + \frac{x^5}{5!} - \frac{x^7}{7!} + \ldots \tag{8.4}$$

$$= \sum_{n=0}^{\infty} (-1)^n \frac{1}{(2n+1)!} x^{2n+1}$$

Expansion of the Binomial Series $f(x) = (a + x)^x$

$$
\begin{aligned}
f(x) &= (a+x)^n & f(0) &= a^n \\
f'(x) &= n(a+x)^{n-1} & f'(0) &= na^{n-1} \\
f''(x) &= n(n-1)(a+x)^{n-2} & f''(0) &= n(n-1)a^{n-2}
\end{aligned}
$$

$$\vdots \qquad\qquad\qquad\qquad \vdots$$

$$f^k(x) = n(n-1) \ldots (n-k+1)(a+x)^{n-k} \qquad f^k(0) = n(n-1) \ldots (n-k+1)a^{n-k}$$

Note that n need not be an integer. Thus the expansion is valid for, e.g. $n = \frac{1}{2}$.

Substituting in equation 8.2 gives

$$(a+x)^n = a^n + na^{n-1}x + \frac{n(n-1)}{2!} a^{n-2}x^2 + \frac{n(n-1) \ldots (n-k+1)}{k!} a^{n-k} x^k + \ldots$$

A useful version of this series is when $a = 1$. We then have

$$(1+x)^n = 1 + nx + \frac{n(n-1)}{2!} x^2 + \frac{n(n-1)(n-2)}{3!} x^3 + \ldots \tag{8.5}$$

Expansion of the Function $f(x) = \dfrac{1}{1-x}$

We know the result already because this is the sum of a geometric series.

$$f(x) = \frac{1}{1-x} \qquad\qquad f(0) = 1$$

$$f'(x) = \frac{1}{(1-x)^2} \qquad\qquad f'(0) = 1$$

$$f''(x) = \frac{1 \times 2}{(1-x)^3} \qquad\qquad f''(0) = 2!$$

$$f'''(x) = \frac{1 \times 2 \times 3}{(1-x)^4} \qquad\qquad f'''(0) = 3!$$

$$\vdots \qquad\qquad\qquad\qquad \vdots$$

$$f^n(x) = \frac{n!}{(1-x)^{n+1}} \qquad\qquad f^n(0) = n!$$

Substituting in equation 8.2 gives the familiar result

$$\frac{1}{1-x} = 1 + x + x^2 + x^3 + \ldots + x^n + \ldots = \sum_{n=0}^{\infty} x^n \qquad (|x| < 1) \qquad (8.6)$$

$$(\text{Absolute value of } x!!)$$

8.3   INTERVAL OF CONVERGENCE OF POWER SERIES

There are functions for which Maclaurin's series converge for values of x within a certain range. This is the case for the geometric series. It is only convergent provided that $-1 < x < 1$. This range is referred to as the *interval of convergence*.

Consider the power series

$$a_0 + a_1 x + a_2 x^2 + a_3 x^3 + \ldots + a_n x^n + a_{n+1} x^{n+1} + \ldots$$

The coefficients are numbers independent of x. The series may converge for certain values of x and diverge for other values.

We wish to find the range of values of x for which the series converges, i.e. the interval of convergence. We form the ratio

$$\frac{a_{n+1} x^{n+1}}{a_n x^n}$$

assuming that all coefficients a_i are non-zero. Now consider its limit:

$$\lim_{n \to \infty} \left| \frac{a_{n+1}}{a_n} x \right| = \frac{|x|}{R}$$

where
$$R = \lim_{n \to \infty} \left| \frac{a_n}{a_{n+1}} \right| \qquad\qquad (8.7a)$$

R is called the *radius of convergence*.

The series is absolutely convergent if $|x| < R$ and divergent if $|x| > R$. Hence a power series is convergent in a definite interval $(-R, R)$ and divergent outside this interval. This is illustrated in figure 8.2.

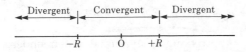

Figure 8.2

Another well-known formula for computing R is due to Cauchy and Hadamard. It is

$$\frac{1}{R} = \lim_{n \to \infty} \sqrt[n]{|a_n|} \qquad (8.7b)$$

The formula is also applicable if some coefficients a_i vanish, e.g. in the trigonometric functions.

EXAMPLE   Consider the power series

$$x - \frac{x^2}{2^2} + \frac{x^3}{3^2} - \frac{x^4}{4^2} + \ldots + (-1)^{n-1} \frac{x^n}{n^2}$$

Using Formula 8.7a we obtain

$$R = \lim_{n \to \infty} \left| \frac{a_n}{a_{n+1}} \right| = \lim_{n \to \infty} \left| \frac{-(n+1)^2}{n^2} \right| = 1$$

Hence the series is convergent if $|x| < 1$ and divergent if $|x| > 1$.

EXAMPLE   Consider the exponential series

$$e^x = \sum_{n=0}^{\infty} \frac{x^n}{n!}$$

Using Formula 8.7a we obtain

$$R = \lim_{n \to \infty} \left| \frac{a_n}{a_{n+1}} \right| = \lim_{n \to \infty} \frac{(n+1)!}{n!} = \lim_{n \to \infty} (n+1) = \infty$$

Hence the series is valid for all values of x; the radius of convergence is ∞.

8.4 APPROXIMATE VALUES OF FUNCTIONS

It is easier to handle a finite number of terms of a series than an infinite number. In a convergent series the terms tend to zero; the evaluation of a power series can therefore be broken off after a certain number of terms. Where we break it off depends on the accuracy required. It is important then to be able to estimate the error.

Consider the series

$$f(x) = a_0 + a_1 x + a_2 x^2 + \ldots + a_n x^n + \ldots$$

Let us divide the series in two parts so that

$$f(x) = \underbrace{a_0 + a_1 x + a_2 x^2 + \ldots + a_n x^n}_{\substack{\text{Approximate polynomial} \\ P_n(x) \text{ of degree } n}} + \underbrace{a_{n+1} x^{n+1} + \ldots}_{\substack{\text{Remainder} \\ R_n(x)}}$$

The first part represents the approximate value of the function $f(x)$, and the second part the *remainder*. If we take the polynomial of degree n as an approximation of the value of the function $f(x)$, it follows that the error is equal to $R_n(x)$, the remainder. This remainder is an infinite series and if we can estimate its magnitude we automatically have an estimate for the value of the error.

To appreciate the behaviour of approximations, let us consider graphically the sine function taking one term, then two terms, and so on, of the power series for $\sin x$.

First approximation, $\sin x \approx x$

Second approximation, $\sin x \approx x - \dfrac{x^3}{6}$

Third approximation, $\sin x \approx x - \dfrac{x^3}{6} + \dfrac{x^5}{120}$

The *first approximation* is represented by a tangent at the position $x = 0$ (figure 8.3a). We can see that the error builds up rapidly after $x \approx \dfrac{\pi}{6}$.

The *second approximation* replaces the sine function by a polynomial of the third degree (figure 8.3b). The range of values of x for which the approximation is sufficiently exact is greater.

The *third approximation* replaces the sine function by a polynomial of the 5th degree (figure 8.3c). The range of values of x for which the approximation is satisfactory is much larger, e.g. consider an extreme case:

$$\sin \frac{\pi}{2} = \sin 90° \approx 1.004\,52, \qquad \text{error} = 0.004\,52$$

In practice, we would have to decide what error we could accept, and this depends very much on the nature of the problem. The above example does illustrate the point that a polynomial can give a good approximation to the value of some other function.

Lagrange succeeded in estimating the error when the first terms of a series are used to calculate the value of a function. He showed that the terms neglected can be represented by the expression

$$R_n(x) = \frac{f^{(n+1)}(\xi)}{(n+1)!} x^{n+1} \tag{8.8}$$

This expression contains the $(n+1)$th derivative of the function at some value of ξ, which lies in the interval $0 < \xi < x$. There exists a value of $\xi = \xi_0$ for which the remainder is a maximum. The error cannot be greater than $R_n(\xi_0)$.

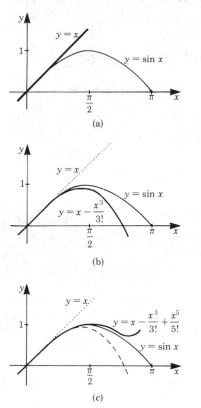

Figure 8.3   (c)

EXAMPLE   Suppose we stop the series for the exponential function after the third term. Then the remainder for $x = 0.5$, say, is

$$R_3(0.5) = e^\xi \frac{(0.5)^4}{4!}$$

As e^x is monotonic increasing, we get the maximum value of the remainder with $\xi_0 = 0.5$:

$$R_3(0.5) = \frac{e^{0.5}(0.5)^4}{24} \approx 0.004$$

The error, in this case, will be less than 0.004 if we stop at the 3rd term of the expansion.

8.5 EXPANSION OF A FUNCTION $f(x)$ AT AN ARBITRARY POSITION

It is often useful to expand a function at a position x_0 which is different from zero. To obtain such an expansion we could proceed as in section 8.2, but instead we introduce a new variable, $u = x - x_0$. Since this auxiliary variable is zero at $x = x_0$, we can expand the function at the position $u = 0$ in terms of ascending powers of u and afterwards express the expansion in terms of x. Hence we proceed as follows:

Since the function is to be expanded at $x = x_0$, we introduce a new variable, $u = x - x_0$.

We resolve in terms of x; $x = u + x_0$ and substitute the expression $u + x_0$ for x in $f(x)$:

$$f(x) = f(u + x_0).$$

We expand at the position $u = 0$ with respect to u to obtain

$$f(x) = f(u_0 + x_0) = f(x_0) + f'(x_0)u + \frac{f''(x_0)}{2!}u^2 + \ldots + \frac{f^{(n)}(x_0)}{n!}u^n + \ldots$$

Now we replace u by $x - x_0$ so that

$$f(x) = f(x_0) + f'(x_0)(x - x_0) + \frac{f''(x_0)}{2!}(x - x_0)^2 + \ldots + \frac{f^{(n)}(x_0)}{n!}(x - x_0)^n + \ldots$$

$$(8.9)$$

This type of series is known as *Taylor's series*.

Note that the geometric meaning of the substitution $u = x - x_0$ is a transformation of coordinates; the variable u has its origin at $x = x_0$. By means of this shift, we are back to the previous situation when we expanded a function at a position where the abscissa was zero.

EXAMPLE   Expand the cosine $f(x) = \cos x$ about the point $x_0 = \frac{\pi}{3}$ or $(60°)$.

Differentiating gives

$$f'(x) = -\sin x, \qquad f''(x) = -\cos x, \qquad f'''(x) = \sin x$$

$$f'\left(\frac{\pi}{3}\right) = -\frac{\sqrt{3}}{2}, \qquad f''\left(\frac{\pi}{3}\right) = -\frac{1}{2}, \qquad f'''\left(\frac{\pi}{3}\right) = \frac{\sqrt{3}}{2}$$

and so on.

Substituting in equation 8.9 gives

$$\cos x = \frac{1}{2} - \left(x - \frac{\pi}{3}\right)\frac{\sqrt{3}}{2} - \left(x - \frac{\pi}{3}\right)^2\frac{1}{4} + \left(x - \frac{\pi}{3}\right)^3\frac{\sqrt{3}}{12} + \ldots$$

Suppose we wish to calculate the value of cosine 61° without using tables. Then

$$\cos 61° = \frac{1}{2} - \frac{\sqrt{3}}{2}\left(\frac{\pi}{180}\right) - \frac{1}{4}\left(\frac{\pi}{180}\right)^2 + \frac{\sqrt{3}}{12}\left(\frac{\pi}{180}\right)^3 + \ldots$$

If we use two terms only

$$\cos 61° \approx \frac{1}{2} - \frac{\sqrt{3}}{2}\left(\frac{\pi}{180}\right) = 0.5000 - 0.015\,11 = 0.484\,89$$

Error $$R_n(x) = \frac{\cos\left(\xi + \dfrac{n+1}{2}\pi\right)}{(n+1)!}\left(x - \frac{\pi}{3}\right)^{n+1}$$

Note: The $(n+1)$ the derivative of $\cos(x)$ is $\cos\left(x + \dfrac{n+1}{2}\pi\right)$. Also, $\dfrac{\pi}{3} < \xi < \dfrac{\pi}{3} + \dfrac{\pi}{180}$.

In this case the error is $\qquad \leqslant \dfrac{1}{2}\left(\dfrac{\pi}{180}\right)^2 = 0.000\,15$

since $$\cos\left(\xi + \frac{n+1}{2}\pi\right) \leqslant 1.$$

(The actual error is 0.000 08.)

EXAMPLE   Expand the function $f(\theta) = (1 - a\sin^2\theta)^{1/2}$.

This expression is important in the study of the slider crank mechanism as used in the car engine.

To expand this function, we could use Maclaurin's or Taylor's series. However, a moment's thought leads us to conclude that we should use the binomial expansion instead. We saw earlier that

$$(1+x)^n = 1 + nx + \frac{n(n-1)}{2!}x^2 + \frac{n(n-1)(n-2)}{3!}x^3 + \dots$$

With $x = -a\sin^2\theta$ and $n = \frac{1}{2}$ we have

$$(1 - a\sin^2\theta)^{1/2} = 1 - \tfrac{1}{2}a\sin^2\theta + \frac{\tfrac{1}{2}(-\tfrac{1}{2})}{2!}a^2\sin^4\theta - \frac{(\tfrac{1}{2})(-\tfrac{1}{2})(-\tfrac{3}{2})}{3!}a^3\sin^6\theta \dots$$

$$= 1 - \tfrac{1}{2}a\sin^2\theta - \tfrac{1}{8}a^2\sin^4\theta - \tfrac{1}{16}a^3\sin^6\theta - \dots$$

8.6 APPLICATIONS OF SERIES

At the beginning of this chapter we mentioned briefly the important applications of series. The numerical values of trigonometric, exponential, logarithmic and many more functions are computed by means of series. The values found in tables were first computed by hand a very long time ago: an extremely tiresome task! Today computers make the task far easier and furthermore, error is reduced.

8.6.1 POLYNOMIALS AS APPROXIMATIONS

The expansion of functions as infinite series has a special significance in calculating approximate values. With a rapidly convergent series and with small values of x we only need to take the first two or three terms of the expansion; in some cases the first term is adequate. If we replace the function by an approximate polynomial, the mathematical expression may be simplified considerably.

EXAMPLE   The atmospheric pressure p is a function of altitude h and is given by

$$p = p_0 e^{-\alpha h}$$

p_0 and α are constants, p_0 being the pressure when $h = 0$. To calculate the pressure difference we have

$$\Delta p = p - p_0 = p_0(e^{-\alpha h} - 1)$$

This expression can be simplified by using an approximation.

Since $e^{-x} = 1 - x + \ldots$, as a first approximation to the pressure difference, it follows that

$$\Delta p \approx p_0(1 - \alpha h - 1) = -p_0 \alpha h$$

Suppose we want to calculate the altitude h when the pressure p is decreased by 1% of the pressure at $h = 0$, i.e.

$$\Delta p / p_0 = \frac{-1}{100}, \qquad \alpha = 0.121 \times 10^{-3} \frac{1}{m}$$

We have

$$h = -\frac{\Delta p}{p_0 \alpha} = \frac{1}{100} \times \frac{1}{0.121 \times 10^{-3}} = 82.64 \, m$$

The error is 0.4 m or 0.49% of the true value.

EXAMPLE   Detour problem. Figure 8.4 shows two possible paths that can be taken when travelling a distance S from A to B, a direct one and an indirect one via C. The problem is to find how much longer is the detour via C than the direct path?

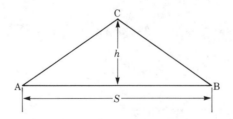

Figure 8.4

Let u be the detour. If h is the height of an assumed equilateral triangle, then, by Pythagoras' theorem, we have

$$u = 2\left(\sqrt{\left(\frac{S}{2}\right)^2 + h^2} - \frac{S}{2}\right)$$

$$u = S\left(\sqrt{1 + \left(\frac{2h}{S}\right)^2} - 1\right)$$

To investigate the behaviour of u as a function of h, it is much simpler to express it by an approximate polynomial. Using the binomial expansion, we have

$$\frac{1+u}{S} = f(h) = \left(1 + \left(\frac{2h}{S}\right)^2\right)^{1/2} = 1 + \frac{1}{2}\left(\frac{2h}{S}\right)^2 - \frac{\frac{1}{2}(1-\frac{1}{2})}{2!}\left(\frac{2h}{S}\right)^4 + \ldots$$

Provided that $h < S$, we can use a first-degree approximation by taking the first two terms of the series:

$$f(h) \approx 1 + \frac{1}{2}\left(\frac{2h}{S}\right)^2$$

Substituting in the equation for u gives

$$u = S\left(1 + \tfrac{1}{2}\left(\frac{2h}{S}\right)^2 - 1\right) = \frac{2h^2}{S}$$

As an example, let $S = 100$ km. The function is shown in figure 8.5. An examination of the graph shows, e.g. that when $h = 5$ km, the detour is only 0.5 km.

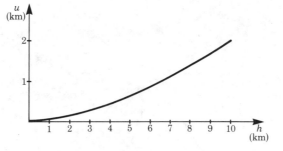

Figure 8.5

EXAMPLE   Obtain a closer approximation for one of the roots of the equation

$$x^4 - 1.5x^3 + 3.7x - 21.554 = 0$$

A rough estimate gave $x = 2.4$.

Let x be a rough approximation for the root of an equation found by trial and error. If the true solution is $x + h$, then, by Taylor's theorem, we have

$$0 = f(x + h) \approx f(x) + h \cdot f'(x)$$

Solving for h gives

$$h \approx -\frac{f(x)}{f'(x)}; \qquad \text{hence} \qquad x - \frac{f(x)}{f'(x)} \qquad \text{is a better approximation.}$$

This is also known as the Newton–Raphson approximation formula (see Chapter 17). Returning to the example, we find

$$f'(x) = 4x^3 - 4.5x^2 + 3.7 \qquad \text{and} \qquad f'(2.4) = 33.076$$

Also $f(2.4) = -0.2324$

It follows that $h = \dfrac{0.2324}{33.076} = 0.007$.

A more accurate approximation is $x = 2.4 + 0.007 = 2.407$.

8.6.2   INTEGRATION OF FUNCTIONS WHEN EXPRESSED AS POWER SERIES

We often encounter integrals whose integrands are complicated functions. This makes their integration extremely difficult or even near impossible. If the function to be integrated can be expressed as a power series, then we can integrate it term by term within the interval of convergence. In this way we can solve practical problems more easily. This is illustrated by the following examples.

EXAMPLE   The function e^{-x^2} is known as the Gaussian bell-shaped curve. It is symmetrical about the y-axis. In statistics and the theory of errors the dispersion of measured values about a mean is described in terms of a function of a similar type. We wish to compute the integral

$$\int_0^x e^{-t^2}\, dt$$

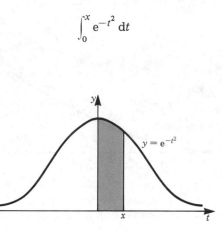

Figure 8.6

This corresponds to the area under the curve between $t = 0$ and $t = x$ (figure 8.6). It is not possible to evaluate this integral as it stands; instead we replace it by a power series. Remember that

$$e^x = 1 + x + \frac{x^2}{2} + \frac{x^3}{6} + \frac{x^4}{24} + \ldots$$

Substituting $-t^2$ for x gives

$$e^{-t^2} = 1 - t^2 + \frac{t^4}{2} - \frac{t^6}{6} + \frac{t^8}{24} - + \ldots$$

Substituting for e^{-t^2} in the integral and integrating term by term gives

$$\int_0^x e^{-t^2}\, dt = x - \frac{x^3}{3} + \frac{x^5}{10} - \frac{x^7}{42} + \frac{x^9}{216} - + \ldots$$

As $x \to \infty$ the integral has a limiting value. This value is given here without proof:

$$\int_0^\infty e^{-t^2}\, dt = \frac{\sqrt{\pi}}{2}$$

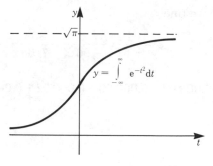

Figure 8.7

The total area under the bell-shaped curve (figure 8.7) is then

$$\int_{-\infty}^{\infty} e^{-t^2}\, dt = \sqrt{\pi}$$

It is useful to normalise the curve so that the area under it is equal to unity, i.e.

$$\int_{-\infty}^{\infty} \frac{e^{-t^2}}{\sqrt{\pi}}\, dt = 1$$

In the next example we will use the following statement which is given without proof.

Multiplying two power series is valid within the interval of convergence if they are both absolutely convergent.

Absolute convergence means that the sum of the absolute values of the summands converges as well.

EXAMPLE   Evaluate
$$\int_0^{0.4} \sqrt{\frac{4-x^2}{4+4x^3}}\, dx$$

First we express the integrand as a product, i.e.

$$\int \sqrt{\frac{4-x^2}{4+4x^3}}\, dx = \int \left(1-\left(\frac{x}{2}\right)^2\right)^{1/2} (1+x^3)^{-1/2}\, dx$$

The binomial series converges for $|x| < 1$. The condition is satisfied in the case of our two functions. The expansions are

$$\left(1-\left(\frac{x}{2}\right)^2\right)^{1/2} = 1 - \frac{1}{8}x^2 - \frac{1}{128}x^4 - \frac{1}{1024}x^6 - \cdots$$

$$(1+x^3)^{-1/2} = 1 - \frac{x^3}{2} + \frac{3}{8}x^6 - + \cdots$$

Multiplying the two series gives, for the integrand I

$$I = 1 - \frac{1}{8}x^2 - \frac{1}{2}x^3 - \frac{1}{128}x^4 + \frac{1}{16}x^5 + \frac{383}{1024}x^6 + \cdots$$

Integrating term by term we find

$$\int_0^{0.4} I \, dx = \left[x - \frac{1}{24}x^3 - \frac{1}{8}x^4 - \frac{1}{640}x^5 + \frac{1}{96}x^6 + \frac{383}{7168}x^7 + \dots \right]_0^{0.4}$$

$$\int_0^{0.4} I \, dx \approx 0.4 - 0.002\,67 - 0.003\,20 - 0.000\,02 + 0.000\,04 + 0.000\,09$$

$$\approx 0.3942$$

8.6.3   EXPANSION IN A SERIES BY INTEGRATING

The expansion of a function in a series can sometimes be achieved by expanding its derivative first and then integrating term by term.

Integrating a convergent power series term by term is valid.

EXAMPLE   Obtain a series for $\tan^{-1}x$.

We know that $\tan^{-1}x = \int \dfrac{dx}{1+x^2}$.

If we expand the integrand in a series and integrate term by term we will obtain a series for $\tan^{-1}x$.

Expanding the integrand by means of the binomial theorem gives

$$\frac{1}{1+x^2} = (1+x^2)^{-1} = 1 - x^2 + x^4 - x^6 + x^8 - \dots$$

which is convergent for $|x| < 1$.

Hence we have

$$\int_0^x \frac{dx}{1+x^2} = \tan^{-1}x = x - \frac{x^3}{3} + \frac{x^5}{5} - \frac{x^7}{7} + \frac{x^9}{9} - \dots \qquad (|x| \leqslant 1)$$

APPENDIX:

COMMONLY USED APPROXIMATE POLYNOMIALS

This table contains a number of typical functions, together with the first terms when expanded as power series. These expansions can be used to obtain approximate values for the functions. The range of values of x for which the approximations are valid are given; these are based on the error being smaller than 1% and 10%, respectively.

TABLE 8.1   APPROXIMATIONS FOR TYPICAL FUNCTIONS

Function	First approximation	Error less than 1% for $x = 0$ to $x =$	Error less than 10% for $x = 0$ to $x =$	Second approximation	Error less than 1% for $x = 0$ to $x =$	Error less than 10% for $x = 0$ to $x =$
$\sin x$	x	0.24	0.74	$x - \dfrac{x^3}{3!}$	1.00	1.66
$\cos x$	$1 - \dfrac{x^2}{2}$	0.66	1.05	$1 - \dfrac{x^2}{2!} + \dfrac{x^4}{4!}$	1.18	1.44
$\tan x$	x	0.17	0.53	$x + \dfrac{x^3}{3}$	0.52	0.91
e^x	$1 + x$	0.14	0.53	$1 + x + \dfrac{x^2}{2}$	0.43	1.10
$\ln(1+x)$ $x > -1$	x	0.02	0.20	$x - \dfrac{x^2}{2}$	0.17	0.58
$\sqrt{1+x}$ $|x| < 1$	$1 + \dfrac{x}{2}$	0.32	1.42	$1 + \dfrac{x}{2} - \dfrac{x^2}{8}$	0.66	1.74
$\dfrac{1}{\sqrt{1+x}}$ $|x| < 1$	$1 - \dfrac{x}{2}$	0.16	0.55	$1 - \dfrac{x}{2} + \dfrac{3}{8}x^2$	0.32	0.73
$\dfrac{1}{1-x}$ $|x| < 1$	$1 + x$	0.10	0.31	$1 + x + x^2$	0.21	0.46
$\dfrac{1}{1-x^2}$ $|x| < 1$	$1 + x^2$	0.31	0.56	$1 + x^2 + x^4$	0.46	0.68

TABLE 8.2   POWER SERIES OF IMPORTANT FUNCTIONS

This table is included for further reference. It contains some expansions which have not been discussed in the text. In some cases negative powers occur.

$$e^x = 1 + \frac{x}{1!} + \frac{x^2}{2!} + \frac{x^3}{3!} + \dots$$

$$\sin x = x - \frac{x^3}{3!} + \frac{x^5}{5!} - \frac{x^7}{7!} + \dots$$

$$\cos x = 1 - \frac{x^2}{2!} + \frac{x^4}{4!} - \frac{x^6}{6!} + \dots$$

$$\tan x = x + \frac{x^3}{3} + \frac{2x^5}{15} + \frac{17x^7}{315} + \frac{62x^9}{2835} + \dots \qquad \left(|x| < \frac{\pi}{2}\right)$$

$$\cot x = \frac{1}{x} - \frac{x}{3} - \frac{x^3}{45} - \frac{2x^5}{945} - \frac{x^7}{4725} - \dots \qquad (0 < |x| < \pi)$$

$$\sec x = 1 + \frac{x^2}{2!} + \frac{5x^4}{4!} + \frac{61x^6}{6!} + \frac{1385x^8}{8!} + \dots \qquad \left(|x| < \frac{\pi}{2}\right)$$

$$\sin^{-1}x = x + \frac{x^3}{2 \times 3} + \frac{3x^5}{2 \times 4 \times 5} + \frac{(3 \times 5)x^7}{2 \times 4 \times 6 \times 7} + \dots \qquad (|x| < 1)$$

$$\cos^{-1}x = \frac{\pi}{2} - \sin^{-1}x$$

$$\tan^{-1}x = x - \frac{x^3}{3} + \frac{x^5}{5} - \frac{x^7}{7} + \frac{x^9}{9} - + \dots \qquad (|x| < 1)$$

$$\cot^{-1}x = \frac{\pi}{2} - \tan^{-1}x$$

$$\sinh x = x + \frac{x^3}{3!} + \frac{x^5}{5!} + \frac{x^7}{7!} + \dots$$

$$\cosh x = 1 + \frac{x^2}{2!} + \frac{x^4}{4!} + \frac{x^6}{6!} + \dots$$

$$\tanh x = x - \frac{x^3}{3} + \frac{2x^5}{15} - \frac{17x^7}{315} + \frac{62x^9}{2835} - \dots \qquad \left(|x| < \frac{\pi}{2}\right)$$

$$\coth x = \frac{1}{x} + \frac{x}{3} - \frac{x^3}{45} + \frac{2x^5}{945} - \frac{x^7}{4725} + \dots \qquad (0 < |x| < \pi)$$

$$\sinh^{-1}x = x - \frac{1 \times x^3}{2 \times 3} + \frac{(1 \times 3)x^5}{2 \times 4 \times 5} - \frac{(1 \times 3 \times 5)x^7}{2 \times 4 \times 6 \times 7} + \dots \qquad (|x| < 1)$$

$$\sinh^{-1}x = (\ln 2)x + \frac{1}{(2 \times 2)x^2} - \frac{1 \times 3}{(2 \times 4 \times 4)x^4} + \frac{1 \times 3 \times 5}{(2 \times 4 \times 6 \times 6)x^6} - \dots$$
$$(|x| > 1)$$

$$\cosh^{-1}x = (\ln 2)x - \frac{1}{(2 \times 2)x^2} - \frac{1 \times 3}{(2 \times 4 \times 4)x^4} - \frac{1 \times 3 \times 5}{(2 \times 4 \times 6 \times 6)x^6} - \dots$$
$$(x > 1)$$

$$\tanh^{-1}x = x + \frac{x^3}{3} + \frac{x^5}{5} + \frac{x^7}{7} + \dots \qquad (|x| < 1)$$

$$\coth^{-1}x = \frac{1}{x} + \frac{1}{3x^3} + \frac{1}{5x^5} + \frac{1}{7x^7} + \dots \qquad (|x| > 1)$$

$$\ln(1 + x) = x - \frac{x^2}{2} + \frac{x^3}{3} - \frac{x^4}{4} + - \dots \qquad (-1 < x \leqslant 1)$$

$$(1 + x)^n = 1 + nx + \frac{n(n-1)}{2!}x^2 + \frac{n(n-1)(n-2)}{3!}x^3$$
$$+ \frac{n(n-1)\dots(n-k+1)x^k}{k!} + \dots \qquad (|x| < 1)$$

This last formula is valid both for integral exponents and for fractional exponents.

Note: There is no expansion for $\cot x$ and for $\coth x$ at $x = 0$ because these functions have a pole at this value. The series expansions were obtained by expanding $x \cot x$ and $x \coth x$ and dividing the result by x.

EXERCISES

8.2 Expansion of a Function in a Power Series

1. Expand the following functions at $x_0 = 0$ in a series up to the first four terms:

   (a) $f(x) = \sqrt{1-x}$ (b) $f(t) = \sin(\omega t + \pi)$

   (c) $f(x) = \ln[(1+x)^5]$ (d) $f(x) = \cos x$

   (e) $f(x) = \tan x$ (f) $f(x) = \cosh x$

8.3 Interval of Convergence of a Power Series

2. Obtain the radius of convergence of the following series:

   (a) $f(x) = \sin x = \displaystyle\sum_{n=0}^{\infty} \frac{(-1)^n}{(2n+1)!} x^{2n+1}$

   (b) $f(x) = \dfrac{1}{1-3x} = \displaystyle\sum_{n=0}^{\infty} 3^n x^n$

8.4 Approximate Value of Functions

3. Sketch in the neighbourhood of $x_0 = 0$ the function $f(x)$ and the graphs of the approximate polynomials $P_1(x)$, $P_2(x)$ and $P_3(x)$.

   (a) $y = \tan x$ (b) $y = \dfrac{x}{4-x}$

8.5 Expansion of a Function at an Arbitrary Position

4. Expand the following functions at $x_0 = \pi$:

   (a) $y = \sin x$ (b) $y = \cos x$

5. Expand the function $f(x) = \ln x$ at $x_0 = 1$.

6. Expand the function $f(x) = \dfrac{4}{1-3x}$ at $x_0 = 2$. Obtain the first four terms.

8.6 Applications of Series

7. Determine the intersection — which lies in the first quadrant — of the functions $e^x - 1$ and $2 \sin x$. Approximate both functions by a polynomial of the third degree, $P_3(x)$.

8. Calculate $\sqrt{42} = \sqrt{36+6}$ to 4 d.p.

9. Replace the function $f(x)$ by an approximate polynomial in the interval $(0, 0.3)$. The error should not exceed 1%.

   (a) $f(x) = \ln(1+x)$ (b) $f(x) = \dfrac{1}{\sqrt{1+x}}$

10. Given the functions $f(x)$, compute approximately (see table 8.1) the value of $f(\frac{1}{4})$. The value obtained should have an accuracy of 10%.

   (a) $f(x) = e^x$ (b) $f(x) = \ln(1+x)$ (c) $f(x) = \sqrt{1+x}$

11. Let the series for the function $f(x) = \sum_{n=0}^{\infty} a_n x^n$ be given. Obtain a series

   expansion for the integral $\int f(x)\,dx$ by integrating the series term by term for

   the following functions:

   (a) $f(x) = \dfrac{1}{1+x} = \sum_{n=0}^{\infty}(-1)^n x^n = 1 - x + x^2 - x^3 + x^4 - \ldots$ $(|x| < 1)$

                                    (geometric series)

   (b) $f(x) = \cos x = \sum_{n=0}^{\infty}(-1)^n \dfrac{x^{2n}}{2n!} = 1 - \dfrac{x^2}{2!} + \dfrac{x^4}{4!} - \dfrac{x^6}{6!} + \ldots$

12. Solve the following integrals using a series expansion:

   (a) $\displaystyle\int_0^{0.58} \sqrt{1+x^2}\,dx$ (b) $\displaystyle\int_0^x \dfrac{\sin t}{t}\,dt$

   (Integral (b) cannot be evaluated by any other method.)

13. (a) Obtain a power series for $\sin^{-1}x$ by first expanding $\dfrac{1}{\sqrt{1-x^2}}$, which is

        the derivative of $\sin^{-1}x$, and, second, integrating term by term.

   (b) Since $\sin^{-1}(1) = \dfrac{\pi}{2}$, by inserting $x = 1$ into the series one obtains a

        series for $\dfrac{\pi}{2}$. Compute the value of this series up to the fifth term and

        compare with the correct numerical value.

CHAPTER 9

Complex numbers

9.1 DEFINITION AND PROPERTIES OF COMPLEX NUMBERS

9.1.1 IMAGINARY NUMBERS

The square of positive as well as negative real numbers is always a positive real number. For example, $3^2 = (-3)^2 = 9$. The root of a positive number is therefore a positive or negative number. We now introduce a new type of number whose square always gives a negative real number: they are called *imaginary numbers*.

> **DEFINITION**   The unit of imaginary numbers is the number j with the property that
>
> $$j^2 = -1 \qquad\qquad (9.1)$$
>
> The imaginary unit j corresponds to 1 for real numbers.

An arbitrary imaginary number is made up of the imaginary unit j and any real number y; thus yj is the general form for an imaginary number.

We know that we cannot extract the root of a negative number when dealing with real numbers; nevertheless, we can factorise the root of a negative number thus:

$$\sqrt{-5} = \sqrt{5(-1)} = \sqrt{5} \times \sqrt{-1}$$

Since $j^2 = -1$, it follows that $j = \sqrt{-1}$; hence $\sqrt{5} \times \sqrt{-1} = j\sqrt{5}$.

The root of a negative number is an imaginary number. Moreover, with $j^2 = -1$, we can simplify higher powers of j.

The ordinary rules of algebra extend to imaginary numbers. In addition one must remember that $j^2 = -1$, i.e.

$$
\begin{aligned}
j^1 &= j \\
j^2 &= -1 \\
j^3 &= j^2 \times j = -j \\
j^4 &= j^2 \times j^2 = 1
\end{aligned}
$$

9.1.2 COMPLEX NUMBERS

The sum z of a real number x and an imaginary number jy is called a *complex number*, complex meaning 'composed'. Thus

$$z = x + jy$$

where x is the real part of z and y is the imaginary part of z.

If we replace j by $-j$, we obtain a different complex number $z*$ given by

$$z* = x - jy$$

$z*$ is called the *complex conjugate* of z.

A complex number has the value zero only if both the real part and the imaginary part are zero.

9.1.3 FIELDS OF APPLICATION

The most obvious property of imaginary numbers is that we can 'extract' the root of a negative number, which means we obtain an expression which we can handle. This property enables us, in principle, to solve equations of any degree. Consider, for instance, the quadratic equation

$$ax^2 + bx + x = 0$$

We have seen that it is unsolvable in terms of real x if $b^2 < 4ac$, i.e. when the radicand $b^2 - 4ac$ becomes negative. The solution in terms of complex x is

$$x_{1,2} = \frac{1}{2a}(-b \pm j\sqrt{4ac - b^2})$$

Complex numbers are important in the solution of differential equations which are dealt with in Chapter 10. They are also a useful concept in electrical engineering, and they are indispensable in the study of quantum physics.

9.1.4 OPERATIONS WITH COMPLEX NUMBERS

When handling complex numbers there are two rules to remember:

The complex number $x + jy$ is zero if and only if $x = 0$ and $y = 0$.

Complex numbers obey the ordinary rules of algebra, in addition $j^2 = -1$.

Addition and Subtraction of Complex Numbers

> **Rule** The sum of complex numbers is obtained by adding real and imaginary part separately.
>
> The difference of complex numbers is obtained by subtracting real and imaginary parts separately.

Let

$$z_1 = x_1 + jy_1$$

$$z_2 = x_2 + jy_2$$

Then their sum is

$$z_1 + z_2 = (x_1 + jy_1) + (x_2 + jy_2)$$
$$= (x_1 + x_2) + j(y_1 + y_2)$$

Their difference is

$$z_1 - z_2 = (x_1 - x_2) + (y_1 - y_2)j$$

Example

$$z_1 = 6 + 7j$$

$$z_2 = 3 + 4j$$

$$z_1 + z_2 = (6 + 7j) + (3 + 4j)$$
$$= 9 + 11j$$

Example

$$z_1 - z_2 = 3 + 3j$$

Product of Complex Numbers

> **Rule** The product z_1, z_2 of two complex numbers is obtained by simple multiplication of the terms, taking into account $j^2 = -1$.

General expression

$$z_1 z_2 = (x_1 + jy_1)(x_2 + jy_2)$$
$$= x_1 x_2 + jx_1 y_2 + jy_1 x_2 + j^2 y_1 y_2$$
$$= (x_1 x_2 - y_1 y_2) + j(x_1 y_2 + x_2 y_1)$$

Example

$$z_1 z_2 = (6 + 7j)(3 + 4j)$$
$$= 18 + 24j + 21j - 28$$
$$= -10 + 45j$$

Division of Complex Numbers

> **Rule** Division of a complex number by another complex number is carried out by multiplying numerator and denominator by the conjugate of the divisor to transform the latter into a real number. Note that the denominator then appears as the sum of two squares.

General expression

$$\frac{z_1}{z_2} = \frac{x_1 + jy_1}{x_2 + jy_2} = \frac{(x_1 + jy_1)(x_2 - jy_2)}{(x_2 + jy_2)(x_2 - jy_2)}$$
$$= \frac{(x_1 x_2 + y_1 y_2) - j(x_1 y_2 - y_1 x_2)}{x_2{}^2 + y_2{}^2}$$

The conjugate of z_2 is $z_2{}^* = x_2 - jy_2$

Example

$$\frac{z_1}{z_2} = \frac{6 + 7j}{3 + 4j} = \frac{(6 + 7j)(3 - 4j)}{(3 + 4j)(3 - 4j)}$$
$$= \frac{-10 - 3j}{25} = \frac{-2}{5} - \frac{3}{25}j$$

9.2 GRAPHICAL REPRESENTATION OF COMPLEX NUMBERS

9.2.1 GAUSS COMPLEX NUMBER PLANE: ARGAND DIAGRAM

The complex number $z = x + jy$ can be represented in an x–y coordinate system by placing the real part along the x-axis and the imaginary part along the y-axis in a similar way to the components of a vector. Figure 9.1 shows this.

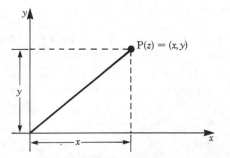

Figure 9.1

We obtain a point $P(z)$ in this plane which corresponds to the complex number z. This plane is called the Gauss number plane, better known as the *Argand diagram*. In this way we have produced a geometrical picture of a complex number.

EXAMPLE   Where is the point in the Argand diagram corresponding to the complex number $z = 4 - 2j$? Figure 9.2 shows the answer.

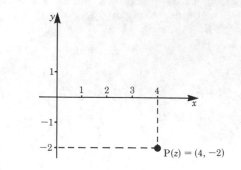

Figure 9.2

Addition

$$z_3 = z_1 + z_2$$

If $z_1 = 6 + 3j$ and $z_2 = 2 + 5j$ we know that the sum $z_3 = 8 + 8j$. This addition may be represented in an Argand diagram, as shown in figure 9.3.

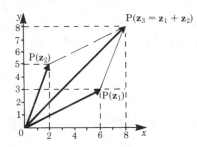

Figure 9.3

To obtain the complex number z_3, we first draw the two complex numbers z_1 and z_2 as vectors as shown. Then we shift vector $\mathbf{z}_2$ parallel to itself in such a way that its tail is made to coincide with the tip of vector $\mathbf{z}_1$. The tip of vector $\mathbf{z}_2$ locates the tip of the required vector or complex number z_3 whose magnitude is given by the length OP. This construction is based on the parallelogram rule for the addition of vectors. The same result is obtained if we shift $\mathbf{z}_1$ instead of $\mathbf{z}_2$. We dealt with vectors in Chapter 3, so this construction should not be new to the reader.

To add more than two complex numbers we draw a polygon by joining the vectors tip to tail. The tip of the closing vector represents the required complex number.

Subtraction

$$z_3 = z_1 - z_2 = z_1 + (-z_2)$$

The problem of the subtraction of two complex numbers can be transformed into one of addition if to one complex number we add the negative of the other.

9.2.2   POLAR FORM OF A COMPLEX NUMBER

Instead of specifying a complex number by means of the coordinates x and y, we could specify it by means of a distance r from the origin of the coordinates and an angle α, as shown in figure 9.4 x and y are the cartesian coordinates, and r and α are the polar coordinates.

From the figure we see that

$$x = r\cos\alpha$$

$$y = r\sin\alpha$$

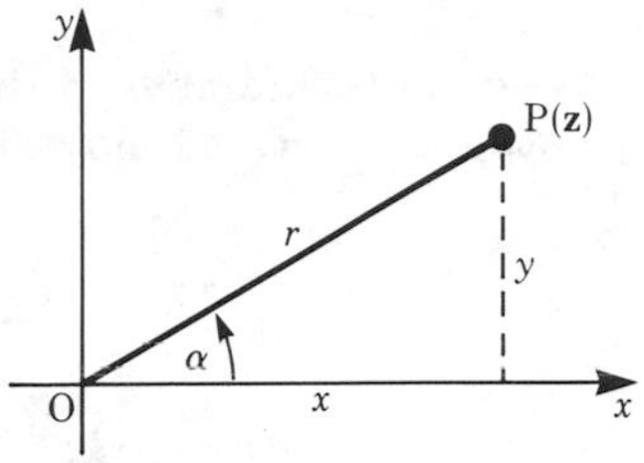

Figure 9.4

Substituting in the expression for the complex number, $z = x + \mathrm{j}y$, gives

$$z = r(\cos\alpha + \mathrm{j}\sin\alpha) \tag{9.2}$$

This expresses a complex number in terms of the trigonometric functions. It follows that the conjugate z^* of the complex number z is

$$z^* = r(\cos\alpha - \mathrm{j}\sin\alpha)$$

If we know r and α we can calculate x and y from the above equations. If on the other hand we know x and y and we want to express a complex number in polar form, it follows from figure 9.4 that

$$r = \sqrt{x^2 + y^2} \qquad \text{taking the positive value only}$$

$$\alpha = \tan^{-1}\left(\frac{y}{x}\right)$$

r is known as the *modulus* and α as the *argument*.

Table 9.1 illustrates the values taken by α according to the sign of x and y.

TABLE 9.1

x	y	$\tan\alpha$	P(z) lies in	α in the range
positive	positive	positive	1st quadrant	0 to $\frac{\pi}{2}$
negative	positive	negative	2nd quadrant	$\frac{\pi}{2}$ to π
negative	negative	positive	3rd quadrant	π to $\frac{3\pi}{2}$
positive	negative	negative	4th quadrant	$\frac{3\pi}{2}$ to 2π

EXAMPLE   Express the complex number $z = 1 - j$ in polar form.

$z = 1 - j$ means that $x = 1$ and $y = -1$.

Hence
$$r = \sqrt{x^2 + y^2} = \sqrt{1 + 1} = \sqrt{2}$$

$$\alpha = \tan^{-1}\left(\frac{-1}{1}\right) = \tan^{-1}(-1) = \frac{3\pi}{4} \quad \text{or} \quad \frac{7\pi}{4}$$

We represent the complex number in the Argand diagram (figure 9.5) and find that it lies in the fourth quadrant. It follows then that

$$\alpha = \frac{7\pi}{4}$$

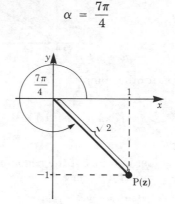

Figure 9.5

Thus the complex number $z = 1 - j$ can be written in polar form:

$$z = \sqrt{2}\left(\cos\frac{7\pi}{4} + j\sin\frac{7\pi}{4}\right)$$

EXAMPLE   Express the complex number $z = 6j$ in polar form.

In this case, $z = x + jy \equiv 0 + 6j$, i.e. $x = 0$. Consequently $r = 6$

$$\tan\alpha = \frac{y}{x} = \frac{6}{0} = \infty, \quad \text{i.e.} \quad \alpha = \frac{\pi}{2} \quad \text{or} \quad \frac{3\pi}{2}$$

To determine α, the complex number is shown in figure 9.6. It follows, therefore, that $\alpha = \frac{\pi}{2}$. In polar form the complex number is

$$z = 6\left(\cos\frac{\pi}{2} + j\sin\frac{\pi}{2}\right)$$

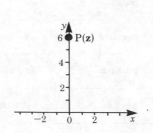

Figure 9.6

9.3 EXPONENTIAL FORM OF COMPLEX NUMBERS

9.3.1 EULER'S FORMULA

It is also possible to express a complex number in the form

$$z = r\,e^{j\alpha} \tag{9.3}$$

As before, r is the modulus of z and α is the argument. We can show that it is equivalent to the polar form of a complex number:

$$z = r(\cos\alpha + j\sin\alpha)$$

In other words we want to prove that

$$r\,e^{j\alpha} = r(\cos\alpha + j\sin\alpha)$$

Dividing by r gives
$$e^{j\alpha} = \cos\alpha + j\sin\alpha$$

In Chapter 8, section 8.2, we showed that the expansion for e^x is

$$e^x = 1 + x + \frac{x^2}{2!} + \frac{x^3}{3!} + \frac{x^4}{4!} + \dots$$

This expansion remains valid if we replace x by $j\alpha$. Remembering that $j^2 = -1$, we obtain

$$e^{j\alpha} = 1 + j\alpha - \frac{\alpha^2}{2!} - j\frac{\alpha^3}{3!} + \frac{\alpha^4}{4!} + j\frac{\alpha^5}{5!} \dots$$

We also showed that
$$\sin\alpha = \alpha - \frac{\alpha^3}{3!} + \frac{\alpha^5}{5!} - \dots$$

Hence
$$j\sin\alpha = j\alpha - j\frac{\alpha^3}{3!} + j\frac{\alpha^5}{5!} - \dots$$

$$\cos\alpha = 1 - \frac{\alpha^2}{2!} + \frac{\alpha^4}{4!} - \dots$$

Comparing these expressions, we obtain

$$\cos\alpha + j\sin\alpha = 1 + j\alpha - \frac{\alpha^2}{2!} - j\frac{\alpha^3}{3!} + \frac{\alpha^4}{4!} + j\frac{\alpha^5}{5!} - \dots = e^{j\alpha}$$

Euler's formula

$$e^{j\alpha} = \cos\alpha + j\sin\alpha \tag{9.4}$$

This enables us to express a complex number in a third way.

A table giving the various forms in which complex numbers can be expressed will be found in the appendix at the end of this chapter.

9.3.2   EXPONENTIAL FORM OF THE SINE AND COSINE FUNCTIONS

The conjugate complex number of $e^{j\alpha}$ is obtained by replacing j by $-j$, so that if $z = e^{j\alpha}$ then $z* = e^{-j\alpha}$.

Now we will try to express the sine and cosine function in terms of $e^{j\alpha}$.

According to Euler's formula, we have

$$e^{j\alpha} = \cos\alpha + j\sin\alpha$$
$$e^{-j\alpha} = \cos\alpha - j\sin\alpha$$

Adding both equations gives

$$\cos\alpha = \frac{1}{2}(e^{j\alpha} + e^{-j\alpha}) = \cosh j\alpha$$

Subtracting both equations gives

$$\sin\alpha = \frac{1}{2j}(e^{j\alpha} - e^{-j\alpha}) = \frac{1}{j}\sinh j\alpha$$

9.3.3   COMPLEX NUMBERS AS POWERS

Given the complex number $z = x + jy$, we wish to calculate the modulus and argument of

$$w = e^z$$

Substituting for z, we have

$$w = e^{(x+jy)} = e^x\, e^{jy} \equiv r\, e^{j\alpha}$$

Comparing the last two expressions, it follows that $r = e^x$ and $jy = j\alpha$. Hence the modulus of w is e^x, and the argument of w is y.

EXAMPLE   If $z = 2 + j\dfrac{\pi}{2}$, calculate $w = e^z$.

$$w = e^z = e^{(2 + j\pi/2)} = e^2\, e^{j\pi/2}$$

Hence the modulus is $r = e^2$ and the argument is $\alpha = \dfrac{\pi}{2}$, and we have

$$w = e^2\left(\cos\frac{\pi}{2} + j\sin\frac{\pi}{2}\right) = j\,e^2$$

The solution is shown in the Argand diagram (figure 9.7).

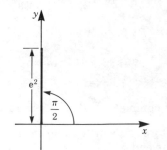

Figure 9.7

EXAMPLE   If $z = -2 + j\dfrac{3\pi}{4}$, calculate $w = e^z$.

$$w = e^{\left(-2 + j\frac{3\pi}{4}\right)} = e^{-2} e^{j\frac{3\pi}{4}}$$

$$r = e^{-2} \quad \text{and} \quad \alpha = \frac{3\pi}{4}$$

Hence

$$w = e^{-2}\left(\cos\frac{3\pi}{4} + j\sin\frac{3\pi}{4}\right)$$

$$= e^{-2}\left(-\frac{1}{\sqrt{2}} + j\frac{1}{\sqrt{2}}\right)$$

This solution is shown in the Argand diagram (figure 9.8).

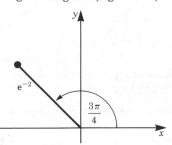

Figure 9.8

Suppose now that z is a function of some parameter t. The simplest case is that of the linear function where a and b are constants.

$$z(t) = at + jbt$$

One important interpretation of t in practice is the time, so that $z(t)$ grows with time, i.e. the real part and the imaginary part grow with time.

Substituting for z in $w = e^z$ gives $w(t) = e^{(at + jbt)} = e^{at} e^{jbt}$

Using Euler's formula, we get $w(t) = e^{at}(\cos bt + j\sin bt)$

To examine the behaviour of this function, we can consider the real and the imaginary parts separately and represent each one graphically as a function of the time t.

The real part of $w(t)$ is $e^{at}\cos bt$. It is the product of an exponential function and a trigonometric function of period $p = \dfrac{2\pi}{b}$.

If a is positive then $w = e^{at}\cos bt$ represents a vibration whose amplitude grows exponentially with time, as shown in figure 9.9.

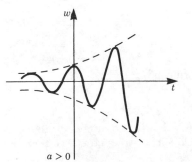

Figure 9.9

If a is negative then $w = e^{at} \cos bt$ represents a vibration whose amplitude decreases exponentially with time; the vibration is said to be damped and is shown below in figure 9.10.

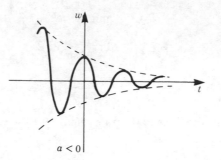

Figure 9.10 $a < 0$

The imaginary part of $w(t)$ is $e^{at} \sin bt$. It is also the product of an exponential function and a trigonometric function whose graphical representation is similar to figure 9.9 or figure 9.10, depending on the sign of a.

The mathematical solution of vibration problems is often simplified by means of complex numbers. Thus, in solving a practical problem, we start by considering the physical situation consisting of real quantities. We then perform all calculations using complex numbers and finally consider and interpret the results of the real and imaginary parts.

9.3.4 MULTIPLICATION AND DIVISION IN EXPONENTIAL FORM

Addition and subtraction of complex numbers are best carried out using the form $z = x + jy$. Multiplication and division, on the other hand are best carried out by expressing the complex numbers either in exponential form or in polar form.

Consider two complex numbers

$$z_1 = r_1 e^{j\alpha_1} \quad \text{and} \quad z_2 = r_2 e^{j\alpha_2}$$

Multiplying gives

$$z = z_1 z_2 = r_1 e^{j\alpha_1} r_2 e^{j\alpha_2} = r_1 r_2 e^{j(\alpha_1 + \alpha_2)} \tag{9.5}$$

Here we use the power rule $a^n a^m = a^{n+m}$.

Dividing gives

$$z = \frac{z_1}{z_2} = \frac{r_1 e^{j\alpha_1}}{r_2 e^{j\alpha_2}} = \frac{r_1}{r_2} e^{j(\alpha_1 - \alpha_2)} \tag{9.6}$$

Here we use the power rule $a^n / a^m = a^{n-m}$.

Rule   To multiply (or divide) complex numbers, we multiply (or divide) the modulii and add (or subtract) the arguments.

9.3.5   **RAISING TO A POWER, EXPONENTIAL FORM**

We have $z^n = (r\,e^{j\alpha})^n = r^n\,e^{jn\alpha}$ (9.7)

Rule   To raise a complex number to a given power we raise the modulus to that power and multiply the argument by that power.

Figure 9.11 shows the points z and z^2 in the complex plane with $r = 2$.

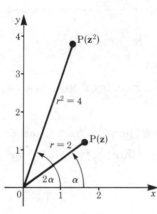

Figure 9.11

In the case of $z^{1/n}$ we have

$$z^{1/n} \; = \; \sqrt[n]{z} \; = \; \sqrt[n]{r\,e^{j\alpha}} \; = \; \sqrt[n]{r}\,e^{j\alpha/n}$$ (9.8)

Rule   To extract the root of a complex number, we find the root of the modulus and divide the argument by the index.

9.3.6   **PERIODICITY OF $r\,e^{j\alpha}$**

We should like to mention the fact, perhaps surprising to the reader, that the complex number

$$z \; = \; r\,e^{j\alpha}$$

is identical to $$z \; = \; r\,e^{j(\alpha + 2\pi)}$$

If we examine figure 9.12 we can see that the same point $P(z)$ is obtained whether the angle is α or $\alpha + 2\pi$. In fact we could equally take the angle to be $\alpha + 4\pi$, $\alpha + 6\pi$, $\alpha - 2\pi$, $\alpha - 4\pi$, etc. Hence, generally

$$r\,e^{j\alpha} = r\,e^{j(\alpha + 2\pi k)}$$

where

$$k = \pm 1,\ \pm 2,\ \pm 3\ \text{etc.}$$

In particular we have

$$1 = e^{j2\pi}$$

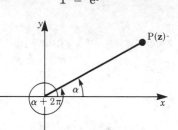

Figure 9.12

9.3.7   TRANSFORMATION OF A COMPLEX NUMBER FROM ONE FORM INTO ANOTHER

Transformation from the Algebraic Form into the Exponential Form

The transformation from $x + jy$ into $r\,e^{j\alpha}$ is based on the relationships derived in section 9.2.2, i.e.

$$r = \sqrt{x^2 + y^2}, \qquad \tan \alpha = \frac{y}{x}$$

EXAMPLE   Convert the complex number $z = -\sqrt{5} + 2j$ to the exponential form.

$$r = \sqrt{(-\sqrt{5})^2 + 2^2} = 3$$

$$\tan \alpha = \frac{2}{-\sqrt{5}} = -0.894$$

The angle is in the second quadrant; therefore $\alpha = 138.19°$ or 0.768π radians.

Transformation of the Exponential Form into the Algebraic Form

Since $z = r\,e^{j\alpha} = r(\cos \alpha + j \sin \alpha) = x + jy$

then $x = r \cos \alpha$

$\quad\ \ y = r \sin \alpha$

EXAMPLE   Convert the expression $z = e^{(0.5 + 1.3j)}$ to the algebraic form.

$$z = e^{(0.5 + 1.3j)} = e^{0.5}\,e^{1.3j}$$

$$x = e^{0.5} \cos 1.3 = 0.441$$

$$y = e^{0.5} \sin 1.3 = 1.589$$

Hence $z = 0.441 + 1.589j$

9.4 OPERATIONS WITH COMPLEX NUMBERS EXPRESSED IN POLAR FORM

In the previous section, the rules for operations with complex numbers have been derived using the exponential form. We will now show that the same rules are obtained if we express the complex numbers in polar form. We make use of the addition theorems for trigonometric functions given in Chapter 2, section 2.7.

9.4.1 MULTIPLICATION AND DIVISION

Let z_1 and z_2 be expressed in polar form so that

$$z_1 = r_1(\cos\alpha_1 + j\sin\alpha_1)$$
$$z_2 = r_2(\cos\alpha_2 + j\sin\alpha_2)$$

Multiplication

$$z_1 z_2 = r_1 r_2(\cos\alpha_1 + j\sin\alpha_1)(\cos\alpha_2 + j\sin\alpha_2)$$
$$= r_1 r_2[(\cos\alpha_1\cos\alpha_2 - \sin\alpha_1\sin\alpha_2) + j(\sin\alpha_1\cos\alpha_2 + \cos\alpha_1\sin\alpha_2)]$$

Using the addition formulae for the sine and cosine functions,

$$z_1 z_2 = r_1 r_2[\cos(\alpha_1 + \alpha_2) + j\sin(\alpha_1 + \alpha_2)] \tag{9.9}$$
$$= r(\cos\alpha + j\sin\alpha)$$

Thus the modulus of the product equals $r_1 r_2$, and the argument is $(\alpha_1 + \alpha_2)$.

This is exactly in accordance with the rule derived in the previous section: to multiply complex numbers we multiply the moduli and add the arguments.

Figure 9.13 illustrates geometrically the multiplication of complex numbers. Draw a triangle OPP_1 similar to the triangle OP_2Q (Q has coordinates $1, 0$), then

$$\frac{z}{z_1} = \frac{z_2}{1}, \qquad \text{hence} \qquad z = z_1 z_2$$

Also $\alpha = $ angle $QOP = \alpha_1 + \alpha_2$

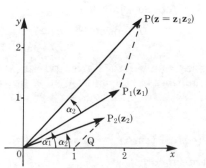

Figure 9.13

Division

Both the numerator and the denominator are multiplied by the conjugate of the divisor.

$$\frac{z_1}{z_2} = \frac{r_1(\cos\alpha_1 + j\sin\alpha_1)}{r_2(\cos\alpha_2 + j\sin\alpha_2)} \frac{r_2(\cos\alpha_2 - j\sin\alpha_2)}{r_2(\cos\alpha_2 - j\sin\alpha_2)}$$

Using the addition formulae for the sine and cosine functions,

$$\frac{z_1}{z_2} = \frac{r_1}{r_2}\left[\cos(\alpha_1 - \alpha_2) + j\sin(\alpha_1 - \alpha_2)\right] \tag{9.10}$$

Thus the modulus of the quotient equals $\dfrac{r_1}{r_2}$, and the argument is $(\alpha_1 - \alpha_2)$.

Again this is in accordance with the rule derived in the previous section: to divide complex numbers we divide the moduli and subtract the arguments.

Figure 9.14 illustrates the division of complex numbers by a reasoning similar to that for multiplication.

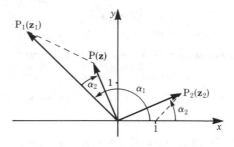

Figure 9.14

9.4.2 RAISING TO A POWER

We saw earlier that $\quad z_1 z_2 = r_1 r_2[\cos(\alpha_1 + \alpha_2) + j\sin(\alpha_1 + \alpha_2)]$

If we now let $z_1 = z_2 = z$, $r_1 = r_2 = r$ and $\alpha_1 = \alpha_2 = \alpha$ it follows that

$$z^2 = r^2(\cos 2\alpha + j\sin 2\alpha)$$

Similarly
$$z^3 = r^2(\cos 2\alpha + j\sin 2\alpha)r(\cos\alpha + j\sin\alpha)$$

$$= r^3(\cos 3\alpha + j\sin 3\alpha)$$

The general expression is $\quad z^n = r^n(\cos n\alpha + j\sin n\alpha) \tag{9.11}$

> **Rule** To raise a complex number to a given power, we raise the modulus to that power and multiply the argument by that power.

By setting $r = 1$ we have

$$(\cos\alpha + j\sin\alpha)^n = \cos n\alpha + j\sin n\alpha$$

This is known as *De Moivre's theorem*.

9.4.3 ROOTS OF A COMPLEX NUMBER

De Moivre's theorem holds true for *positive*, *negative* and *fractional powers*. We can, therefore, use this fact to determine all the distinct roots of any number.

Since $x + jy = r(\cos \alpha + j \sin \alpha)$, then, by De Moivre's theorem, it follows that

$$\sqrt[n]{x + jy} = \sqrt[n]{r}\left(\cos \frac{\alpha}{n} + j \sin \frac{\alpha}{n}\right)$$

However, using this equation, we obtain one root only. In order to obtain all the roots we must consider the fact that the cosine and sine functions are periodic functions of period 2π radians or $360°$. Thus we can write

$$(\cos \alpha + j \sin \alpha)^n = [\cos(\alpha + 2\pi k) + j \sin(\alpha + 2\pi k)]^n$$

$$= \cos(n\alpha + 2\pi nk) + j \sin(n\alpha + 2\pi nk)$$

where $k = 0, \pm1, \pm2, \pm3, \ldots$

When raising a complex number to an integral power there is no ambiguity: the result is independent of periodicity. But extracting the roots of a complex number means raising it to a fractional power. Now periodicity becomes important, and we have

$$\sqrt[n]{x + jy} = \sqrt[n]{r}\left[\cos\left(\frac{\alpha}{n} + \frac{2\pi}{n}k\right) + j \sin\left(\frac{k}{n} + \frac{2\pi}{n}k\right)\right] \qquad (9.12)$$

where $k = 0, \pm1, \pm2, \ldots$

> **Rule** The *n*th roots of a complex number are obtained by extracting the *n*th root of the modulus and dividing the argument by *n*. Due to the periodicity of the trigonometric functions, there are *n* solutions.

By giving k the values $0, 1, 2, 3, \ldots, (n-1)$, we obtain the n different roots of a complex number; for example,

$k = 0$, $\qquad\qquad z_1 = \sqrt[n]{r}\left(\cos\frac{\alpha}{n} + j \sin\frac{\alpha}{n}\right)$

$k = 1$, $\qquad\qquad z_2 = \sqrt[n]{r}\left[\cos\left(\frac{\alpha}{n} + \frac{2\pi}{n}\right) + j \sin\left(\frac{\alpha}{n} + \frac{2\pi}{n}\right)\right]$

$k = 2$, $\qquad\qquad z_3 = \sqrt[n]{r}\left[\cos\left(\frac{\alpha}{n} + \frac{4\pi}{n}\right) + j \sin\left(\frac{\alpha}{n} + \frac{4\pi}{n}\right)\right]$

and so on. With $k = n$, we would obtain the same value as with $k = 0$; also we would not obtain any new values for $k > n$ or $k = -1, -2, -3, \ldots$. The root corresponding to $k = 0$ is called the *principal value*.

EXAMPLE   Calculate the four roots of

$$z^4 = \cos\frac{2\pi}{3} + j\sin\frac{2\pi}{3}$$

In this case $r = 1$ and $n = 4$; hence

$$z^4 = \cos\left(\frac{\pi}{6} + \frac{2\pi}{4}k\right) + j\sin\left(\frac{\pi}{6} + \frac{2\pi}{4}k\right)$$

$$= \cos\left(\frac{\pi}{6} + \frac{\pi}{2}k\right) + j\sin\left(\frac{\pi}{6} + \frac{\pi}{2}k\right)$$

The roots are

$k = 0,$ $z_1 = \cos\dfrac{\pi}{6} + j\sin\dfrac{\pi}{6} = \dfrac{1}{2}\sqrt{3} + \dfrac{1}{2}j$

$k = 1,$ $z_2 = \cos\dfrac{2\pi}{3} + j\sin\dfrac{2\pi}{3} = -\dfrac{1}{2} + \dfrac{\sqrt{3}}{2}j$

$k = 2,$ $z_3 = \cos\dfrac{7\pi}{6} + j\sin\dfrac{7\pi}{6} = -\dfrac{\sqrt{3}}{2} - \dfrac{1}{2}j$

$k = 3,$ $z_4 = \cos\dfrac{10\pi}{6} + j\sin\dfrac{10\pi}{6} = \dfrac{1}{2} - \dfrac{\sqrt{3}}{2}j$

The roots z_1 and z_3, as well as z_2 and z_4, are opposite to each other; moreover, we can see that the argument, starting from the principal value, is successively increased by $\dfrac{\pi}{2}$. Figure 9.15 shows all the values on a circle of radius 1. They form a square.

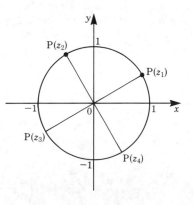

Figure 9.15

All the nth roots of a complex number of modulus 1 have modulus 1. When depicted in an Argand diagram they form the vertices of a regular n-sided polygon inscribed in a unit circle.

APPENDIX

Summary of operations with complex numbers

Designation	Formulae
Imaginary unit j Imaginary number η	$j^2 = -1$ $\eta = jy$ (y real)
Complex number z in arithmetic form Complex conjugate	$z = x + jy \quad (x, y \text{ real})$ x = real part y = imaginary part $z^* = x - jy$
Complex numbers in polar form Transformation $(x, y) \leftrightarrow (r, \alpha)$	$z = r(\cos\alpha + j\sin\alpha)$ $\left. \begin{array}{l} x = r\cos\alpha \\ y = r\sin\alpha \end{array} \right\} \quad \left. \begin{array}{l} r = \sqrt{x^2 + y^2} \\ \tan\alpha = \dfrac{y}{x} \end{array} \right\}$
Complex number in exponential form Euler's formula	$z = r\,e^{j\alpha}$ $e^{j\alpha} = \cos\alpha + j\sin\alpha$
Exponential form for cosine and sine functions	$\cos\alpha = \dfrac{1}{2}(e^{j\alpha} + e^{-j\alpha}) = \cosh j\alpha$ $\sin\alpha = \dfrac{1}{2j}(e^{j\alpha} - e^{-j\alpha}) = \dfrac{1}{j}\sinh j\alpha$
Periodicity of complex numbers	$z = r\,e^{j\alpha}$ $\quad = r\,e^{j(\alpha + 2k\pi)} \quad (k = \pm 1, \pm 2, \pm 3, \ldots)$
Multiplication and division in exponential form Raising to a power and extracting roots in exponential form	$z_1 = r_1 e^{j\alpha_1},\ z_2 = r_2 e^{j\alpha_2}$ $z_1 z_2 = r_1 r_2\, e^{j(\alpha_1 + \alpha_2)}$ $\dfrac{z_1}{z_2} = \dfrac{r_1}{r_2}\, e^{j(\alpha_1 - \alpha_2)}$ $z = r\,e^{j\alpha}$ $z^n = r^n e^{jn\alpha}$ $\sqrt[n]{z} = \sqrt[n]{r}\; e^{j[(\alpha + 2\pi k)/n]} \quad (k = 0, \pm 1, \pm 2, \ldots)$
Multiplication and division in polar form Raising to a power and extracting roots in polar form	$z_1 = r_1(\cos\alpha_1 + j\sin\alpha_1)$ $z_2 = r_2(\cos\alpha_2 + j\sin\alpha_2)$ $z_1 z_2 = r_1 r_2[\cos(\alpha_1 + \alpha_2) + j\sin(\alpha_1 + \alpha_2)]$ $\dfrac{z_1}{z_2} = \dfrac{r_1}{r_2}[\cos(\alpha_1 - \alpha_2) + j\sin(\alpha_1 - \alpha_2)]$ $z = r(\cos\alpha + j\sin\alpha)$ $z^n = r^n[\cos n\alpha + j\sin n\alpha]$ $\sqrt[n]{z} = \sqrt[n]{r}\left[\cos\left(\dfrac{\alpha}{n} + \dfrac{2\pi k}{n}\right) + j\sin\left(\dfrac{\alpha}{n} + \dfrac{2\pi k}{n}\right)\right]$ $(k = 0, \pm 1, \pm 2, \ldots)$

EXERCISES

9.1   Definition and Properties of Complex Numbers

1. Express the following in terms of j:

    (a) $\sqrt{4-7}$ (b) $\sqrt{-144}$ (c) $\dfrac{\sqrt{5}}{\sqrt{-4}}$ (d) $\sqrt{4(-25)}$

2. Compute

    (a) j^8 (b) j^{15} (c) j^{45} (d) $(-j)^3$

3. Evaluate

    (a) $\sqrt{-48}+\sqrt{-75}-\sqrt{-27}$ (b) $\sqrt{-12}-\sqrt{-8}+\sqrt{-0.6}$

    (c) $\sqrt{-3}\sqrt{-3}$ (d) $\sqrt{-a}\sqrt{+b}$

    (e) $5j^3\,2j^6$ (f) $(-j)^3 j^2$

    (g) $8j/2j$ (h) $1/j^3$

    (i) $6j/j^7\sqrt{3}$ (j) $\dfrac{1}{j^5}+\dfrac{1}{j^7}$

    (k) $\sqrt{b-a}\,\sqrt{a-b}$ (l) $\dfrac{\sqrt{-3}\sqrt{12}}{j\sqrt{-a^2}}$

4. Determine the imaginary part of z:

    (a) $z = 3+7j$ (b) $z = 15j-4$

5. Determine the conjugate complex number z^* of z:

    (a) $z = 5+2j$ (b) $z = \frac{1}{2}-\sqrt{3}j$

6. Evaluate the (complex) roots of the following quadratic equations:

    (a) $x^2+4x+13 = 0$ (b) $x^2+\frac{3}{2}x+\frac{25}{16} = 0$

7. Calculate the sum z_1+z_2:

    (a) $z_1 = 3-2j$ (b) $z_1 = \frac{3}{4}+\frac{3}{4}j$

    $z_2 = 7+5j$ $z_2 = \frac{3}{4}-\frac{3}{4}j$

8. Compute $w = z_1-z_2+z_3{}^*$:

    (a) $z_1 = 5-2j$ (b) $z_1 = 4-3.5j$

    $z_2 = 2-3j$ $z_2 = 3+2j$

    $z_3 = -4+6j$ $z_3 = 7.5j$

9. Compute the product $w = z_1 z_2$:

    (a) $z_1 = 1+j$ (b) $z_1 = 3-2j$

    $z_2 = 1-j$ $z_2 = 5+4j$

10. Determine

    (a) $(16+j\sqrt{2})/2\sqrt{2}$ (b) $(4-j\sqrt{3})/2j$

    (c) $(2+3j)/(2-4j)$ (d) $1/(1+j)$

    (e) $\dfrac{1+j}{1-j}-\dfrac{1-j}{1+j}$ (f) $\dfrac{(5+j\sqrt{3})(5-j\sqrt{3})}{2-j\sqrt{3}}$

11. Convert the following sums into products:

(a) $4x^2 + 9y^2$ (b) $a + b$

9.2   Graphical Representation of Complex Numbers

12. Plot each point z_i and $-z_i{}^*$ in the complex number plane:

(a) $z_1 = -1 - j$ (b) $z_2 = 3 + 2j$ (c) $z_3 = 5 + 3j$

(d) $z_4 = \frac{3}{2}j$ (e) $z_5 = -3 + \frac{1}{2}j$ (f) $z_6 = \sqrt{2}$

13. Using figure 9.16, determine the real and imaginary parts of each point $z_1, z_2, \ldots, z_6$.

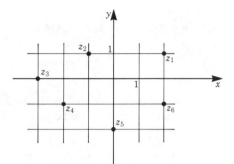

Figure 9.16

14. Convert the complex number $z = x + jy$ to the polar form $z = r(\cos \alpha + j \sin \alpha)$:

(a) $z = j - 1$ (b) $z = -(1 + j)$

15. Transform the complex number $z = r(\cos \alpha + j \sin \alpha)$ into the form $z = x + jy$:

(a) $z = 5\left(\cos \dfrac{\pi}{3} - j \sin \dfrac{\pi}{3}\right)$ (b) $z = 4(\cos 225° + j \sin 225°)$

16. Compute $z_1 z_2$:

(a) $z_1 = 2(\cos 15° + j \sin 15°)$ (b) $z_1 = \sqrt{5}\,(\cos 80° + j \sin 80°)$

 $z_2 = 3(\cos 45° + j \sin 45°)$ $z_2 = \sqrt{5}\,(\cos 40° + j \sin 40°)$

17. Calculate z_1/z_2:

(a) $z_1 = \cos 70° + j \sin 70°$ (b) $z_1 = 4$

 $z_2 = \cos 25° + j \sin 25°$ $z_2 = 4(\cos 30° + j \sin 30°)$

                                             (*Hint*: $4 = 4(\cos 360° + j \sin 360°)$)

18. What is meant geometrically by the multiplication (or division) of a complex number by $-j$?

19. Calculate

(a) $(1 - j)^5$ (b) $(\frac{1}{2} - j\frac{1}{2}\sqrt{3})^3$

20. (a) Prove that $(\cos 50° - j\sin 50°)^4 = \cos 200° - j\sin 200°$.

(b) State De Moivre's theorem.

21. Calculate all the roots of

(a) $\sqrt{-5 + 12j}$ (b) $\sqrt[4]{\cos 60° + j\sin 60°}$

9.3 Exponential Form of Complex Numbers

22. Using Euler's formula, compute $\cos\alpha$ and $\sin\alpha$ and convert to the algebraic form:

(a) $e^{j\pi/2}$ (b) $e^{j\pi/3}$

23. Let the values for $e^{j\alpha}$ and $e^{-j\alpha}$ be given. Compute the values of α, $\cos\alpha$ and $\sin\alpha$:

(a) $e^{j\alpha} = 1$ (b) $e^{j\alpha} = -1$ (c) $e^{j\alpha} = -j$ (d) $e^{j\alpha} = \frac{1}{2}\sqrt{3} + \frac{j}{2}$

$e^{-j\alpha} = 1$ $e^{-j\alpha} = -1$ $e^{-j\alpha} = j$ $e^{-j\alpha} = \frac{1}{2}\sqrt{3} - \frac{j}{2}$

24. Given the complex number $z = x + jy$, $w = e^z$ is then a new complex number. Put it in the form $w = r\,e^{j\alpha}$ and compute r and α if

(a) $z = 3 + 2j$ (b) $z = 2 - \dfrac{j}{2}$

25. Transform the complex number $w = e^z$ into the form $w = u + jv$ if

(a) $z = \dfrac{1}{2} + j\pi$ (b) $z = \dfrac{3}{2} - j\pi$ (c) $z = -1 - j\dfrac{3\pi}{2}$ (d) $z = 3 - j$

26. Let the complex quantity z be a linear function of the parameter t (for example time), i.e. $z(t) = at + jbt$ $(0 \leqslant t \leqslant \infty)$. Given

(a) $z(t) = -t + j2\pi t$,

(b) $z(t) = 2t - j\frac{3}{2}t$,

(i) what is the real part, $\text{Re}\,[w(t)]$, of $w(t) = e^{z(t)}$?
(ii) what is the period of $\text{Re}\,[w(t)]$?
(iii) what is the amplitude of the function $w(t)$ at time $t = 2$?

27. Compute the product $z_1 z_2$:

(a) $z_1 = 2e^{j\pi/2}$ (b) $z_1 = \frac{1}{2}e^{j\pi/4}$
$z_2 = \frac{1}{2}e^{j\pi/2}$ $z_2 = \frac{3}{2}e^{-j3\pi/4}$

28. Calculate for the pairs of numbers z_1, z_2 in the previous exercise the quotient $z_1{}^*/z_2$.

29. (a) Given $z = 2e^{j\pi/5}$, calculate z^5.
(b) Given $z = \frac{1}{2}e^{j\pi/4}$, calculate z^3.

30. (a) Given $z = 32e^{j10\pi}$, calculate $z^{1/5}$.
(b) Given $z = \frac{1}{16}e^{j6\pi}$, calculate $z^{1/4}$.

31. Given $z = r\,e^{j\alpha}$, what in each of the following cases is the angle α for which $0 \leqslant \alpha \leqslant 2\pi$?

(a) $z = 3e^{j\,7\pi}$
(b) $z = \frac{1}{2}e^{j\,14\pi/3}$

32. Put the following complex numbers into exponential form:

(a) $5 - 5j$
(b) $15 - 13j$

33. (a) Put the expression $z = 2.5e^{j43°30'}$ into the form $x + jy$.

(b) Calculate $e^{j146°}\,e^{-j82°}$ and express the result in the form $z = x + jy$.

9.4   Operations with Complex Numbers Expressed in Polar Form

34. (a) Determine the real and imaginary parts of $\dfrac{(1+j)^2}{\sqrt{2}(1-j)}$

(b) What is the polar form of this complex number?

35. Put $z = -2\,(\cos 30° - j \sin 30°)$:

(a) into the form $x + jy$,

(b) into exponential form.

CHAPTER 10

Differential equations

10.1 CONCEPT AND CLASSIFICATION OF DIFFERENTIAL EQUATIONS

Many natural laws in engineering and science are formulated by equations involving derivatives or differentials of physical quantities.

An example is Newton's axiom in mechanics, which states

$$\text{Force} = \text{mass} \times \text{acceleration}$$

The acceleration is the second derivative of the displacement x with respect to the time t. The law can be written as

$$F = m\ddot{x}(t)$$

The force F may be constant or a function of the displacement, a function of the velocity v or a function of some other parameter of the system. We are interested in the displacement as a function of time, i.e.

$$x = x(t)$$

To find it, we must solve Newton's equation.

An equation containing one or more derivatives is called a *differential equation*. In what follows we will use the term DE for short.

Let us consider a concrete example. The motion of a body of mass m falling freely (see figure 10.1) is described by the DE

$$m\ddot{x} = -mg$$

or $$\ddot{x} = -g$$

(Air resistance is neglected; g, the acceleration due to gravity, is $9.81\,\text{m/s}^2$.) We require the displacement $x(t)$, which is determined by the DE given above.

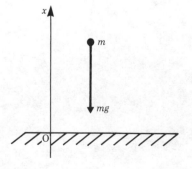

Figure 10.1

Later we will show how to solve such an equation systematically, but for the time being we will merely quote the result.

The position of a body at an instant of time t which satisfies the DE $\ddot{x} = -g$ is given by

$$x(t) = -\tfrac{1}{2}gt^2 + C_1 t + C_2$$

C_1 and C_2 are arbitrary constants. You can verify the correctness of this solution for yourself by differentiating it twice. Thus a DE serves to determine a required function, unlike an algebraic equation which determines numbers.

In general, the DE of a function $y(x)$ may contain one or more derivatives of that function, as well as the function itself and the independent variable x.

Examples of DEs are

$$y'' + x^2 y' + y^2 + \sin x = 0$$

$$y'' + x = 0$$

$$e^x y' - 3x = 0$$

Among the great number of possible types of DE encountered in engineering and science the most important ones are the linear DEs of the first and of the second order, with constant coefficients.

But first we must define the terms *order of a DE* and *linear DE*.

Order of a DE

DEFINITION   The *order* of a DE is defined by the highest derivative contained therein. Thus an nth order DE contains an nth derivative.

Examples are

$$y' + ax = 0, \qquad \text{which is of the first order}$$

and $\qquad y'' + 7y = 0, \qquad$ which is of the second order.

Linear DE

DEFINITION   If the function y and its derivatives $(y', y'', \ldots)$ in a DE are all to the first power and if no products like yy', $y'y'''$ etc. occur, then the DE is *linear*.

Examples are

$$y'' + 7y + \sin x = 0 \quad \text{and} \quad 5y' = xy \quad \text{which are linear DEs.}$$

$$y'' + y^2 = 0 \quad \text{and} \quad (y'')^2 = x^2 y \quad \text{which are non-linear DEs.}$$

Linear DE with Constant Coefficients

DEFINITION   The DE

$$a_2 y'' + a_1 y' + a_0 y = f(x)$$

where $a_2 \neq 0$ and a_2, a_1 and a_0 are arbitrary real constants, is called a *second-order linear* differential equation with *constant coefficients* since all a_j are constants.

Given the linear DE with constant coefficients

$$a_2 y'' + a_1 y' + a_0 y = f(x)$$

We must distinguish between two cases:

$$f(x) = 0 \quad \text{and} \quad f(x) \neq 0$$

If $f(x) = 0$, then the DE is referred to as a *homogeneous* DE. If $f(x) \neq 0$, then it is referred to as a *non-homogeneous* DE.

A homogeneous DE is $my'' + \gamma y' + ky = 0$

A non-homogeneous DE is $my'' + \gamma y' + ky = \sin \omega x$

If in the DE in the definition above, $a_2 = 0$ and $a_1 \neq 0$, the equation becomes $a_1 y' + a_0 y = f(x)$, which is a *first-order linear* DE with constant coefficients. (a_1 and a_0 are real numbers.)

The following equations are examples of first-order linear DEs:

$$y' - gt = 0$$

$$y' - xy = 0$$

Every function which satisfies a DE is called a *solution* of that DE. The purpose of this chapter is to deal with the problem of finding solutions of DEs. Before proceeding further, let us consider the solution of the equation $y'' = -g$. The following equations are possible solutions of this DE, as can easily be verified by inserting in the DE:

$$y_1 = -\tfrac{1}{2}gx^2 + C_1 x + C_2$$

$$y_2 = -\tfrac{1}{2}gx^2 + C_2$$

$$y_3 = -\tfrac{1}{2}gx^2 + C_1 x$$

and $$y_4 = -\tfrac{1}{2}gx^2$$

The solutions y_2, y_3 and y_4 are obviously special cases of the solution y_1. They are obtained when constants are set to zero. We are allowed to give C_1 and C_2 any value we like, e.g.

$$C_1 = -1, \qquad C_2 = 5$$

Hence the solution $y = -\frac{1}{2}gx^2 - x + 5$ is another solution of the DE $y'' = -g$. Thus we have made it clear that y_1 is a solution of the DE, no matter what values we assume for C_1 and C_2. This implies that the solution of the DE is not uniquely determined. The constants which appear in the solution and which we can choose freely are called *integration constants*. The solution is referred to as the *general solution* before the constants are evaluated.

The number of constants which appear in the solution of a DE is determined by the following lemma.

Lemma 10.1   The general solution of a first-order DE contains exactly one undetermined integration constant. The general solution of a second-order DE contains exactly two integration constants, which can be chosen independently of each other.

This statement follows from the fact that a first-order DE requires one integration and hence one constant of integration, while a second-order DE requires two integrations, and hence two constants of integration.

A *special solution* of the DE is obtained by assigning particular values to the constants in the general solution. The special solution is called a *particular solution* or a *particular integral*.

In the example above, the second, third and fourth solutions are particular solutions of the general solution, i.e. of the first solution $(C_1 = 0,\ C_2 = 0,$ and $C_1 = C_2 = 0,$ respectively).

We are, above all, interested in the general solution, since it contains all the particular solutions. A particular solution is obtained if additional conditions are imposed. These conditions are referred to as *boundary conditions*.

There is a similarity with the problem of integration. An indefinite integral is a general solution, while the definite integral is the particular solution when certain conditions are imposed, such as the limits of integration.

The constants in the general solution of a DE are chosen in such a way as to satisfy the boundary conditions. The problem in science and engineering is that of obtaining a particular solution by fixing boundary conditions in order to solve a particular case.

We will now develop methods for solving first- and second-order DEs with constant coefficients.

10.2 PRELIMINARY REMARKS

It has already been mentioned that a special case of the linear second-order DE is obtained by setting $a_2 = 0$ in

$$a_2 y'' + a_1 y' + a_0 y = f(x)$$

The result is the linear first-order DE

$$a_1 y' + a_0 y = f(x)$$

For this reason, we will derive the solution for the second-order DE and only refer briefly to its application to the first-order DE. The main reason for doing this is that in engineering and science many of the problems we meet lead to second-order DEs.

Finding a solution for the non-homogeneous second-order DE is made easier by the following lemma.

> **Lemma 10.2**   Consider the non-homogeneous DE
>
> $$a_2 y'' + a_1 y' + a_0 y = f(x)$$
>
> Let y_c be the general solution of the homogeneous equation
>
> $$a_2 y'' + a_1 y' + a_0 y = 0$$
>
> y_c is also called the *complementary function*.
>
> Let y_p be a particular solution of the non-homogeneous DE
>
> $$a_2 y'' + a_1 y' + a_0 y = f(x)$$
>
> Then the general solution of the DE is given by
>
> $$y = y_c + y_p \qquad (10.1)$$

PROOF   We will first show that $y = y_c + y_p$ is a solution of the DE.

According to the assumptions we have made for the homogeneous DE,

$$a_2 y_c'' + a_1 y_c' + a_0 y_c = 0 \qquad [1]$$

For the non-homogeneous DE we have

$$a_2 y_p'' + a_1 y_p' + a_0 y_p = f(x) \qquad [2]$$

Substituting $y = y_c + y_p$ in the non-homogeneous equation gives

$$a_2 (y_c + y_p)'' + a_1 (y_c + y_p)' + a_0 (y_c + y_p) = f(x)$$

Rearranging gives

$$(a_2 y_c'' + a_1 y_c' + a_0 y_c) + (a_2 y_p'' + a_1 y_p' + a_0 y_p) = f(x)$$

But the first bracket is zero, according to equation 1, and the second bracket is in accordance with equation 2. It follows that $y = y_c + y_p$ is a solution of the non-homogeneous DE. Furthermore, since we assumed that y_c is the general solution of the homogeneous DE, it contains two arbitrary constants (cf. lemma 10.1). Hence the solution $y = y_c + y_p$ also contains two arbitrary constants which can be chosen independently of each other: it is the general solution.

According to lemma 10.2, the general, or complete, solution of

$$a_2 y'' + a_1 y' + a_0 y \ = \ f(x)$$

can be achieved in three steps:

Step 1   Find the complementary function y_c of the homogeneous equation.

Step 2   Find a particular integral y_p of the non-homogeneous equation.

Step 3   Add both solutions to obtain the general solution of the non-homogeneous equation:

$$y \ = \ y_c + y_p$$

To solve DEs, engineers and scientists will often look up solutions from a collection of solutions and will only try to find solutions for themselves when such a collection is not at hand. Even in these circumstances they will not necessarily follow a systematic procedure that is always successful; instead they will try to find a solution and then use the principle of verification to prove that it is valid. If the assumed solution is found not to be valid, it is modified and the process repeated until a valid solution is found. To guess successfully requires experience which the learner, obviously, does not possess; in the following section we will therefore consider systematic methods of solution.

10.3   GENERAL SOLUTION OF FIRST- AND SECOND-ORDER DEs WITH CONSTANT COEFFICIENTS

10.3.1   HOMOGENEOUS LINEAR DE

In this section, we derive a method for finding solutions of first- and second-order homogeneous DEs with constant coefficients. The method is always successful.

Homogeneous First-Order DE

We will consider briefly the first-order DE with constant coefficients

$$a_1 y' + a_0 y \ = \ 0$$

Rearranging the equation gives $a_1 \dfrac{\mathrm{d}y}{\mathrm{d}x} \ = \ -a_0 y$

or by 'separating the variables': $\dfrac{\mathrm{d}y}{y} \ = \ -\dfrac{a_0}{a_1}\,\mathrm{d}x$

Integrating both sides gives $\ln y \ = \ -\dfrac{a_0}{a_1}x + \text{constant}$

For convenience, we can write $\ln C$ for the constant. The solution is

$$y \ = \ C\,e^{r_1 x}, \qquad \text{where} \qquad r_1 \ = \ -\dfrac{a_0}{a_1}$$

This type of equation is frequently encountered in, e.g. the decay of radioactive substances, the tension of a belt round a pulley, the discharge of a capacitor in an electric circuit.

Note: If it is possible to write any DE with only x terms on one side and only y terms on the other, the solution can be obtained by straightforward integration of both sides. This is called *separation of variables*.

Homogeneous Second-Order DE

We now seek a general solution of the homogeneous second-order DE with constant coefficients, i.e.

$$a_2 y'' + a_1 y' + a_0 y = 0$$

The solution of this equation will contain two arbitrary constants, corresponding to two different solutions y_1 and y_2. The solutions must be *linearly independent*, i.e. they cannot be represented by $y_1 = C y_2$, where C is some constant, for all values of x in the interval considered. The following lemma will be useful for finding the general solution.

> **Lemma 10.3**   If the homogeneous linear DE
>
> $$a_2 y'' + a_1 y' + a_0 y = 0$$
>
> has two different solutions y_1 and y_2, then the following expression is also a solution of the DE:
>
> $$y = C_1 y_1 + C_2 y_2$$
>
> C_1 and C_2 may be real or complex quantities. This expression is the general solution of the DE.

PROOF   We assume that

$$a_2 y_1'' + a_1 y_1' + a_0 y_1 = 0 \qquad [1]$$

$$a_2 y_2'' + a_1 y_2' + a_0 y_2 = 0 \qquad [2]$$

Substituting $y = C_1 y_1 + C_2 y_2$ in the DE gives

$$a_2(C_1 y_1 + C_2 y_2)'' + a_1(C_1 y_1 + C_2 y_2)' + a_0(C_1 y_1 + C_2 y_2) = 0$$

Rearranging the terms gives

$$C_1(a_2 y_1'' + a_1 y_1' + a_0 y_1) + C_2(a_2 y_2'' + a_1 y_2 + a_0 y_2) = 0$$

By equations (1) and (2), both expressions in brackets are identically zero. Hence we have proved that

$$y = C_1 y_1 + C_2 y_2$$

is a solution of the DE. It is the general solution since it contains two arbitrary constants.

We must find two linearly independent solutions y_1 and y_2. Guided by the results for the first-order DE, we assume that the second-order DE is solved by functions of the type $y = e^{rx}$. The admissible values for the unknown r are to be determined. Table 10.1 shows the systematic procedure and an example.

TABLE 10.1

Systematic procedure for the solution of the homogeneous second-order DE	Example

Let the equation be

$$a_2 y'' + a_1 y' + a_0 y = 0 \qquad\qquad y'' + 3y' + 2y = 0$$

Let $y = e^{rx}$ be a solution of the DE. Substituting for

$$y' = r e^{rx} \qquad\qquad y = e^{rx}, \qquad y' = \frac{dy}{dx} = r e^{rx}$$

and
$$y'' = r^2 e^{rx} \qquad\qquad y'' = \frac{d^2y}{dx^2} = r^2 e^{rx}$$

gives
$$a_2 r^2 e^{rx} + a_1 r e^{rx} + a_0 e^{rx} = 0$$

We can factorise e^{rx}:

$$e^{rx}(a_2 r^2 + a_1 r + a_0) = 0 \qquad\qquad e^{rx}(r^2 + 3r + 2) = 0$$

Since $e^{rx} \neq 0$, the expression in the bracket must be zero:

$$a_2 r^2 + a_1 r + a_0 = 0 \qquad\qquad r^2 + 3r + 2 = 0$$

This is a quadratic in r. It is called the *auxiliary equation* of the DE. Its roots are

$$r_{1,2} = \frac{-a_1 \pm \sqrt{a_1^2 - 4a_2 a_0}}{2a_2} \qquad\qquad r_1 = -1, \qquad r_2 = -2$$

Provided that r_1 and r_2 are different, the general solution of the DE is

$$y = C_1 e^{r_1 x} + C_2 e^{r_2 x} \qquad\qquad y = C_1 e^{-x} + C_2 e^{-2x}$$

The roots of the auxiliary equation will depend on the values of the constants a_2, a_1 and a_0. We must therefore examine these roots carefully. There are, in fact, three cases to distinguish.

Let us recapitulate. Given the DE $a_2 y'' + a_1 y' + a_0 y = 0$

we found the solution to be $\quad y = C_1 e^{r_1 x} + C_2 e^{r_2 x}$ $\qquad\qquad$ (10.2)

with $\qquad\qquad\qquad\qquad r_{1,2} = \dfrac{-a_1 \pm \sqrt{a_1^2 - 4a_2 a_0}}{2a_2}$

Case 1: The expression $a_1^2 - 4a_2 a_0$ is positive.

Here the roots are real and unequal.

EXAMPLE   Solve $2y'' + 7y' + 3y = 0$.

The auxiliary equation is $2r^2 + 7r + 3 = 0$.

The roots are $r_1 = -0.5$, $r_2 = -3$, and the general solution is

$$y = C_1 e^{-0.5x} + C_2 e^{-3x}$$

The solutions are combinations of exponential functions. A detailed discussion of this type of solution with respect to applications can be found in section 10.4.2.

Case 2: The expression $a_1^2 - 4a_2 a_0$ is zero.
Here the two roots are equal and so far we know only one solution, namely e^{rx}:

$$r = -\frac{a_1}{2a_2}$$

We need to find a second solution. Let us assume that it is of the type $y = u\,e^{rx}$, where u is some function of x. Differentiating we find

$$y' = u'\,e^{rx} + ru\,e^{rx} \qquad y'' = u''\,e^{rx} + 2ru'\,e^{rx} + r^2 u\,e^{rx}$$

Substituting in the DE gives $\left(\text{use } r = -\dfrac{a_1}{2a_2}\right)$ $a_2 u''\,e^{rx} = 0$

The DE is satisfied only if $u'' = 0$, i.e. if u is a linear function

$$u = C_1 + C_2 x$$

Then the solution of the equation is

$$y = (C_1 + C_2 x)\,e^{rx} \tag{10.3}$$

As it contains two arbitrary constants, it is the general solution.

Case 3: The expression $a_1^2 - 4a_2 a_0$ is negative.
Here the roots r_1 and r_2 are complex conjugates. As we are concerned with real solutions we must show how complex roots lead to real solutions.

To simplify, let the roots be denoted by

$$r_1 = a + jb \qquad \text{and} \qquad r_2 = a - jb$$

where $$a = -\frac{a_1}{2a_2} \qquad \text{and} \qquad b = \frac{1}{2a_2}\sqrt{4a_2 a_0 - a_1^2}$$

The general solution of the DE is then

$$y = C_1 e^{(a+jb)x} + C_2 e^{(a-jb)x}$$
$$= e^{ax}(C_1 e^{jbx} + C_2 e^{-jbx})$$

From Euler's formula (equation 9.4 in Chapter 9, section 9.3.1)

$$e^{\pm jx} = \cos x \pm j\sin x$$

Substituting for the complex exponential gives

$$y = e^{ax}[(C_1 + C_2)\cos bx + j(C_1 - C_2)\sin bx]$$

or $$y = e^{ax}(A\cos bx + B\sin bx) \tag{10.4}$$

where $A = C_1 + C_2$ and $B = j(C_1 - C_2)$.

To demonstrate that we can obtain a real solution from this general complex solution, we will consider the following lemma.

Lemma 10.4   Let the solution of the homogeneous DE

$$a_2 y'' + a_1 y' + a_0 y = 0$$

be a complex function y of the real variable x so that

$$y = y_1(x) + jy_2(x)$$

The constants a_2, a_1 and a_0 are assumed to be real.

Then the real part y_1 and the imaginary part y_2 are particular solutions, and the general real valued solution is given by

$$y = C_1 y_1 + C_2 y_2 \qquad (10.5)$$

with real constants C_1 and C_2.

PROOF   According to our assumption, we have

$$a_2(y_1 + jy_2)'' + a_1(y_1 + jy_2)' + a_0(y_1 + jy_2) = 0$$

Collecting real and imaginary parts gives

$$(a_2 y_1'' + a_1 y_1' + a_0 y_1) + j(a_2 y_2'' + a_1 y_2' + a_0 y_2) = 0$$

But a complex number is exactly equal to zero if the real and the imaginary parts are zero at the same time. Hence

$$a_2 y_1'' + a_1 y_1' + a_0 y_1 = 0$$

and

$$a_2 y_2'' + a_1 y_2' + a_0 y_2 = 0$$

From this it follows that both y_1 and y_2 are solutions of the DE and, according to lemma 10.3, the general solution is

$$y = C_1 y_1 + C_2 y_2$$

thus proving the lemma.

We can now state that if the auxiliary equation of the homogeneous DE has conjugate complex roots $r_1 = a + jb$ and $r_2 = a - jb$ there is a real solution given by

$$y = e^{ax}(A \cos bx + B \sin bx)$$

EXAMPLE   Solve $y'' + 4y' + 13y = 0$.

The auxiliary equation is $r^2 + 4r + 13 = 0$,

whose roots are $r_1 = -2 + 3j$ and $r_2 = -2 - 3j$.

By the above, the general solution is

$$Y = e^{-2x}(C_1 \cos 3x + C_2 \sin 3x)$$

Summary   The solution of the homogeneous second-order DE

$$a_2 y'' + a_1 y' + a_0 y = 0$$

with constant coefficients may be summarised by the following steps.

Set up the auxiliary equation.

$$y'' \text{ is replaced by } r^2$$
$$y' \text{ is replaced by } r$$
$$y \text{ is replaced by } 1$$

The auxiliary equation is $a_2 r^2 + a_1 r + a_0 = 0$.

Calculate the roots r_1 and r_2 of the auxiliary equation:

$$r_{1,2} = \frac{-a_1 \pm \sqrt{a_1^2 - 4a_2 a_0}}{2a_2}$$

Obtain the general solution according to the following three possible cases.

Case 1   If $r_1 \neq r_2$ are real and unequal roots

$$y = C_1 e^{r_1 x} + C_2 e^{r_2 x} \tag{10.2}$$

Case 2   If $r_1 = r_2$ are equal roots

$$y = e^{r_1 x}(C_1 + C_2 x) \tag{10.3}$$

Case 3   If r_1 and r_2 are complex roots with

$$r_1 = a + jb \qquad \text{and} \qquad r_2 = a - jb$$
$$y = e^{ax}(C_1 \cos bx + C_2 \sin bx) \tag{10.4}$$

10.3.2   NON-HOMOGENEOUS LINEAR DE

According to lemma 10.2, the complete solution of the non-homogeneous DE is the sum of the complementary function and a particular integral. We have learned how to find the complementary solution, i.e. the solution of the homogeneous equation. We must find methods for obtaining a particular solution. One method, called the variation of parameters, which always yields a solution is discussed later in this section. The only problem with this method is that it is longwinded. Consequently we often tend to find a way of guessing a particular solution. You may think this is unsatisfactory, but with practice you will soon appreciate its value.

EXAMPLE   Find a particular integral of the DE

$$y'' + y = 5$$

One such particular integral is

$$y_p = 5$$

since $y_p'' = 0$ and $y_p = 5$ satisfy the DE.

Generally, if the right-hand side of the non-homogeneous DE is a constant, so that

$$a_2 y'' + a_1 y' + a_0 y = C \qquad (a_0 \neq 0),$$

then a particular integral is $y_p = \dfrac{C}{a_0}$, since $y_p' = 0$ and $y_p'' = 0$.

Solution by Substitution or by Trial

Given the DE $a_2 y'' + a_1 y' + a_0 y = f(x)$. We wish to obtain particular solutions for typical functions $f(x)$, the right-hand side of this non-homogeneous equation.

The most important cases encountered in practice are those where $f(x)$ is of the type $C e^{\lambda x}$, $C \sin ax$, or $C \cos ax$ or of the polynomial type.

If the function $f(x)$ is the sum of two or more types, a particular solution is found for each term separately and then these solutions are added. Note that the DE is linear!

Polynomial function $f(x) = a + bx + cx^2 + \ldots$ in which $a, b, c, \ldots$ are constants.

The only functions whose differential coefficients are positive integral powers of the variable x are themselves positive integral powers of x. Hence, for a particular integral, we assume

$$y_p = A + Bx + Cx^2 + \ldots .$$

The degree of the function assumed for y_p must equal the degree of $f(x)$, and no powers of x can be omitted, even if the RHS of the DE does not contain all powers. Substituting y_p and its derivatives in the DE and comparing coefficients of the different powers of x gives equations for the coefficients $A, B, C, \ldots .$

EXAMPLE   Find a particular integral of the DE

$$y'' - 3y' + 2y = 3 - 2x^2$$

Since the RHS is a quadratic, we assume

$$y_p = A + Bx + Cx^2$$

Hence $y_p' = B + 2Cx$ and $y_p'' = 2C$

Substituting in the DE gives

$$2C - 3(B + 2Cx) + 2(A + Bx + Cx^2) = 3 - 2x^2$$

Comparing coefficients we find

for x^2, $2C = -2, \quad C = -1$

for x, $-6C + 2B = 0, \quad B = -3$

constant terms, $2C - 3B + 2A = 3 \quad A = -2$

A particular integral is

$$y_p = -2 - 3x - x^2$$

Exponential function $f(x) = C e^{\lambda x}$.

We have seen that differentiating an exponential gives an exponential. Hence we assume for a particular solution that

$$y_p = A e^{\lambda x}$$

Substituting in the DE, we find

$$(a_2 \lambda^2 + a_1 \lambda + a_0) A e^{\lambda x} = C e^{\lambda x}$$

The unknown factor A is then given by

$$A = \frac{C}{a_2\lambda^2 + a_1\lambda + a_0}$$

If, however, $e^{\lambda x}$ happens to be a term of the complementary function the method fails, since $a_2\lambda^2 + a_1\lambda + a_0 = 0$. In this case, we can substitute $y_p = Ax\,e^{\lambda x}$.

Should this fail because $x\,e^{\lambda x}$ is a term of the complementary function, then we assume $y_p = Ax^2 e^{\lambda x}$, and so on.

EXAMPLE   Find a particular integral of

$$y'' - 4y' + 3y = 5e^{-3x}$$

The roots of the auxiliary equation are 3 and 1. Thus e^{-3x} is not a term of the complementary function; hence we assume

$$y_p = A\,e^{-3x}$$
$$y_p' = -3A\,e^{-3x}$$
$$y_p'' = 9A\,e^{-3x}$$

Substituting in the DE gives

$$[9 - 4(-3) + 3]A\,e^{-3x} = 5e^{-3x}$$

so that $A = \frac{5}{24}$

A particular integral is

$$y_p = \frac{5}{24}e^{-3x}$$

The complete solution is

$$y = C_1 e^{3x} + C_2 e^x + \frac{5}{24}e^{-3x}$$

EXAMPLE   Suppose that the RHS of the previous example was $5e^x$. As e^x is a term of the complementary function, we assume

$$y_p = Ax\,e^x$$
$$y_p' = Ax\,e^x + A\,e^x = A(x\,e^x + e^x)$$
$$y_p'' = Ax\,e^x + A\,e^x + A\,e^x = A(x\,e^x + 2\,e^x)$$

Substituting in the DE, we have

$$(x + 2 - 4x - 4 + 3x)A\,e^x = 5e^x \qquad \text{or} \qquad -2A = 5$$

Hence $A = -\frac{5}{2}$

A particular integral is

$$y_p = -\frac{5}{2}x\,e^x$$

The complete solution is

$$y = C_1 e^{3x} + C_2 e^x - \frac{5}{2}x\,e^x$$

Trigonometric function $f(x) = R_1 \sin ax + R_2 \cos ax$.

The differential coefficients of sine and cosine functions are trigonometric functions also. We therefore assume for the particular integral that

$$y_p = A \sin ax + B \cos ax$$

We then calculate the derivatives, substitute in the DE and compare the coefficients of sine and cosine in order to obtain equations for A and B.

If the complementary function contains terms of the same form, i.e. $\sin ax$, $\cos ax$, the method fails and, as for type 2, we substitute

$$y = Ax \sin ax + Bx \cos ax$$

EXAMPLE   Solve $y'' - 3y' + 2y = 7 \sin 4x$.

The roots of the auxiliary equation are $r_1 = 1$, $\quad r_2 = 2$

The complementary function is

$$y_c = C_1 e^x + C_2 e^{2x}$$

To find a particular integral, we assume that

$$y_p = A \sin 4x + B \cos 4x$$
$$y_p' = 4A \cos 4x - 4B \sin 4x$$
$$y_p'' = -16A \sin 4x - 16B \cos 4x$$

The DE becomes

$$-16A \sin 4x - 16B \cos 4x - 12A \cos 4x + 12B \sin 4x + 2A \sin 4x + 2B \cos 4x = 7 \sin 4x$$

We compare the coefficients of $\sin 4x$ and $\cos 4x$:

$$-14A + 12B = 7$$
$$-14B - 12A = 0$$

Hence $\qquad A = \dfrac{-49}{170} \quad$ and $\quad B = \dfrac{21}{85}.$

The general solution is

$$y = C_1 e^x + C_2 e^{2x} - \tfrac{49}{170} \sin 4x + \tfrac{21}{85} \cos 4x$$

EXAMPLE   Solve $y'' + 9y = \sin 3x$.

The roots of the auxiliary equations are

$$r_1 = 3j, \qquad r_2 = -3j$$

The complementary function is

$$y_c = C_1 \cos 3x + C_2 \sin 3x$$

Since $f(x) = \sin 3x$ is a term of the complementary function, we assume for a particular integral that

$$y_p = Ax \sin 3x + Bx \cos 3x$$

Thus $\qquad y_p' = 3Ax \cos 3x + A \sin 3x - 3Bx \sin 3x + B \cos 3x$

and $\qquad y_p'' = -9Ax \sin 3x + 6A \cos 3x - 9Bx \cos 3x - 6B \sin 3x$

The differential equation becomes

$$-9Ax \sin 3x + 6A \cos 3x - 9Bx \cos 3x - 6B \sin 3x + 9Ax \sin 3x + 9Bx \cos 3x = \sin 3x$$

Comparing the coefficients of $\sin 3x$ and $\cos 3x$, we find

$$B = -\tfrac{1}{6}, \qquad A = 0$$

The complete solution is

$$y = C_1 \cos 3x + C_2 \sin 3x - \tfrac{1}{6}x \cos 3x$$

Method of Variation of Parameters

Let us consider the linear non-homogeneous DE with constant coefficients. No assumptions are made about the type of $f(x)$.

$$a_2 y'' + a_1 y' + a_0 y = f(x) \tag{1}$$

Let y_1 and y_2 be independent solutions of the homogeneous equation. We then know that the complementary function is $y_c = C_1 y_1 + C_2 y_2$.

We need to find a particular integral y_p. We assume that it is of the following form:

$$y_p = V_1 y_1 + V_2 y_2 \tag{2}$$

$V_1(x)$ and $V_2(x)$ are two functions of x to be determined. Hence we require two equations for the two unknowns, V_1 and V_2.

Substituting equation 2 in equation 1 gives one equation which must be satisfied by V_1 and V_2. We then try to find another equation which will simplify the calculation of V_1 and V_2. Although this equation may be chosen arbitrarily, it must not contradict the first equation. Differentiating equation 2, we find

$$y_p' = (V_1 y_1' + V_2 y_2') + (V_1' y_1 + V_2' y_2)$$

y_p' can be simplified by choosing the second equation for the unknowns V_1 and V_2 to be

$$V_1' y_1 + V_2' y_2 = 0 \tag{3}$$

Hence

$$y_p' = V_1 y_1' + V_2 y_2' \tag{4}$$

and

$$y_p'' = V_1 y_1'' + V_2 y_2'' + V_1' y_1' + V_2' y_2' \tag{5}$$

Substituting equations 4 and 5 in equation 1 and rearranging, we have

$$V_1(a_2 y_1'' + a_1 y_1' + a_0 y_1) + V_2(a_2 y_2'' + a_1 y_2' + a_0 y_2) + a_2(V_1' y_1' + V_2' y_2') = f(x)$$

Since y_1 and y_2 satisfy the homogeneous DE, the expressions in the first two brackets vanish. Hence we have

$$a_2(V_1' y_1' + V_2' y_2') = f(x) \tag{6}$$

Equations 3 and 6 are really the only ones that concern us. It is these that we have to solve in order to find V_1' and V_2'. We now obtain

$$V_1' = \frac{-f(x) y_2}{a_2(y_1 y_2' - y_1' y_2)}, \qquad V_2' = \frac{f(x) y_1}{a_2(y_1 y_2' - y_1' y_2)} \tag{10.6a}$$

Since y_1 and y_2 are independent solutions of the homogeneous equation, the denominator does not vanish identically.

If we denote the expressions on the right-hand sides by $g_1(x)$ and $g_2(x)$, then V_1 and V_2 are obtained by integration:

$$V_1 = \int g_1(x)\, dx, \qquad V_2 = \int g_2(x)\, dx \qquad (10.6b)$$

You may have noticed that this method is somewhat lengthy.

This is the reason for attempting the trial solution approach first.

EXAMPLE   Solve $y'' - y = 4e^x$.

The complementary function is

$$y_c = C_1 e^x + C_2 e^{-x}$$

i.e.
$$y_1 = e^x, \qquad y_2 = e^{-x}$$

For a particular integral let

$$y_p = V_1 e^x + V_2 e^{-x}$$

First, we compute the denominator of the integrands:

$$y_1 y_2' - y_1' y_2 = e^x(-e^{-x}) - e^x e^{-x} = -2$$

Second, we compute the parameters V_1 and V_2 by integration:

$$V_1 = -\int \frac{f(x) y_2}{-2}\, dx = \frac{1}{2} \int 4\, e^x e^{-x}\, dx = 2x$$

$$V_2 = \int \frac{f(x) y_1}{-2}\, dx = -\frac{1}{2} \int 4\, e^{2x}\, dx = -e^{2x}$$

Third, we can write y_p explicitly:

$$y_p = 2x\, e^x - e^{2x} e^{-x} = e^x(2x - 1)$$

The complete solution of the DE is

$$y = C_1 e^x + C_2 e^{-x} + (2x - 1) e^x$$

10.4 BOUNDARY VALUE PROBLEMS

10.4.1 FIRST-ORDER DEs

Let us consider the equation $a_1 y' + a_0 y = 0$. The auxiliary equation is $r_1 = -\dfrac{a_0}{a_1}$. The solution is

$$y = C\, e^{r_1 x}$$

Since C can take on any value, there is an infinite number of solutions. But we often know the value of the function or its derivative at a particular point. For example, we might state that for a body in motion its velocity is v_0 at time $t = 0$. Such a condition is referred to as a *boundary condition* or *initial condition*, and this fixes the value of the constant C. The general solution becomes a *particular solution* because it satisfies a preassigned condition. According to lemma 10.2 a first-order DE contains one arbitrary constant. This constant is determined by one boundary condition.

EXAMPLE   Solve $y' + 3y = 0$, so that when $x = 0$, $y = 2$ (i.e. the solution is to contain the point $x = 0$, $y = 2$).

The general solution is $y = C e^{-3x}$

Substituting the boundary condition, we have

$$2 = C e^0 = C; \quad \text{hence} \quad C = 2$$

Consequently, the particular solution satisfying the boundary condition is

$$y = 2 e^{-3x}$$

10.4.2   SECOND-ORDER DEs

The general solution of a second-order DE has two arbitrary constants. We therefore require two boundary conditions to calculate their values. These conditions may take various forms. For example, the solution might have to pass through two points in the x-y plane, or it might have to pass through one point and have a certain slope at another. These conditions could be stated thus:

at $x = x_1$, $y = y_1$ and at $x = x_2$, $y = y_2$

or   at $x = x_1$, $y = y_1$ and at $x = x_2$, $y' = y_2'$

EXAMPLE   Figure 10.2 shows a cantilever beam of length L supporting a load W at the free end. The DE is given by

$$EIy'' - M = 0$$

where the bending moment M at a section XX is Wx. The product EI is constant and is a property of the beam material and its cross-sectional dimensions.

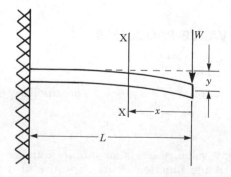

Figure 10.2

Solve the DE, given that

when $x = L$, $y = 0$

and when $x = L$, $y' = 0$ } boundary conditions

The DE is
$$EIy'' = Wx$$

This second-order DE can be solved directly by integrating twice: thus

$$EIy' = \frac{Wx^2}{2} + C_1 \qquad [1]$$

and

$$EIy = \frac{Wx^3}{6} + C_1x + C_2 \qquad [2]$$

Let us now consider the boundary conditions. Since $y' = 0$ when $x = L$, we have

$$C_1 = -\frac{WL^2}{2}$$

Since $y = 0$ when $x = L$, we have

$$C_2 = -\frac{WL^3}{3}$$

The desired solution is

$$y = \frac{1}{EI}\left(\frac{Wx^3}{6} - \frac{WL^2x}{2} + \frac{WL^3}{3}\right)$$

or

$$y = \frac{W}{EI}\left(\frac{x^3}{6} - \frac{L^2x}{2} + \frac{L^3}{3}\right)$$

EXAMPLE   Solve $y'' + y = 0$, given that $y(0) = 0$ and $y(\pi) = 1$ (boundary conditions).

The roots of the auxiliary equation are $r_1 = j$ and $r_2 = -j$. The general solution is

$$y = C_1\sin x + C_2\cos x$$

Substituting the first boundary condition in this equation, we have

$$0 = C_1\sin 0 + C_2\cos 0 = C_1 \times 0 + C_2$$

Hence $C_2 = 0$

The second boundary condition stipulates that

$$1 = C_1\sin \pi + C_2\cos \pi = C_2 \times 0 - C_2$$

Hence $C_2 = -1$

In this example, the two given boundary conditions contradict each other: they cannot both be satisfied if the solution is expected to be a differentiable function. No differentiable solution exists.

10.5 SOME APPLICATIONS OF DEs

10.5.1 RADIOACTIVE DECAY

Let $N(t)$ be the number of radioactive atoms present at time t. We assume that the rate of decay with time is proportional to the number of atoms remaining, i.e.

$$\frac{\mathrm{d}N(t)}{\mathrm{d}t} \propto N(t)$$

If we introduce a factor of proportionality k, bearing in mind that the number of atoms is decreasing with time, the DE is

$$\frac{\mathrm{d}}{\mathrm{d}t}N(t) = -kN(t) \qquad (k > 0)$$

or
$$\dot{N} + kN = 0$$

This is a homogeneous first-order DE whose solution is

$$N = C\,e^{-kt}$$

To fix C, we need to choose some initial value. For example, let N_0 be the number of atoms at time $t = t_0 = 0$. Then

$$N_0 = C\,e^{-kt_0} = C\,e^0 = C$$

The particular solution is
$$N = N_0\,e^{-kt}$$

10.5.2 THE HARMONIC OSCILLATOR

Free Undamped Oscillations

Figure 10.3 shows a mass m on a spring of stiffness k (load per unit elongation). If the mass is pulled down by an amount x from the equilibrium position, the spring will exert a restoring force trying to bring back the mass towards that position.

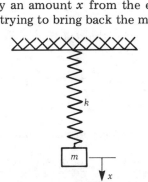

Figure 10.3

By Newton's second law of motion,

$$m\,\ddot{x}(t) = -kx(t)$$

$$\ddot{x} + \omega_n^2 x = 0, \qquad \omega_n^2 = \frac{k}{m}$$

$$\omega_n = \text{natural frequency}$$

This is a linear second-order DE. The auxiliary equation is

$$r^2 + \omega_n^2 = 0$$

The roots are $r_1 = j\omega_n$ and $r_2 = -j\omega_n$.

The general solution is (cf. section 10.3.1, case 3)

$$x = C_1 \cos \omega_n t + C_2 \sin \omega_n t$$

We need two boundary conditions to determine the values of C_1 and C_2.

For example, the boundary conditions of an oscillation are

$$x = 0 \quad \text{at} \quad t = 0 \qquad \text{(position at the instant } t = 0)$$

$$\dot{x} = v_0 \quad \text{at} \quad t = 0 \qquad \text{(velocity at the instant } t = 0)$$

Substituting the first condition in the DE above gives

$$0 = C_1 \cos 0 + C_2 \sin 0$$

Hence $\qquad\qquad\qquad C_1 = 0$

Substituting the second boundary condition gives

$$\dot{x} = v_0 = -\omega_n C_1 \sin 0 + \omega_n C_2 \cos 0 = \omega_n C_2$$

Hence $C_2 = \dfrac{v_0}{\omega_n}$

The particular solution is

$$x = \frac{v_0}{\omega_n} \sin \omega_n t$$

showing that the motion of the mass is oscillatory at a frequency of ω_n rad/s and of constant amplitude v_0/ω_n.

The general solution of the DE is a superposition of two trigonometric functions with the same period (figure 10.4):

$$x(t) = C_1 \cos \omega_n t + C_2 \sin \omega_n t$$

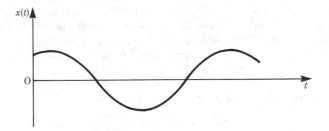

Figure 10.4

According to the superposition formula in Chapter 1, section 1.5.6, $x(t)$ can be expressed in the form

$$x(t) = C \cos (\omega_n t - \alpha)$$

where $\qquad\qquad\qquad C = \sqrt{C_1^2 + C_2^2}$

and $\qquad\qquad\qquad \alpha = \tan^{-1}\left(\dfrac{C_2}{C_1}\right)$

Thus, if we start with

$$x = C \cos(\omega_n t - \alpha)$$

as the general solution, we have two unknown constants, C and α. These are determined by the boundary conditions as before.

Since $x = 0$ at $t = 0$ and $\dot{x} = v_0$ at $t = 0$, we have

$$0 = C \cos(-\alpha) = C \cos \alpha$$

Since $C \neq 0$, it follows that $\alpha = \pi/2$.

Differentiating x gives

$$\dot{x} = -\omega_n C \sin(\omega_n t - \alpha)$$

Inserting the second boundary condition gives

$$v_0 = \omega_n C$$

Therefore

$$C = \frac{v_0}{\omega_n}$$

The particular solution is

$$x = \frac{v_0}{\omega_n} \cos\left(\omega_n t - \frac{\pi}{2}\right) = \frac{v_0}{\omega_n} \sin \omega_n t$$

which is identical to the previous solution.

Finally, we could also have chosen the solution

$$x = C_0 \sin(\omega_n t + \alpha_0)$$

You should verify this for yourself.

Damped Harmonic Oscillator

The harmonic oscillator considered above is an ideal case. In reality, friction is present in all systems in the form of dry friction, viscous friction and internal friction between the molecules in a material. Friction in whatever form slows down motion because it dissipates energy in the form of heat which cannot be recovered. No matter how small the friction is in a system (such as our spring–mass system) oscillations will eventually die out. The effect of friction is known as *damping*.

The friction or damping force is given in some cases by

$$F = -c\dot{x}$$

where c is a friction or damping coefficient, $\dot{x}$ is the velocity and the minus sign indicates that the force acts in a direction opposite to the motion. By Newton's second law, the equation of motion for our spring–mass system becomes

$$m\ddot{x} + c\dot{x} + kx = 0$$

This is the DE of motion for free oscillations or vibrations, meaning that there are no external forces acting on the system.

The auxiliary equation is

$$mr^2 + cr + k = 0$$

whose roots are

$$r_{1,2} = \frac{-c}{2m} \pm \frac{\sqrt{c^2 - 4mk}}{2m} = -a \pm b$$

As we saw in section 10.3.1, there are three cases to consider, these depend on the value of $c^2 - 4mk$, i.e.

$$c^2 - 4mk > 0, \qquad c^2 - 4mk < 0, \qquad c^2 - 4mk = 0$$

Case 1: $c^2 - 4mk > 0$.

This means that the roots are real and unequal. In this case the general solution is

$$x = C_1 e^{r_1 t} + C_2 e^{r_2 t}$$
$$= e^{-at}[C_1 e^{bt} + C_2 e^{-bt}]$$

This corresponds to an *overdamped system*, and its response from a given initial displacement is shown in figure 10.5. No oscillations are present. The system will return to the equilibrium position slowly.

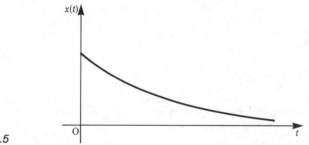

Figure 10.5

Case 2: $c^2 - 4mk = 0$.

The roots are equal, i.e. $r_1 = r_2 = -a$. The general solution is

$$x = (C_1 + C_2 t)e^{-at}$$

The system will return to the equilibrium position more quickly than the system in case 1 but again there will be no oscillations. It is referred to as *critical* or *aperiodic* and the damping is called *critical damping*. Its response from a given initial displacement and initial velocity is shown in figure 10.6.

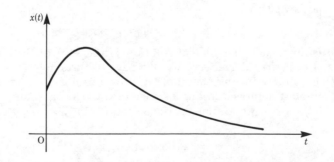

Figure 10.6

Case 3: $c^2 - 4mk < 0$.

The roots in this case are complex conjugate, i.e. $r_1 = -a + jb$, $r_2 = -a - jb$, with $a > 0$. The general solution is

$$x = e^{-at}[C_1 e^{jbt} + C_2 e^{-jbt}]$$

or
$$x = e^{-at}[C_1(\cos bt + j \sin bt) + C_2(\cos bt - j \sin bt)]$$

$$= e^{-at}(A \cos bt + B \sin bt)$$

where $A = C_1 + C_2$ and $B = j(C_1 - C_2)$ and A and B are arbitrary.

We should point out that although C_1 and C_2 may be complex, A and B are not necessarily complex. As we are dealing with a real physical problem, the solution must be real, hence A and B must be real, which means that C_1 and C_2 must be complex conjugate numbers.

The displacement x may be put in another form thus:

$$x = C e^{-at} \cos(bt - \alpha)$$

An examination of this function shows that the system will oscillate, but the oscillations will die out due to the exponential factor. Its response from a given initial displacement and velocity is shown in figure 10.7. It is a *damped* oscillation.

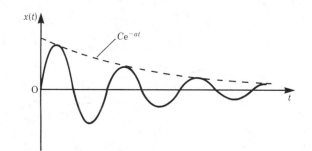

Figure 10.7

Forced Oscillations

The damped oscillator shown in figure 10.8 is now subjected to an exciting force given by $F_0 \cos \omega t$. F_0 is constant and ω is the frequency of excitation, or forcing frequency.

Newton's second law gives

$$m\ddot{x} + c\dot{x} + kx = F_0 \cos \omega t$$

According to lemma 10.2, the general solution is the sum of the complementary function x_c and the particular integral x_p. We have just examined the three possible solutions of the homogeneous equation; it now remains to find the particular integral. The simplest approach in this instance is to use a trial solution, as discussed in section 10.3.2. Hence we assume an oscillation at the frequency of the exciting force:

$$x_p = x_0 \cos(\omega t - \alpha_1)$$

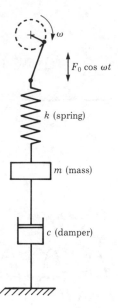

$F_0 \cos \omega t$

k (spring)

m (mass)

c (damper)

Figure 10.8

Substituting in the DE and comparing coefficients, we find

$$x_0 = \frac{F_0}{\sqrt{(k-m\omega^2)^2 + c^2\omega^2}}$$

and

$$\tan \alpha_1 = \frac{\omega c}{k-m\omega^2}$$

The general solution is

$$x = x_c + x_p, \qquad x_c = \text{complementary function}$$

i.e.

$$x = x_c + \frac{F_0}{\sqrt{(k-m\omega^2)^2 + c^2\omega^2}} \cos(\omega t - \alpha_1)$$

If there is damping in the system, the complementary function will die out after a certain time (known as the *transient phase*) and the motion will be given by

$$x = \frac{F_0}{\sqrt{(k-m\omega^2)^2 + c^2\omega^2}} \cos(\omega t - \alpha_1)$$

The system will oscillate at the frequency ω of the excitation. This phase of the motion is called the *steady state*. Figure 10.9 shows the complementary function x_c, the particular integral x_p, and the response of the system from the instant the excitation is applied, i.e. $x = x_c + x_p$.

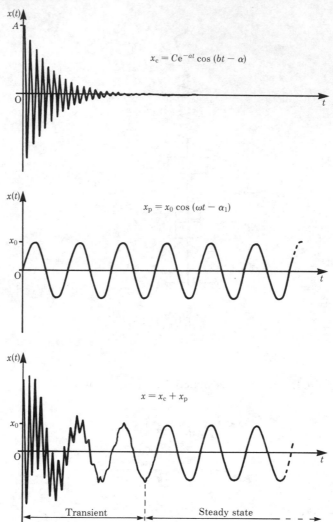

Figure 10.9

From a practical point of view, the amplitude x_0 of the steady state is most important. It depends on the excitation frequency ω. If we vary ω we will reach a value which will make x_0 a maximum. This condition is referred to as *resonance* because ω corresponds to the natural frequency ω_n of the system. This maximum is obtained by setting

$$\frac{dx_0}{d\omega} = 0$$

which gives

$$\omega = \omega_d = \sqrt{\omega_n^2 - \frac{c^2}{2m^2}}$$

where $\omega_n^2 = \dfrac{k}{m}$ is the undamped natural frequency, and ω_d is the damped natural frequency of the system.

If the system is undamped, then $c = 0$. We see that the excitation frequency corresponds to the undamped natural frequency and the amplitude grows beyond all bounds because the denominator in $x_0 = \dfrac{F_0}{k - m\omega^2}$ vanishes. This situation is shown in figure 10.10a which shows the amplitude of the steady state as a function of the excitation frequency.

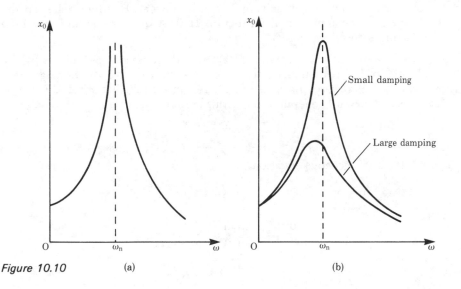

Figure 10.10   (a)   (b)

In practice the amplitude is reduced due to the presence of damping, no matter how small, as shown in figure 10.10b. The greater the damping the smaller the amplitude. With a small amount of damping the amplitude at resonance can be very large and engineers avoid this situation.

The following sections offer, in a concise fashion, some further methods of solving certain types of DE. In sections 10.7.3 and 10.7.4, concepts which have not yet been introduced will be referred to, namely 'partial derivative', 'total differential' and 'partial DE'. It may be advisable to skip the rest of this chapter during a first course and to return to it when the need arises.

10.6   GENERAL LINEAR FIRST-ORDER DEs

10.6.1   SOLUTION BY VARIATION OF THE CONSTANT

The DEs discussed so far have *constant* coefficients, but that is only a special case of what we shall start to discuss now. In this section, we are concerned with linear first-order equations. First order means that no higher derivatives other than y' appear; linear means that no powers of y and y' and no products like yy' appear. The general form is thus

$$p(x)\,y' + q(x)\,y = f(x)$$

The coefficients p and q are arbitrary functions of x and the following are examples:

$$y' + \frac{y}{x} = 4x^2, \quad \text{with } p(x) = 1, \; q(x) = \frac{1}{x}, \; f(x) = 4x^2$$

$$\sqrt{x}\,y' - y = 1, \quad \text{with } p(x) = \sqrt{x}, \; q(x) = -1, \; f(x) = 1$$

We will now derive a method for solving first-order linear DEs which relies on the method of variation of parameters from section 10.3. A quicker method, using the integrating factor, is described in section 10.6.2.

We have seen how to solve systematically a DE with constant coefficients by the method of variation of parameters. Step 1 requires us to solve the homogeneous equation, and step 2 to vary the constant. A general linear first-order DE can be solved by a straightforward generalisation of this method.

$$p(x)\,y' + q(x)\,y = f(x)$$

Step 1 Solve the homogeneous equation

$$p(x)\frac{\mathrm{d}y}{\mathrm{d}x} + q(x)\,y = 0$$

$$\frac{\mathrm{d}y}{y} = -\frac{q(x)}{p(x)}\,\mathrm{d}x$$

$$\int\frac{\mathrm{d}y}{y} = \ln|y| = -\int\frac{q(x)}{p(x)}\,\mathrm{d}x + C_1$$

Hence
$$y = C\,\mathrm{e}^{-\int\frac{q(x)}{p(x)}\,\mathrm{d}x} \tag{10.7}$$

The function $\mathrm{e}^{\int(q/p)\,\mathrm{d}x} = I(x)$ is called *integrating factor*, for reasons that will soon become clear. In some other references, the integrating factor is abbreviated as IF. $[I(x)]^{-1}$ is a particular solution of the homogeneous equation.

In order to solve the non-homogeneous equation, we will now vary the constant C.

Step 2 Let the constant C become a function $v(x)$. Assume that

$$y = \frac{v(x)}{I(x)} \quad \text{solves the given equation, i.e.} \quad y = v(x)\,\mathrm{e}^{-\int\frac{q(x)}{p(x)}\,\mathrm{d}x}$$

Compute y':
$$y' = \frac{v'(x)}{I(x)} - \frac{q(x)}{p(x)}\frac{v(x)}{I(x)} = \frac{1}{I(x)}\left(v'(x) - \frac{q(x)}{p(x)}v(x)\right)$$

Inserting this into the original equation gives

$$\frac{1}{I(x)}p(x)\,v'(x) = f(x)$$

This equation allows us to compute $v(x)$. Thus

$$v(x) = \int v'(x)\,\mathrm{d}x = \int I(x)\frac{f(x)}{p(x)}\,\mathrm{d}x$$

The solution of the equation $p(x)y' + q(x)y = f(x)$ reads

$$y(x) = \frac{1}{I(x)} \int I(x) \frac{f(x)}{p(x)} \, dx$$

We note in passing that the *general* solution of any first-order DE must contain one free parameter. In the case under consideration, this is the constant which arises in the last integration.

EXAMPLE $y' + \dfrac{y}{x} = 4x^2$

Step 1 The homogeneous equation reads

$$y' + \frac{y}{x} = 0$$

Its solution is

$$\frac{dy}{y} = -\frac{dx}{x}$$

$$\ln |y| = -\ln |x| + C_1$$

$$y = \frac{C}{x}$$

Step 2 Variation of the constant $C = v(x)$.

Assume
$$y = \frac{v(x)}{x}, \qquad y' = \frac{v'(x)}{x} - \frac{v(x)}{x^2}$$

Inserting into the original equation gives

$$\text{LHS} = y' + \frac{y}{x} = \frac{v'(x)}{x} - \frac{v(x)}{x^2} + \frac{v(x)}{x^2} = \frac{v'(x)}{x}$$

$$\text{RHS} = 4x^2$$

Thus
$$\frac{v'(x)}{x} = 4x^2$$

and
$$v(x) = \int 4x^3 \, dx = x^4 + C$$

The general solution of the given equation reads

$$y(x) = x^3 + \frac{C}{x}$$

Let us convince ourselves that this claim is correct (direct computation):

$$y' = 3x^2 - \frac{C}{x^2}, \qquad y' + \frac{y}{x} = 3x^2 - \frac{C}{x^2} + x^2 + \frac{C}{x^2} = 4x^2$$

The DE is indeed solved by $y(x)$. As the solution contains one free parameter (namely C) we can be certain that it is the *general* solution.

10.6.2   A STRAIGHTFORWARD METHOD INVOLVING THE INTEGRATING FACTOR

Remember that the integrating factor is the reciprocal of a particular solution of the *homogeneous* equation

$$I(x) = e^{\int(q(x)/p(x))\,dx}$$

In other words, $\dfrac{C}{I(x)}$ solves $p(x)y' + q(x)y = 0$.

The name integrating factor is justified by the following observation. If the given non-homogeneous DE is multiplied through by the integrating factor, then the LHS can be expressed almost as an ordinary derivative:

$$I(x)p(x)y' + I(x)q(x)y = I(x)f(x)$$

Observe that

$$[I(x)y]' = I(x)'y + I(x)y' = \frac{q(x)}{p(x)}I(x)y + I(x)y'$$

Thus

$$p(x)[I(x)y]' = I(x)f(x)$$

The solution of the equation $p(x)y' + q(x)y = 0$ is

$$y(x) = \frac{1}{I(x)}\int\frac{I(x)}{p(x)}f(x)\,dx \qquad (10.8)$$

where $I(x) = e^{\int(q(x)/p(x))\,dx}$

For the sake of clarity, let us list the steps necessary for solving a linear first-order DE, $p(x)y' + q(x)y = f(x)$, using this result.

As a preliminary step, identify $p(x)$, $q(x)$ and $f(x)$.

Step 1 Solve the integral

$$\int\frac{q(x)}{p(x)}\,dx$$

and write down the integrating factor

$$I(x) = e^{\int(q(x)/p(x))\,dx}$$

Step 2 Solve the integral

$$\int\frac{I(x)}{p(x)}f(x)\,dx$$

and write down the general solution. The necessary constant emerges because of the last integration.

$$y(x) = \frac{1}{I(x)}\int\frac{I(x)}{p(x)}f(x)\,dx$$

EXAMPLE   Solve $\sqrt{x}\,y' - y = 1$.

$$p(x) = \sqrt{x}, \qquad q(x) = -1, \qquad f(x) = 1$$

Step 1 $$-\int\frac{dx}{\sqrt{x}} = -2\sqrt{x}, \qquad I(x) = e^{-2\sqrt{x}}$$

Step 2 $y(x) = e^{2\sqrt{x}} \int \dfrac{e^{-2\sqrt{x}}}{\sqrt{x}} \, dx = e^{2\sqrt{x}}(-e^{-2\sqrt{x}} + C) = C\,e^{2\sqrt{x}} - 1$

The following example shows that the method just described can also be used for DEs with constant coefficients.

EXAMPLE   The DE for the current i in an electrical circuit consisting of an inductor L and a resistor R in series is given by

$$\frac{di}{dt} + \frac{R}{L}\, i = \frac{E}{L} \sin \omega t$$

where $E \sin \omega t$ is the voltage applied to the circuit. Solve the equation and discuss the solution.

Step 1 $I(x) = e^{\int (R/L)\, dt} = e^{(R/L)t}$

Step 2 $i(t) = e^{-(R/L)t} \left(\dfrac{E}{L} \int e^{(R/L)t} \sin \omega t \, dt + c \right)$

The integral has to be evaluated by parts. (Remember that $\int u \, dv = uv - \int v \, du$.)

This integral is, leaving out the constant

$$\frac{L\, e^{(R/L)t}}{R^2 + \omega^2 L^2} (R \sin \omega t - \omega L \cos \omega t)$$

Step 3   The solution of the DE is

$$i = \frac{E}{\sqrt{R^2 + \omega^2 L^2}} (R \sin \omega t - \omega L \cos \omega t) + C\, e^{-(R/L)t}$$

or $i = \dfrac{E}{\sqrt{R^2 + \omega^2 L^2}} \sin(\omega t - \alpha) + C\, e^{-(R/L)t}$

where $\alpha = \tan^{-1}\left(\dfrac{\omega L}{R}\right)$

As t increases, the last term decreases and the current i tends to a steady periodic value.

We conclude this section with a word of warning. The process of first determining the integrating factor and then the general solution of a linear first-order DE is guaranteed to work *in principle* but not always in practice! The snag lies in the annoying fact that a given integral may not have an elementary solution. Thus it may well prove to be unavoidable to resort to numerical methods, even in cases of presumably innocuous DEs.

10.7   SOME REMARKS ON GENERAL FIRST-ORDER DEs

*   ### 10.7.1   BERNOULLI'S EQUATIONS

The general *Bernoulli DE* for arbitrary n is

$$y' + q(x)\, y = f(x)\, y^n   \qquad (n \neq 1)$$

Note that n may also be negative, but it must not be unity. A Bernoulli DE can be converted to a first-order linear equation by means of the substitution

$$u = y^{1-n} = \frac{1}{y^{n-1}} \tag{10.9}$$

Then $\dfrac{u'}{1-n} y^n = y'$, $uy^n = y$; hence Bernoulli's equation becomes

$$\frac{1}{1-n} u'y^n + q(x)uy^n = f(x)y^n$$

Dividing by y^n gives

$$\frac{1}{1-n} u' + q(x)u = f(x)$$

This is a linear first-order DE. Its solution has been shown in the preceding section.

EXAMPLE $y' - xy = -y^3 e^{-x^2}$

This is a Bernoulli-type equation with $n = 3$. We put $u = y^{-2}$, $y = uy^3$.

Hence $y' = -\dfrac{1}{2} u'y^3$

Inserting these into the given DE gives

$$-\frac{1}{2} u'y^3 - xuy^3 = -y^3 e^{-x^2}, \qquad u' + 2xu = 2e^{-x^2}$$

This is a *linear* first-order DE for the function u.

The integrating factor is $I(x) = e^{x^2}$. The solution for u is

$$u(x) = e^{-x^2} 2 \int e^{x^2} e^{-x^2} \, dx = 2x e^{-x^2} + C e^{-x^2}$$

After substituting this into $y = u^{-1/2}$, we obtain the solution in its final form:

$$y = \frac{1}{\sqrt{u}} = \frac{1}{\sqrt{2x+C}} e^{x^2/2}$$

10.7.2   SEPARATION OF VARIABLES

If the equation is neither linear nor of the Bernoulli type, then we may still be able to solve it using only elementary tools. The simplest case is when the equation can be rewritten with only y terms on the LHS and only x terms on the RHS. The DE is said to have *separable variables* when it can be written in one of the following equivalent forms:

$$p(y)y' + q(x) = 0$$
$$p(y)\,dy = -q(x)\,dx$$

The solution of such an equation is obtained by simple integration:

$$\int p(y)\,dy = -\int q(x)\,dx = C$$

EXAMPLE   The variables in the following equation can be separated:

$$y'x^3 = 2y^2$$

Dividing by x^3y^2, we obtain

$$\frac{1}{y^2}\,y' = \frac{2}{x^3}, \qquad \text{i.e.} \qquad \frac{1}{y^2}\,dy = \frac{2}{x^3}\,dx$$

This is an equation of the type required with $p(y) = \dfrac{1}{y^2}$ and $q(x) = -\dfrac{2}{x^3}$.

Now, straightforward integration gives

$$\frac{1}{y} = \frac{1}{x^2} + C$$

and hence

$$y = \frac{x^2}{Cx^2 + 1}$$

10.7.3 EXACT EQUATIONS

If, in a given DE, the variables cannot be separated, there is still a chance of finding an easy way to solve the equation. We must, however, refer to the basic concepts of partial derivative and total differential which are covered in Chapter 12. Logically speaking, it should be read beforehand.

DEFINITION   Let $p(x,y)\,dy + q(x,y)\,dx = 0$.

If the following condition holds then the DE is said to be *exact*:

$$\frac{\partial p}{\partial x} = \frac{\partial q}{\partial y}$$

EXAMPLE $2xyy' + y^2 = x^2$

This can be rewritten as

$$2xy\,dy + (y^2 - x^2)\,dx = 0$$

We identify $p(x,y)$ and $q(x,y)$, so that

$$p(x,y) = 2xy, \qquad q(x,y) = y^2 - x^2$$

Let us now check whether the condition of exactness holds:

$$\frac{\partial p}{\partial x} = 2y, \qquad \frac{\partial p}{\partial y} = 2y$$

Hence the given equation is exact.

The LHS of an exact DE can be considered as the total differential of some function $F(x,y)$:

$$p\,dy + q\,dx = dF = \frac{\partial F}{\partial y}\,dy + \frac{\partial F}{\partial x}\,dx = 0$$

The equation is therefore solved by the functions $y(x)$ which are defined implicitly by $F(x,y) = C = $ constant. The obvious question is how a suitable function $F(x,y)$ may be found for any given exact DE. We will describe in general terms a method for finding such a function. Later we will refer back to the example just given.

The starting point can be either one of the following equations:

$$\frac{\partial F}{\partial y} = p(x,y), \qquad \frac{\partial F}{\partial x} = q(x,y)$$

Let us choose the first one.

Step 1 From this first equation we find, by integration, that $F = \int p(x,y)\,dy + C$. The constant of integration C is an, as yet, undetermined function of x only, i.e. $C = v(x)$. The reason is that any such function vanishes if the partial derivative $\frac{\partial}{\partial y}$ is taken.

Step 2 In order to determine $v(x)$, insert F into the equation $\frac{\partial F}{\partial x} = q(x,y)$. This yields a differential equation for $v(x)$, i.e.

$$\frac{\partial F}{\partial x} = \frac{\partial}{\partial x} \int p(x,y)\,dy + \frac{d}{dx}\,v(x) = q(x,y)$$

Note that since $v(x)$ is a function of x only, the partial derivative $\frac{\partial}{\partial x}v(x)$ equals the usual derivative $\frac{d}{dx}\,v(x)$.

$$v(x) = \int \left[q(x,y) - \frac{\partial}{\partial x} \int p(x,y)\,dy \right] dx$$

Step 3 Insert the result of the last integration into the equation for F.

Exact DE $p\,dy + q\,dx = 0$. It is solved by functions $y(x)$ which are given implicitly by $F(x,y) = C = $ constant.

The function $F(x,y)$ can be obtained in either one of two ways.

If we start with the equation $\frac{\partial F}{\partial y} = p$, then the formula reads

$$F = \int p\,dy + \int \left[q - \frac{\partial}{\partial x} \int p\,dy \right] dx \tag{10.10a}$$

If we choose to start with the equation $\frac{\partial F}{\partial x} = q$ then the formula reads

$$F = \int q\,dx + \int \left[p - \frac{\partial}{\partial y} \int q\,dx \right] dy \tag{10.10b}$$

We will now solve the equation given in the previous example; it has already been proved to be exact. The equation is

$$2xyy' + y^2 = x^2$$

Step 1
$$p(x,y) = 2xy$$

$$F = \int 2xy \, dy = xy^2 + v(x)$$

Step 2
$$q(x,y) = y^2 - x^2 = \frac{\partial F}{\partial x} = y^2 + \frac{d}{dx} v(x)$$

$$v(x) = -\int x^2 \, dx = -\frac{x^3}{3}$$

Step 3
$$F(x,y) = xy^2 - \frac{x^3}{3}$$

Therefore, the general solution of the given DE is

$$xy^2 - \frac{x^3}{3} = C, \quad \text{i.e.} \quad y^2 = \frac{x^2}{3} + \frac{C}{x}$$

It is not hard to verify that these functions do indeed solve the equation.

EXAMPLE   Solve the DE

$$y \frac{dy}{dx} + x = \pm\sqrt{x^2 + y^2}$$

Even though this is not an exact equation, the notion of exactness aids us in finding solutions.

Rearranging the equation we have

$$\frac{x \, dx + y \, dy}{\pm\sqrt{x^2 + y^2}} = dx$$

Inspection reveals that the LHS is a differential, i.e.

$$d(\pm\sqrt{x^2 + y^2}) = dx$$

By integration, we find

$$\pm\sqrt{x^2 + y^2} = x + C$$

Squaring gives

$$y^2 = 2Cx + C^2$$

This is the equation of a parabola.

* 10.7.4   **THE INTEGRATING FACTOR – GENERAL CASE**

If the given DE is not exact, then sometimes it is possible to turn it into an exact equation by multiplying it by a suitable function $\mu(x,y)$. This function is also called an *integrating factor*.

EXAMPLE   Suppose the given DE is

$$(xy-1)y'+y^2 = 0$$

If it were exact then we would have

$$\frac{\partial p}{\partial x} = \frac{\partial q}{\partial y}$$

Now, $p = xy - 1$ and $q = y^2$, so

$$\frac{\partial p}{\partial x} = y \neq \frac{\partial q}{\partial y} = 2y$$

It does, however, become exact if it is multiplied by $\mu = \dfrac{1}{y}$.

The DE $\left(x - \dfrac{1}{y}\right)y' + y = 0$ is exact.

PROOF $p = x - \dfrac{1}{y}$ and $q = y$, so

$$\frac{\partial p}{\partial x} = 1 = \frac{\partial q}{\partial y}$$

This equation can therefore be treated as described in section 10.7.3.

How can an integrating factor $\mu(x,y)$ be found in the general case? In order to provide the answer, we must formulate the problem mathematically.

Given: $p(x,y)y' + q(x,y) = 0$ with $\dfrac{\partial p}{\partial x} \neq \dfrac{\partial q}{\partial y}$

Wanted: $\mu(x,y)$ such that $\dfrac{\partial(p\mu)}{\partial x} = \dfrac{\partial(q\mu)}{\partial y}$

If an integrating factor μ exists, then it must satisfy the following condition which derives from the equation above and from the product rule:

$$\frac{\partial p}{\partial x}\mu + p\frac{\partial \mu}{\partial x} = \frac{\partial q}{\partial y}\mu + q\frac{\partial \mu}{\partial y}$$

This is a partial DE, and it would seem less easy to solve a partial DE when it is not possible first to solve the ordinary DE. However, we do not need the general solution of the partial DE: *any* non-zero particular solution μ will suffice.

Even though we can offer no general advice on how to find the integrating factor μ, there are two important special cases in which μ can be readily obtained. Below we state these without proof.

Special case 1

If $\dfrac{1}{p}\left(\dfrac{\partial q}{\partial y} - \dfrac{\partial p}{\partial x}\right) = f(x)$ is a function of x only, then $\mu = e^{\int f(x)\,dx}$.

Special case 2

If $\dfrac{1}{q}\left(\dfrac{\partial p}{\partial y}-\dfrac{\partial q}{\partial x}\right)=g(y)$ is a function of y only, then $\mu=e^{\int g(y)\,dy}$.

EXAMPLE   Let us return to the equation encountered in the last example, i.e.

$$(xy-1)y'+y^2 = 0$$

Both $\dfrac{\partial p}{\partial x}$ and $\dfrac{\partial q}{\partial y}$ are functions of y, as is q. Therefore from special case 2, we get

$$g(y) = -\frac{1}{y}, \qquad \int g(y)\,dy = -\ln|y|+C$$

The function $\mu(x,y)=\mu(y)=e^{-\ln|y|}=\dfrac{1}{|y|}$ is an integrating factor. μ has already been used above as an integrating factor.

Finally, we can solve the equation:

$$\left(x-\frac{1}{y}\right)y'+y = 0$$

It is exact, and we must now find a function F such that

$$\frac{\partial F}{\partial y} = x-\frac{1}{y} \qquad \text{and} \qquad \frac{\partial F}{\partial x} = y$$

The solution is obtained by the method outlined in section 10.7.3. It reads

$$F(x,y) = xy-\ln|y|-C$$

Therefore, the general solution of the given equation is the class of functions which is given in implicit form by

$$xy-\ln|y| = C$$

Let us note in passing that the method outlined in this section is a generalisation of the technique covered in section 10.6.2. You are invited to prove for yourself that if a linear first-order DE is treated as proposed here, then $\mu(x,y)=\mu(x)=I(x)$. *Hint*: Use the normalised form of the equation, which means that it has been divided by the first coefficient: $y'+\dfrac{q(x)}{p(x)}y=\dfrac{f(x)}{p(x)}$.

It is, admittedly, not altogether satisfactory that we are unable here to present a more general procedure for finding the integrating factor μ. For a more extensive treatment, the reader is referred to the standard treatises on DEs.

10.8   SIMULTANEOUS DEs

Often problems arise involving several dependent variables. They give rise to a set of differential equations. The number of equations corresponds to the number of dependent variables.

We will restrict ourselves to the solution of simultaneous first- and second-order DEs with constant coefficients and two dependent variables. Before considering practical problems, let us look at the form of the equations.

Let x and y be the dependent variables and t the independent variable. These quantities are related by means of a set of simultaneous DEs such as

$$\frac{dy}{dt} + ay + bx = f(t)$$

$$\frac{d^2y}{dt^2} + A\frac{dy}{dt} + B\frac{dx}{dt} + Cy = g(t)$$

where a, b, A, B and C are constants and $f(t)$ and $g(t)$ are functions of the independent variable t only.

To illustrate the method of solution, consider the examples that follow.

EXAMPLE   Solve

$$\frac{dx}{dt} + 5x - 3y = 0 \qquad\qquad [1]$$

$$\frac{dy}{dt} + 15x - 7y = 0 \qquad\qquad [2]$$

First method If we want to solve for x first we must eliminate y and $\dfrac{dy}{dt}$ from these equations.

Differentiating equation 1 with respect to t gives

$$\frac{d^2x}{dt^2} + 5\frac{dx}{dt} - 3\frac{dy}{dt} = 0$$

Inserting the expression for $\dfrac{dy}{dt}$ from equation 2 and the expression for y from equation 1 we obtain

$$\frac{d^2x}{dt^2} - 2\frac{dx}{dt} + 10x = 0$$

We know how to solve this equation. Its solution is

$$x = e^t(A\cos 3t + B\sin 3t)$$

We can now obtain the solution for y from equation 1 or equation 2:

$$y = e^t[(2A + B)\cos 3t + (2B - A)\sin 3t)]$$

You should note that there are only two arbitrary constants.

Second method We saw earlier that to solve a DE we assumed a solution e^{rt} and found the values of r which would satisfy the equation. There is no reason why we cannot use the same method for simultaneous DEs.

Hence let $x = a\,e^{rt}$, $y = b\,e^{rt}$.

It follows that $\dfrac{dx}{dt} = ra\,e^{rt}$, $\dfrac{dy}{dt} = rb\,e^{rt}$.

Substituting in equations 1 and 2 we have

$$[(r + 5)a - 3b]\,e^{rt} = 0$$

$$[15a + (r - 7)b]\,e^{rt} = 0$$

Since $e^{rt} \neq 0$, it follows that

$$(r+5)a - 3b = 0 \qquad\qquad [3]$$

$$15a + (r-7)b = 0 \qquad\qquad [4]$$

To calculate the value of r we must eliminate a and b. Hence

$$(r+5)(r-7) + 45 = 0$$

or

$$r^2 - 2r + 10 = 0$$

This is the auxiliary equation we have met before. Its roots are

$$r_1 = 1 + 3j \qquad \text{and} \qquad r_2 = 1 - 3j$$

x is then given by

$$x = e^t(a_1 e^{3jt} + a_2 e^{-3jt})$$

or

$$x = e^t(A_1 \cos 3t + A_2 \sin 3t)$$

Similarly

$$y = e^t(b_1 e^{3jt} + b_2 e^{-3jt})$$

or

$$y = e^t(B_1 \cos 3t + B_2 \sin 3t)$$

b_1 is connected with a_1, and b_2 with a_2 by equation 3 or equation 4, i.e.

$$b_1 = \frac{r_1 + 5}{3} a_1 \qquad b_2 = \frac{r_2 + 5}{3} a_2$$

We now consider two further examples taken from electrical and mechanical engineering.

EXAMPLE   Two electrical circuits are coupled magnetically. Each circuit consists of an inductor and a resistor; a voltage is applied to one of the circuits. Applying Kirchhoff's law to each circuit, the equations relating the two currents i_1 and i_2 (measured in amps) are

$$L_1 \frac{di_1}{dt} + M \frac{di_2}{dt} + R_1 i_1 = E_1$$

$$L_2 \frac{di_2}{dt} + M \frac{di_1}{dt} + R_2 i_2 = 0$$

L_1, L_2 are the values of the inductors, in henries. R_1, R_2 are the values of the resistors, in ohms. M is the coefficient of mutual inductance, in henries. E_1 is the applied voltage, assumed constant in this instance, in volts. The independent variable t is the time in this case.

Proceeding as in the previous example, we assume an exponential solution for the complementary function, so that

$$i_1 = A e^{rt}, \qquad i_2 = B e^{rt}$$

Substituting in the DE we find

$$(L_1 r + R_1)A + MrB = 0$$

$$MrA + (L_2 r + R_2)B = 0$$

liminating A and B from these two equations gives the auxiliary equation, i.e.

$$(L_1 r + R_1)(L_2 r + R_2) - M^2 r^2 = 0$$

i.e. $$(L_1 L_2 - M^2)r^2 + (L_1 R_2 + L_2 R_1)r + R_1 R_2 = 0$$

The form of the solution will depend on the nature of the roots of this equation. We saw in section 10.3.1 that the roots can be (1) real and unequal, (2) real and equal or (3) complex conjugate.

If case 3 applies, for example, the solution is

$$i_1 = e^{-at}(A_1 \cos bt + B_1 \sin bt) + \frac{E_1}{R_1}$$

$\frac{E_1}{R_1}$ is the particular integral and $i_2 = e^{-at}(A_2 \cos bt + B_2 \sin bt)$. Remember that A_2 and A_1, B_2 and B_1 are related.

EXAMPLE   This example concerns the calculation of the translational natural frequencies of a particular two-storey building whose idealised mathematical model leads to the following DEs:

$$15\,000\ddot{x}_1 + 31 \times 10^7 x_1 - 6 \times 10^7 x_2 = 0$$
$$8500\ddot{x}_2 + 6 \times 10^7 x_2 - 6 \times 10^7 x_1 = 0$$

where x_1 and x_2 are the displacements of each floor under free vibration conditions. The dot notation, as you will remember, refers to differentiation with respect to time t.

The method of solution in this case is approached in a different way from that of the previous examples because of the vibratory nature of the problem.

We could assume an exponential solution as before.

Instead, let

$$x_1 = A_1 \cos \omega_n t, \qquad\qquad x_2 = A_2 \cos \omega_n t$$

Therefore

$$\ddot{x}_1 = -\omega_n^2 A_1 \cos \omega_n t, \qquad \ddot{x}_2 = -\omega_n^2 A_2 \cos \omega_n t$$

ω_n is the natural frequency. We are using this method of solution because a vibration can be represented by a sine or cosine function. We have already discussed some aspects of vibrations in section 10.5.2.

Substituting in the differential equations we have, after dividing by 10^4

$$(-1.5\omega_n^2 + 31 \times 10^3)A_1 - 6 \times 10^3 A_2 = 0$$
$$-6 \times 10^3 A_1 + (-0.85\omega_n^2 + 6 \times 10^3)A_2 = 0$$

Since we are concerned with the values of the natural frequencies ω_n, we eliminate A_1 and A_2 from these two equations. Hence

$$(31 \times 10^3 - 1.5\omega_n^2)(6 \times 10^3 - 0.85\omega_n^2) - 36 \times 10^6 = 0$$

Expanding and collecting terms gives

$$1.275\omega_n{}^4 - 35.35 \times 10^3\,\omega_n{}^2 + 150 \times 10^6 \;=\; 0$$

known as the frequency equation, whose roots are $\omega_1{}^2 = 22\,495.7$, $\omega_2{}^2 = 5229.75$. Therefore the two natural frequencies of this building are 150 rad/s or 23.87 Hz and 72.32 rad/s or 11.5 Hz.

As a matter of interest, this was a real problem! Dangerous vibrations were observed when the building was commissioned due to a fan operating on the first floor at a frequency of 12 Hz.

10.9 HIGHER-ORDER DEs INTERPRETED AS SYSTEMS OF FIRST-ORDER SIMULTANEOUS DEs

Any DE of order n can be transformed into a system of n simultaneous first-order DEs. In fact, this is no more than a way of rephrasing the problem: it does not lead us any closer to a solution. However, it can be quite useful in a number of circumstances. If we wish to solve a DE by numerical methods, we find that first-order DEs are much easier to handle than higher-order DEs. We will return to this point in Chapter 17.

Let us start with a linear second-order DE

$$py'' + qy' + ry \;=\; f(x)$$

The fundamental idea is to introduce a new function, $u = y'$. The given equation is then equivalent to the following pair of first-order simultaneous DEs:

$$pu' + qu + ry \;=\; f(x)$$

$$y' \;=\; u$$

The general case is treated quite similarly. Given a DE of order n, we introduce $n-1$ new functions, $u_1 = y'$, $u_2 = u_1' = y''$, $u_3 = u_2' = y'''$,.... By inserting the us for all higher-order derivatives of y, a first-order DE (for the n functions, $u_1, \ldots, u_{n-1}, y$) is obtained. In conjunction with the defining equations for the us, we have a system of n simultaneous first-order DEs.

10.10 SOME ADVICE ON INTRACTABLE DEs

The sorts of differential equations discussed in this chapter can give only a glimpse of the subtle ideas involved in this topic. Sometimes a judicious change of variables may provide the answer, but each equation requires individual attention. Should you be confronted with severe problems, two routes can be taken.

On the one hand, the remedy *might* be provided by more powerful theoretical means. A very worthwhile subject, which could not be included in great detail in this book, is the theory of Laplace transforms. This is outlined in the following chapter.

On the other hand, if you are only interested in numerical data, then computer methods can provide the answer swiftly and reliably. Some elementary algorithms for solving DEs are presented in Chapter 17.

In any case, the first step when encountering a new DE is to classify it according to the criteria: What is its order? Is it linear? What types are its coefficients (constant or variable)? Is it an ordinary DE or a partial DE? Only then will you be able to use other sources of information efficiently.

EXERCISES

10.1   Concept and Classification of Differential Equations

1. Which of the following are linear first- and second-order DEs with constant coefficients?

    (a) $y' + x^2 y = 2x$ (b) $5y'' - 2y' - 4x = 3y$

    (c) $y^4 + 2y'' + 3y' = 0$ (d) $\sin xy'' - y = 0$

    (e) $y'' - x^5 = 2$ (f) $2y'' - y' + \frac{3}{2}y = 0$

2. Which of the following are homogeneous and non-homogeneous DEs and what is the order in each case?

    (a) $y'' + ax = 0$ (b) $\frac{5}{4}y'' + \frac{2}{3}y' = \frac{1}{2}y$

    (c) $2y' = 3y$ (d) $\frac{3}{10}y'' + \frac{2}{5}y' + \frac{1}{6}y - \sin x = 0$

    (e) $3y'' + y' = 2y$

10.3   General Solutions of First and Second Order DEs with Constant Coefficients

3. Solve the following DEs. In the case of complex roots give the real solution.

    (a) $2y'' - 12y' + 10y = 0$ (b) $4y'' - 12y' + 9y = 0$

    (c) $y'' + 2y' + 5y = 0$ (d) $y'' - \frac{1}{2}y' + \frac{5}{8}y = 0$

    (e) $\frac{1}{4}y'' + \frac{1}{2}y' - 2y = 0$ (f) $5y'' - 2y' + y = 0$

4. Solve the following DEs:

    (a) $2y' + 8y = 0$ (b) $\frac{1}{5}y' = 6y$ (c) $3y' = 6y$

5. Obtain the general solution of the following second-order DEs:

    (a) $S''(t) = 2t$ (b) $x''(t) = -\omega^2 \cos \omega t$

6. Given the following non-homogeneous DEs, obtain the particular integral using a trial solution.

    (a) $y'' + y' + y = 2x + 3$ (b) $y'' + 4y' + 2y = 2x + 3$

7. Obtain the general solution of the following non-homogeneous DEs:

    (a) $7y'' - 4y' - 3y = 6$ (b) $y'' - 10y' + 9y = 9x$

    (c) $3y'' - y' - 4y = x^2$ (d) $y'' + 2y' + 5y = \cos 2x$

8. A particular integral $y_p(x)$ of the following non-homogeneous DE is known. Check that it is a solution and obtain the general solution of the DE.
    $$\tfrac{1}{2}y'' - 3y' + \tfrac{5}{2}y = \tfrac{3}{4}x^2 - 1, \qquad y_p(x) = \tfrac{3}{10}x^2 + \tfrac{18}{25}x + \tfrac{43}{125}$$

10.4   Boundary Value Problems

9. Solve the following DEs:

    (a) $\frac{1}{2}y' + 2y = 0$ (given that $y(0) = 3$)

    (b) $\frac{4}{7}y' - \frac{6}{5}y = 0$ (given that $y(10) = 1$)

10. The DE $\frac{1}{3}y' - \frac{2}{3}y = 0$ has for its general solution $y(x) = C e^{2x}$. Calculate the value of the constant if

(a) $y(0) = 0$

(b) $y(0) = -2$

(c) $y(-1) = 1$

(d) $y'(-1) = 2 e^{-2}$

11. Solve $y'' + 4y = 0$ for the following boundary conditions:

(a) $y(0) = 0$, $y\left(\frac{\pi}{4}\right) = 1$

(b) $y\left(\frac{\pi}{2}\right) = -1$, $y'\left(\frac{\pi}{2}\right) = 1$

(c) $y(0) = 0$

$y'(0) = 1$

(d) $y\left(\frac{\pi}{4}\right) = a$

$y''(0) = b$

12. Solve $y'' + y = 2y'$, given that $y(0) = 1$ and $y(1) = 0$.

10.6   General Linear First Order DE

13. Solve the following first-order linear DEs:

(a) $xy' = 2y - x$

(b) $y' = \frac{y}{x} + x$

(c) $y' + y \tan x = \sin 2x$

(d) $xy' + (1+x)y = x e^{-x}$

14. Verify that the following DEs can be brought into the form of Bernoulli-type equations and solve them.

(a) $y' + xy = xy^3$

(b) $y' - \frac{2}{x^2 - 1} y = -y^2$

(c) $x^2 y^2 y' + xy^3 = 1$

(d) $yy' + \frac{y^2}{x} + x + 1 = 0$

15. In the following DEs, the variables can be separated. Solve

(a) $y' = e^{(x - 2y)}$

(b) $y' + xy' + \frac{x}{y} = 0$

(c) $xy' + (\ln x)y^2 = 0$

(d) $(y')^2 + x e^x y' + x e^x = 1$

16. Verify that the following are exact DEs. Find F and solve.

(a) $\frac{2y}{x} y' + \left(4 - \frac{y^2}{x^2}\right) = 0$

(b) $(1 - x e^{-y})y' + e^{-y} = 0$

(c) $(2y - x^2 \sin 2y)y' + 2x \cos^2 y = 0$

(d) $(2x - 3)y' + 3x^2 + 2y = 0$

17. You will remember that the integrating factor $\mu(x, y)$ for a DE $p(x, y)y' + q(x, y) = 0$ is easy to find in special cases. Solve the following equations by finding an integrating factor, μ, and then solving the exact equation.

(a) $\sin y y' - \cos y = -e^{2x}$

(b) $(e^y - x)y' + 1 = 0$

18. Solve the following simultaneous DEs:

(a) $x' - 7x + y = 0$

$y' - 2x - 5y = 0$

(b) $x' + y' + 2x + y = 0$

$y' + 5x + 3y = 0$

CHAPTER 11

* Laplace transforms

11.1 INTRODUCTION

In Chapter 10 we learned how to solve certain differential equations of the first and second order. We now consider a special technique for the solution of such ordinary differential equations known as the *Laplace transform*. It was first introduced by the French mathematician P.S. de Laplace in about 1780. The main advantage of the method is that it transforms the DE into an algebraic equation which, in many cases, can be readily solved. The solution of the original DE is then arrived at by obtaining the inverse transforms which usually consist of the ratio of two polynomials. The transforms and their inverses can be derived or obtained by consulting a table of transforms. We shall build up such a table of the functions frequently met in practice. The method is particularly useful in the solution of DEs whose boundary conditions are specified at a particular point and it is extensively used in the study of electrical networks, mechanical vibrations, impact, acoustics, structural problems, control systems, and in many other fields.

It is also used to solve linear DEs of any order, linear DEs with variable coefficients, linear partial DEs with constant coefficients, difference equations and integral equations.

In this chapter, however, we shall restrict ourselves to an introduction to the technique, and solve first- and second-order DEs with constant coefficients.

11.2 THE LAPLACE TRANSFORM DEFINITION

The Laplace transform $\mathscr{L}[F(t)]$ of a function $F(t)$ for values of $t > 0$ is defined as

$$\mathscr{L}[F(t)] = \int_0^\infty e^{-st} F(t)\, dt = f(s) \qquad (11.1)$$

This means that we take the given function, multiply it by e^{-st}, and integrate between the limits $t = 0$ to $t = \infty$. s is a number which may be complex but whose real part is positive and sufficiently large to ensure that the integral is convergent. Since the value of the integral depends on s, the Laplace transform is a function of s.

It is customary to write the Laplace transform of a function in shorthand form thus:

If $y(t)$ is the function, its transform is written $\bar{y}(s)$ or simply $\bar{y}$. For simplicity, we will also use the latter notation.

Proceeding in the opposite direction, i.e. finding the original function $y(t)$ from a given Laplace transform $\bar{y}(s)$, is called finding the *inverse transform*.

Before we can appreciate the usefulness of the Laplace transform, we need to derive the transforms of some of the more common functions encountered in physical problems.

These functions are

(1) A, a constant.

(2) e^{at}, an exponential function with the constant a real or complex.

(3) $\sin \omega t$, $\cos \omega t$, periodic functions where ω is usually a frequency.

(4) At, a linearly increasing function where, in practice, t is usually the time. This function is known as a ramp.

(5) $t \sin \omega t$, $t \cos \omega t$, periodic functions whose amplitudes increase linearly with the independent variable t.

(6) $e^{at} \sin \omega t$, $e^{at} \cos \omega t$, an exponentially increasing or decreasing oscillation, depending on whether a is positive or negative.

We also need to know the transforms of the derivatives of these functions:

$$\frac{d^n y}{dt^n}, \qquad n = 1, 2, \ldots$$

11.3 LAPLACE TRANSFORM OF STANDARD FUNCTIONS

We will now derive the Laplace transforms of the functions mentioned above.

When evaluating the transforms, i.e. solving the integrals, the quantity s is regarded as a constant.

(1) $y(t) = A$, a constant

$$\bar{y}(s) = \int_0^\infty A e^{-st} \, dt = A \int_0^\infty e^{-st} = A \left[\frac{e^{-st}}{-s} \right]_0^\infty = \frac{A}{s} \qquad (11.2)$$

(2) $y(t) = e^{at}$, with a real or complex

$$\bar{y}(s) = \int_0^\infty e^{-st} e^{at} \, dt = \int_0^\infty e^{-(s-a)t} \, dt = \left[\frac{e^{-(s-a)t}}{-(s-a)} \right]_0^\infty = \frac{1}{s-a} \qquad (11.3)$$

Note that $s >$ real part of a for the integral to be convergent.

(3) $y(t) = \sin \omega t$ and $y(t) = \cos \omega t$

The simplest way of obtaining these transforms is to make use of (2).

Remember: $y(t) = \sin \omega t = \frac{1}{2j} (e^{j\omega t} - e^{-j\omega t})$

Hence, from (2), we have

$$\bar{y}(s) = \frac{1}{2j}\left(\frac{1}{s-j\omega} - \frac{1}{s+j\omega}\right) = \frac{\omega}{s^2+\omega^2} \qquad (11.4a)$$

We obtain the Laplace transform of the cosine function in the same way.

$$y(t) = \cos\omega t = \frac{1}{2}(e^{j\omega t} + e^{-j\omega t})$$

$$\bar{y}(s) = \frac{s}{s^2+\omega^2} \qquad (11.4b)$$

(4) $y(t) = At$

$$\bar{y}(s) = \int_0^\infty At\,e^{-st}\,dt = A\int_0^\infty t\,e^{-st}\,dt$$

$$= A\left(-\left[\frac{t}{s}\,e^{-st}\right]_0^\infty + \frac{1}{s}\int_0^\infty e^{-st}\,dt\right) = \frac{A}{s^2} \qquad (11.5)$$

when integrating by parts.

The first term is zero since e^{-st} decreases more rapidly than t increases as $t\to\infty$.

Before proceeding further with transforms of functions, we will consider some important theorems, the first enables us to extend the list of transforms.

THEOREM I: THE SHIFT THEOREM

If $y(t)$ is a function and $\bar{y}(s)$ its transform, and a is any real or complex number, then $\bar{y}(s+a)$ is the Laplace transform of $e^{-at}y(t)$.   (11.6)

PROOF   The Laplace transform of $e^{-at}y(t)$ is

$$\int_0^\infty e^{-st}\,e^{-at}\,y(t)\,dt = \int_0^\infty e^{-(s+a)t}\,y(t)\,dt = \bar{y}(s+a)$$

Thus we see that we simply replace s by $(s+a)$ wherever s occurs in the transform of $y(t)$. If a is negative, then we can show that s is replaced by $(s-a)$.

As an example, let us find the transform of

$$y(t) = e^{-at}\sin\omega t$$

The transform of the sine function is given by $\bar{y}(s) = \dfrac{\omega}{s^2+\omega^2}$

Applying the shift theorem gives

$$\bar{y}(s) = \frac{\omega}{(s+a)^2+\omega^2} \qquad (11.7a)$$

Similarly, if $y(t) = e^{-at}\cos\omega t$, the shift theorem gives

$$\bar{y}(s) = \frac{s+a}{(s+a)^2+\omega^2} \qquad (11.7b)$$

EXAMPLE   Obtain the Laplace transform of $y(t) = 3\,e^{5t} \cos 10t$.

$$a = -5, \qquad \omega = 10$$

$$\bar{y}(s) = 3\frac{s-5}{(s-5)^2 + 10^2} = \frac{3(s-5)}{s^2 - 10s + 125}$$

Let us now continue with the derivation of the Laplace transforms of frequently used functions.

(5) $\qquad\qquad y(t) = t \sin \omega t$

$$\bar{y}(s) = \int_0^\infty e^{-st} t \sin \omega t \, dt = \frac{2\omega s}{(s^2 + \omega^2)^2} \qquad\qquad (11.8)$$

Even though the result is already stated, we wish to know how it is arrived at. Looking at the integral, you will realise that the task of evaluating it is not a straightforward one. The following theorem will be of considerable help.

THEOREM II: TRANSFORM OF PRODUCTS $t\,y(t)$

If $y(t)$ is a function and $\bar{y}(s)$ its transform, then the transform of the new function $ty(t)$ is

$$\mathscr{L}[t\,y(t)] = -\frac{d}{ds}[\bar{y}(s)] \qquad\qquad (11.9)$$

PROOF

$$\frac{d}{ds}[\bar{y}(s)] = \frac{d}{ds}\left(\int_0^\infty e^{-st} y(t)\, dt\right) = -\int_0^\infty e^{-st} t\, y(t)\, dt = -\mathscr{L}[t\,y(t)]$$

This is due to the fact that we can differentiate under the integral sign with respect to a parameter.

Let us go back to $y(t) = t \sin \omega t$. The transform of the sine function is known to be

$$\mathscr{L}(\sin \omega t) = \frac{\omega}{s^2 + \omega^2}$$

Hence, by theorem II, the transform $\bar{y}(s)$ of $t \sin \omega t$ is

$$\mathscr{L}(t \sin \omega t) = -\frac{d}{ds}\left(\frac{\omega}{s^2 + \omega^2}\right) = \frac{2\omega s}{(s^2 + \omega^2)^2}$$

Similarly, the transform of $t \cos \omega t$ is

$$\mathscr{L}(t \cos \omega t) = -\frac{d}{ds}\left(\frac{s}{s^2 + \omega^2}\right) = \frac{s^2 - \omega^2}{(s^2 + \omega^2)^2} \qquad\qquad (11.10)$$

We can extend the use of theorem II. For example,

$$\mathscr{L}\,(t^2 \cos \omega t) = \mathscr{L}\,[t(t \cos \omega t)] = -\frac{\mathrm{d}}{\mathrm{d}s}\left\{\mathscr{L}(t \cos \omega t)\right\}$$

$$= -\frac{\mathrm{d}}{\mathrm{d}s}\left(\frac{s^2 - \omega^2}{(s^2 + \omega^2)^2}\right) = \frac{2s(s^2 - \omega^2)}{(s^2 + \omega^2)^3}$$

If $f(t)$ is a function and $\bar{f}(s)$ its transform, then the transform of the new function is $y(t) = t^n f(t)$

$$\bar{y}(s) = (-1)^n \frac{\mathrm{d}^n}{\mathrm{d}s^n}[\bar{f}(s)] \qquad\qquad (11.11)$$

(6) $y(t) = t^n$, where n is a positive integer.

$$\bar{y}(s) = \int_0^\infty \mathrm{e}^{-st}\, t^n \,\mathrm{d}t = \frac{n!}{s^{n+1}}$$

As before, the result is already stated but the proof is missing yet. We start by writing the function t^n as a product:

$$y(t) = t^{n-1} t$$

The transform of the function t is known to be $\dfrac{1}{s^2}$

Now, using the general result (equation 11.11) we find

$$\bar{y}(s) = (-1)^{n-1}\frac{\mathrm{d}^{n-1}}{\mathrm{d}s^{n-1}}\frac{1}{s^2} = (-1)^{n-1}(-2)(-3)\ldots(-n)\frac{1}{s^{n+1}} = \frac{n!}{s^{n+1}}$$

For example, if $y(t) = t^2$, then $\bar{y}(s) = \dfrac{2}{s^3}$.

For the sake of completeness, we will mention another theorem which, in fact, has already been used implicitly.

THEOREM III: LINEARITY

Linearity of the Laplace transform

Let $y(t)$ be a combination of functions:

$$y(t) = Af(t) + Bg(t) \qquad (A, B \text{ are constants})$$

Then the Laplace transform is the corresponding combination of the transformed functions:

$$\bar{y}(t) = A\bar{f}(t) + B\bar{g}(t) \qquad\qquad (11.12)$$

or $\qquad\qquad \mathscr{L}\,[Af(t) + Bg(t)] = A\mathscr{L}\,[f(t)] + B\mathscr{L}\,[g(t)]$

The proof is obvious. It follows from the linearity of the integral.

As a particular case, note that a constant factor is preserved by the Laplace transform:

$$\mathscr{L}[Af(t)] = A\mathscr{L}[f(t)]$$

For example, the transform of sin ωt is $\dfrac{\omega}{s^2 + \omega^2}$ and the transform of t is $\dfrac{1}{s^2}$

Therefore, the transform of $-6 \sin \omega t + t$ is $\dfrac{-6\omega}{s^2 + \omega^2} + \dfrac{1}{s^2}$

THEOREM IV: TRANSFORMS OF DERIVATIVES

First Derivative of a Function $y(t)$

By definition $\mathscr{L}\left[\dfrac{d}{dt}y(t)\right] = \displaystyle\int_0^\infty e^{-st}\dfrac{dy}{dt}\,dt$

We find, putting $\dfrac{dy}{dt} = \dot{y}$ i.e. using a dot to denote the derivative with respect to time that

$$\int_0^\infty e^{-st}\dot{y}\,dt = \left[e^{-st}y\right]_0^\infty - \int_0^\infty y(-s\,e^{-st})\,dt = -y(0) + s\bar{y} = s\bar{y} - y(0)$$

This result holds for those functions for which $e^{-t}y(t) \to 0$ as $t \to \infty$.

$$\mathscr{L}\left[\dfrac{d}{dt}y(t)\right] = s\,\bar{y}(s) - y(0) \qquad\qquad (11.13)$$

where $y(0)$ is the value of the function at $t = 0$ (initial value or initial condition).

Second Derivative of a Function $y(t)$

$$\mathscr{L}\left(\dfrac{d^2}{dt^2}y(t)\right) = \int_0^\infty e^{-st}\ddot{y}\,dt = \left[\dot{y}\,e^{-st}\right]_0^\infty + s\int_0^\infty e^{-st}\dot{y}\,dt = -\dot{y}(0) - s\,y(0) + s^2\bar{y}$$

$$\mathscr{L}[\ddot{y}(t)] = s^2\bar{y} - s\,y(0) - \dot{y}(0)$$

where $\dot{y}(0)$ is the value of the first derivative at $t = 0$.

By repeating the process we can show that

$$\mathscr{L}[\dddot{y}(t)] = s^3\bar{y} - s^2 y(0) - s\,\dot{y}(0) - \ddot{y}(0)$$

where $\ddot{y}(0)$ is the value of the second derivative at $t = 0$.

The following notation for the values of the function $y(t)$ and its derivatives at $t = 0$ is commonly used:

$y_0 = y(0)$ is its value at $t = 0$,

$y_1 = \dot{y}(0)$ is the value of the first derivative at $t = 0$,

$y_2 = \ddot{y}(0)$ is the value of the second derivative at $t = 0$,

$\vdots$

$y_n = y^{(n)}(0)$ is the value of the nth derivative at $t = 0$.

For example, the Laplace transform of the 4th derivative will be

$$\mathscr{L}\,[y^{(4)}(t)] \;=\; s^4\bar{y} - s^3 y_0 - s^2 y_1 - s y_2 - y_3$$

Transforms of Derivatives

$$\mathscr{L}\,[y^{(n)}(t)] \;=\; s^n\bar{y} - \sum_{i=0}^{n-1} s^{n-i-1} y_i \qquad (11.14)$$

A table of transforms, and a table of inverse transforms, will be found in the appendix to this chapter.

11.4 SOLUTION OF LINEAR DEs WITH CONSTANT COEFFICIENTS

Suppose we have to solve the DE

$$\frac{d^2 y}{dt^2} + A\,\frac{dy}{dt} + By \;=\; f(t)$$

with initial conditions

$$y = y_0, \qquad \frac{dy}{dt} = y_1 \qquad \text{at} \qquad t = 0$$

If we multiply the equation throughout by e^{-st} and integrate each term for $t = 0$ to $t = \infty$, we in fact replace each term by its Laplace transform. In doing so we transform the DE into an algebraic equation in terms of the parameter s.

Using the table of transforms, we find

$$s^2\bar{y} - s y_0 - y_1 + A(s\bar{y} - y_0) + B\bar{y} \;=\; \bar{f}$$

Solving for $\bar{y}$ gives

$$\bar{y} \;=\; \frac{\bar{f} + s y_0 + A y_0 + y_1}{s^2 + As + B} \qquad (11.15)$$

All we need do now is look up the inverse transform. The reader will notice that we do not have to find the values of arbitrary constants; but we may have to express $\bar{y}$ as a partial fraction or in a form from which the inverse can be found easily.

EXAMPLE   Solve the equation $\dfrac{dy}{dt} + 4y = e^{-2t}$

given that $y = 5$ when $t = 0$, i.e. $y_0 = 5$.

The transformed equation is $s\bar{y} - y_0 + 4\bar{y} = \dfrac{1}{s+2}$

Solving for $\bar{y}$ gives

$$\bar{y} = \frac{5s+11}{(s+2)(s+4)} = \frac{1}{2(s+2)} + \frac{9}{2(s+4)}$$

From the table, we can look up the inverse transform. Hence the solution is

$$y = \frac{1}{2}e^{-2t} + \frac{9}{2}e^{-4t}$$

EXAMPLE   Solve the equation $\ddot{y} + 5\dot{y} + 4y = 0$

given that $y = 0$ and $\dot{y} = 3$ at $t = 0$.

From the table we find

$$\underbrace{s^2\bar{y} - sy_0 - y_1}_{\mathscr{L}(\ddot{y})} + \underbrace{5(s\bar{y} - y_0)}_{\mathscr{L}(5\dot{y})} + \underbrace{4\bar{y}}_{\mathscr{L}(4y)} = 0$$

The initial conditions are $y_0 = 0$, $y_1 = 3$.

Hence

$$s^2\bar{y} - 3 + 5s\bar{y} + 4\bar{y} = 0$$

or
$$\bar{y}(s^2 + 5s + 4) = 3$$

Solving for $\bar{y}$ gives

$$\bar{y} = \frac{3}{s^2 + 5s + 4} = \frac{3}{(s+4)(s+1)} = \frac{1}{(s+1)} - \frac{1}{(s+4)}$$

From the table we get

$$y = e^{-t} - e^{-4t}$$

EXAMPLE   Solve the equation

$$\ddot{y} + 8\dot{y} + 17y = 0 \qquad \text{if} \qquad y = 0, \qquad \dot{y} = 3 \qquad \text{at} \qquad t = 0$$

The initial values are $y_0 = 0$, $y_1 = 3$. Hence the transformed equation is

$$s^2\bar{y} - 3 + 8sy + 17y = 0$$

Solving for $\bar{y}$ gives $\bar{y} = \dfrac{3}{s^2 + 8s + 17} = \dfrac{3}{(s+4)^2 + 1}$

From the table we find

$$y = 3e^{-4t}\sin t$$

EXAMPLE   Solve the equation $\ddot{y} + 6y = t$

The initial conditions are $y = 0$ and $\dot{y} = 1$ at $t = 0$.

The transformed equation is

$$s^2\bar{y} - 1 + 6\bar{y} = \frac{1}{s^2}$$

Therefore

$$\bar{y}(s^2 + 6) = \frac{1}{s^2} + 1 = \frac{1 + s^2}{s^2}$$

Solving for $\bar{y}$ gives

$$\bar{y} = \frac{s^2 + 1}{s^2(s^2 + 6)} = \frac{5}{6(s^2 + 6)} + \frac{1}{6s^2}$$

From the table of inverse transforms at the end of this chapter we find

$$y = \frac{1}{6}t + \frac{5}{6} \times \frac{1}{\sqrt{6}}\sin\sqrt{6}t = \frac{1}{6}\left(t + \frac{5}{\sqrt{6}}\sin\sqrt{6}t\right)$$

11.5 SOLUTION OF SIMULTANEOUS DEs WITH CONSTANT COEFFICIENTS

In science and engineering, we frequently encounter systems which give rise to simultaneous differential equations, e.g. an electrical network consisting of two loops, or a double spring–mass system. We will now illustrate their solution by means of the Laplace transform technique.

If $x(t)$ and $y(t)$ are two functions of the independent variable t, their transforms are denoted by $\bar{x}(s)$ and $\bar{y}(s)$, respectively, or simply $\bar{x}$ and $\bar{y}$.

EXAMPLE   Solve the equations

$$3\dot{x} + 2x + \dot{y} = 1$$
$$\dot{x} + 4\dot{y} + 3y = 0$$

The initial conditions are $x = 0$ and $y = 0$ at $t = 0$

Transforming the equations gives

$$3(s\bar{x} - x_0) + 2\bar{x} + s\bar{y} - y_0 = \frac{1}{s}$$
$$s\bar{x} - x_0 + 4(s\bar{y} - y_0) + 3\bar{y} = 0$$

but $x_0 = 0$ and $y_0 = 0$

Hence we obtain a pair of simultaneous equations in $\bar{x}$ and $\bar{y}$:

$$\bar{x}(3s + 2) + \bar{y}s = \frac{1}{s}$$
$$\bar{x}s + \bar{y}(4s + 3) = 0$$

Solving for $\bar{x}$ gives

$$\bar{x} = \frac{(4s+3)}{s(s+1)(11s+6)} = \frac{1}{2s} - \frac{1}{5}\frac{1}{(s+1)} - \frac{3}{10(s+6/11)}$$

Hence

$$x = \frac{1}{2} - \frac{1}{5}e^{-t} - \frac{3}{10}e^{-6t/11}$$

Solving for $\bar{y}$ gives

$$\bar{y} = \frac{-1}{(s+1)(11s+6)} = \frac{1}{5}\left(\frac{1}{s+1} - \frac{1}{s+6/11}\right)$$

Hence

$$y = \frac{1}{5}(e^{-t} - e^{-6t/11})$$

EXAMPLE   Solve the differential equations

$$\ddot{x} + 2x - \dot{y} = 1$$

$$\dot{x} + \ddot{y} + 2y = 0$$

The initial conditions are $x = 1$ and $\dot{x} = y = \dot{y} = 0$ at $t = 0$.

Transforming the equations gives

$$(s^2 + 2)\bar{x} - s\bar{y} = \frac{1}{s} + sx_0 = \frac{1}{s} + s$$

$$s\bar{x} + (s^2 + 2)\bar{y} = x_0 = 1$$

Solving these two simultaneous algebraic equations in $\bar{x}$ and $\bar{y}$ we find

$$\bar{x} = \frac{s^4 + 4s^2 + 2}{s(s^2 + 1)(s^2 + 4)} = \frac{1}{2s} + \frac{s}{3(s^2 + 1)} + \frac{s}{6(s^2 + 4)}$$

Looking up the table of inverse transforms at the end of this chapter we have

$$x = \frac{1}{2} + \frac{1}{3}\cos t + \frac{1}{6}\cos 2t$$

Also

$$\bar{y} = \frac{1}{(s^2 + 1)(s^2 + 4)} = \frac{1}{3}\left(\frac{1}{s^2 + 1} - \frac{1}{s^2 + 4}\right)$$

and from the table of inverse transforms we get

$$y = \frac{1}{3}\sin t - \frac{1}{6}\sin 2t$$

APPENDIX

TABLE OF LAPLACE TRANSFORMS

$F(t) = y(t)$	$\mathscr{L}\,[F(t)] = \bar{y}(s)$
A	$\dfrac{A}{s}$
e^{at}	$\dfrac{1}{s-a}$
$t^n\ (n = 0, 1, 2, 3, \ldots)$	$\dfrac{n!}{s^{n+1}}$
$\sin \omega t$	$\dfrac{\omega}{s^2 + \omega^2}$
$\cos \omega t$	$\dfrac{s}{s^2 + \omega^2}$
$t \sin \omega t$	$\dfrac{2\omega s}{(s^2 + \omega^2)^2}$
$t \cos \omega t$	$\dfrac{s^2 - \omega^2}{(s^2 + \omega^2)^2}$
$\sinh \omega t$	$\dfrac{\omega}{s^2 - \omega^2}$
$\cosh \omega t$	$\dfrac{s}{s^2 - \omega^2}$
$t \sinh \omega t$	$\dfrac{2\omega s}{(s^2 - \omega^2)^2}$
$t \cosh \omega t$	$\dfrac{s^2 + \omega^2}{(s^2 - \omega^2)^2}$
$e^{-at} y(t)$	$\bar{y}(s + a)$
$t^n y(t)$	$(-1)^n \dfrac{\mathrm{d}^n}{\mathrm{d}s^n}\,\bar{y}(s)$
$\dfrac{y(t)}{t}$	$\displaystyle\int_s^\infty \bar{y}(s)\,\mathrm{d}s, \quad \text{if} \quad \lim_{t \to 0}\left(\dfrac{Fy(t)}{t}\right) \text{exists}$
$\dfrac{\sin \omega t}{t}$	$\tan^{-1} \dfrac{\omega}{s}$
$\dot{y}(t)$	$s\bar{y} - y_0$
$\ddot{y}(t)$	$s^2 \bar{y} - sy_0 - y_1$
$\dddot{y}(t)$	$s^3 \bar{y} - s^2 y_0 - sy_1 - y_2$
$\dfrac{\mathrm{d}^n}{\mathrm{d}t^n}\,y(t)$	$\left. s^n \bar{y}(s) - \displaystyle\sum_{i=0}^{n-1} s^{n-i-1}\,\dfrac{\mathrm{d}^i}{\mathrm{d}t^i}\,y(t)\right|_0$
$\displaystyle\int_0^t y(t)\,\mathrm{d}t$	$\dfrac{\bar{y}(s)}{s}$
$2k\,e^{\alpha t} \cos(\omega t + \theta)$	$\dfrac{k\,e^{\mathrm{j}\theta}}{s - \alpha - \mathrm{j}\omega} + \dfrac{k\,e^{\mathrm{j}\theta}}{\sin \alpha + \mathrm{j}\omega}$

TABLE OF INVERSE LAPLACE TRANSFORMS *CONTINUED*

$f(s) = \bar{y}(s)$	$\mathscr{L}^{-1}[f(s)] = y(t)$
$\dfrac{A}{s}$	A
$\dfrac{1}{s^n}$	$\dfrac{t^{n+1}}{(n-1)!}$
$\dfrac{1}{s-a}$	e^{at}
$\dfrac{1}{(s-a)^n}$	$\dfrac{t^{n-1}e^{at}}{(n-1)!}$
$\dfrac{1}{(s-a)(s-b)}$	$\dfrac{1}{a-b}(e^{at}-e^{bt})$
$\dfrac{s}{(s-a)(s-b)}$	$\dfrac{1}{a-b}(a\,e^{at}-b\,e^{bt})$
$\dfrac{1}{s^2+\omega^2}$	$\dfrac{1}{\omega}\sin\omega t$
$\dfrac{s}{s^2+\omega^2}$	$\cos\omega t$
$\dfrac{1}{(s-a)^2+\omega^2}$	$\dfrac{1}{\omega}e^{at}\sin\omega t$
$\dfrac{s-a}{(s-a)^2+\omega^2}$	$e^{at}\cos\omega t$
$\dfrac{1}{s(s^2+\omega^2)}$	$\dfrac{1}{\omega^2}(1-\cos\omega t)$
$\dfrac{1}{s^2(s^2+\omega^2)}$	$\dfrac{1}{\omega^3}(\omega t-\sin\omega t)$
$\dfrac{1}{(s^2+\omega^2)^2}$	$\dfrac{1}{2\omega^3}(\sin\omega t-\omega t\cos\omega t)$
$\dfrac{s}{(s^2+\omega^2)^2}$	$\dfrac{t}{2\omega}\sin\omega t$
$\dfrac{s^2}{(s^2+\omega^2)^2}$	$\dfrac{1}{2\omega}(\sin\omega t+\omega t\cos\omega t)$
$\dfrac{s}{(s^2+\omega_1{}^2)(s^2+\omega_2{}^2)}, \quad \omega_1{}^2\neq\omega_2{}^2$	$\dfrac{1}{\omega_2{}^2-\omega_1{}^2}(\cos\omega_1 t-\cos\omega_2 t)$
$\dfrac{1}{s^2-\omega^2}$	$\dfrac{1}{\omega}\sinh\omega t$
$\dfrac{s}{s^2-\omega^2}$	$\cosh\omega t$

EXERCISES

11.3   Laplace Transform of Standard Functions

1. Obtain the Laplace transforms for the following functions:

(a) $\frac{1}{4}t^3$ (b) $5e^{-2t}$ (c) $4\cos 3t$ (d) $\sin^2 t$

2. Obtain the inverse transforms for the following:

(a) $\dfrac{1}{4s^2+1}$ (b) $\dfrac{1}{s(s+4)}$ (c) $\dfrac{2}{s(s^2+9)}$

(d) $\dfrac{6}{1-s^2}$ (e) $\dfrac{1}{s^2(s^2+1)}$ (f) $\dfrac{4}{s(s^2-6s+8)}$

11.4   Solution of Linear DEs with Constant Coefficients

3. Solve the following differential equations:

(a) $\ddot{y}+5\dot{y}+4y=0$ (initial conditions: $y=0$, $\dot{y}=2$ at $t=0$)

(b) $\ddot{y}+9y=\sin 2t$ (initial conditions: $y=1$, $\dot{y}=-1$ at $t=0$)

(c) $\dot{y}+2y=\cos t$ (initial conditions: $y=1$ at $t=0$)

4. If $\ddot{y}-3\dot{y}+2y=4$ and $y=2$, $\dot{y}=3$ at $t=0$, show that $\bar{y}=\dfrac{2s^2-3s+4}{s(s-1)(s-2)}$, and hence find the solution for y.

5. Given $\dddot{y}+\ddot{y}=e^t+t+1$ with the initial conditions $y=0$, $\dot{y}=0$, $\ddot{y}=0$ at $t=0$, obtain y.

11.5   Solution of Simultaneous DEs with Constant Coefficients

6. Solve the following simultaneous equations for y:

$$\dot{y}+2\dot{x}+y-x=25$$

$$2\dot{y}+x=25e^t$$

Initial conditions: $y=0$, $x=25$ at $t=0$.

7. Solve for y and x given

$$4\dot{x}-\dot{y}+x=1$$

$$4\dot{x}-4\dot{y}-y=0$$

Initial conditions: $x=0$, $y=0$ at $t=0$

8. An electrical circuit consists of a capacitor, C farads, and an inductor, L henries, in series, to which a voltage $E\sin\omega t$ is applied.

If Q is the charge on the capacitor in coulombs show that

$$\bar{Q}=\frac{E}{L(\omega^2-1/LC)}\left[\frac{\omega}{s^2+1/LC}-\frac{\omega}{s^2+\omega^2}\right], \qquad \omega^2 LC\neq 1$$

and hence calculate Q given that $C=50\times 10^{-6}F$, $L=0.1\,\mathrm{H}$, $\omega=500\,\mathrm{rad/s}$, $E=2\,\mathrm{V}$ and $Q=\dot{Q}=0$ at $t=0$.

Functions of several variables; partial differentiation; and total differentiation

12.1 INTRODUCTION

So far we have dealt with functions of only a single variable, such as x, t etc. But functions of more than one variable also occur frequently in engineering and science.

EXAMPLE A voltage V is applied to a circuit having a resistance R, as shown in figure 12.1.

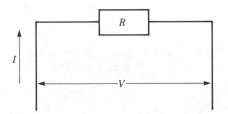

Figure 12.1

What is the value of the current which flows in the circuit?

According to Ohm's law, the current I depends on the resistance R and the applied voltage V, i.e.

$$I = \frac{V}{R}$$

Hence $I = I(V, R)$, which is a function of two variables.

EXAMPLE A gas is trapped inside a cylinder of volume V. The gas pressure on the cylinder walls and the piston is p and the temperature is T (see figure 12.2). The following relationship between volume, pressure and temperature holds true for 1 mol (6.02×10^{23} gas molecules):

$$pV = RT$$

where R, the gas constant, is approximately $8.314 \, \mathrm{J\,K^{-1}\,mol^{-1}}$. The equation can be rewritten as

$$p = R\frac{T}{V}$$

This means that the pressure is a function of two variables, i.e. $p = p(V, T)$.

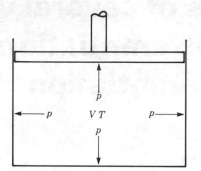

Figure 12.2

12.2 FUNCTIONS OF SEVERAL VARIABLES

Let us now leave the physical examples and consider the mathematical concept. If z is a function of two variables, x and y, then the relationship is usually expressed in the form

$$z = f(x, y)$$

Remember that, geometrically, a function y of one variable x $(y = f(x))$ represents a curve in the x–y plane, as shown in figure 12.3.

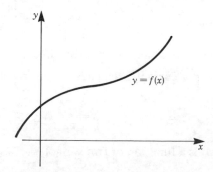

Figure 12.3

Similarly, a function $z = f(x, y)$ of two independent variables x and y can be thought of as representing a surface in three-dimensional space.

A geometrical picture of the function $z = f(x, y)$ can be obtained in two different ways.

12.2.1   REPRESENTING THE SURFACE BY ESTABLISHING A TABLE OF Z-VALUES

By giving x and y a particular value we obtain a value for z by substitution in
$$z = f(x, y)$$

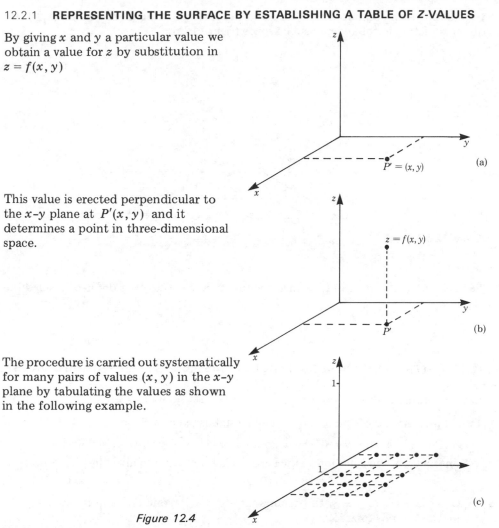

This value is erected perpendicular to the x-y plane at $P'(x, y)$ and it determines a point in three-dimensional space.

The procedure is carried out systematically for many pairs of values (x, y) in the x-y plane by tabulating the values as shown in the following example.

Figure 12.4

EXAMPLE   Values for the function

$$z = \frac{1}{1 + x^2 + y^2}$$

are given in table 12.1.

TABLE 12.1

y \ x	0	1	2	3
0	1	$\frac{1}{2}$	$\frac{1}{5}$	$\frac{1}{10}$
1	$\frac{1}{2}$	$\frac{1}{3}$	$\frac{1}{6}$	$\frac{1}{11}$
2	$\frac{1}{5}$	$\frac{1}{6}$	$\frac{1}{9}$	$\frac{1}{14}$
3	$\frac{1}{10}$	$\frac{1}{11}$	$\frac{1}{14}$	$\frac{1}{19}$

By plotting each computed value of z and the corresponding pair of values (x, y), we obtain a picture in three-dimensional space (figure 12.5).

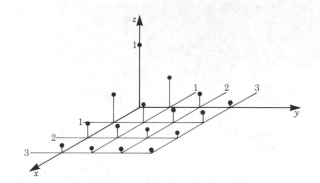

Figure 12.5

The set of values of (x, y) for which the function $z = f(x, y)$ is defined is called the *domain*.

12.2.2   REPRESENTING THE SURFACE BY ESTABLISHING INTERSECTING CURVES

Let us return to the function $z = \dfrac{1}{1 + x^2 + y^2}$. Its domain is the entire x-y plane. Two characteristics of the function can be established at a glance:

(1) For $x = 0$ and $y = 0$ the denominator $1 + x^2 + y^2$ has its smallest value. Consequently the function z (the surface) has a maximum given by

$$f(0, 0) = 1$$

(2) As $x \to \infty$ or $y \to \infty$ the denominator grows beyond all bounds and the function z tends to zero.

Of course, these two characteristics are not sufficient to sketch the surface. Generally speaking, the shape of surfaces is more difficult to determine than that of curves. Nevertheless, we can obtain a true picture of the function if we proceed systematically by dividing the task into parts. The basic idea is to investigate the influence of each variable separately on the shape of the surface by assuming that one of the two variables is constant.

If we regard y as being constant $(y = y_0)$ and vary x, then we obtain z-values which depend only on one variable.

For example, if we set $y = 0$ in the above function we have

$$z(x) = \frac{1}{1 + x^2}$$

This represents an intersecting curve between the surface $z = f(x, y)$ and the x-z plane at $y = 0$ (figure 12.6).

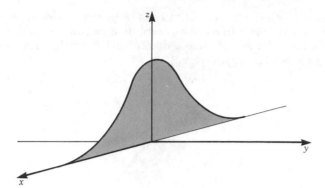

Figure 12.6

For an arbitrary value $y = y_0$, we have

$$z(x) = \frac{1}{1 + x^2 + y_0{}^2}$$

This represents an intersecting curve between the surface $z = f(x, y)$ and a plane parallel to the x–z plane shifted by an amount y_0 along the y-axis (figure 12.7).

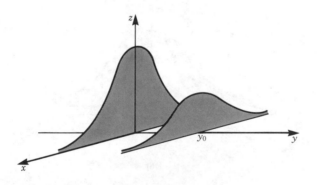

Figure 12.7

Similarly, we obtain a second group of curves by setting x to a constant $(x = x_0)$.

For example, if we set $x = 0$ we have

$$z(y) = \frac{1}{1 + y^2}$$

For an arbitrary value $x = x_0$ we have

$$z(y) = \frac{1}{1 + x_0{}^2 + y^2}$$

Both $z(y)$ curves are shown in figure 12.8a.

By plotting both types of curve in one diagram we obtain a better picture. In this case the graph shows a symmetrical 'hill' (figure 12.8b). (Note that the values in a given row or column of table 12.1 are the values of an intersecting curve.)

The sketch becomes clearer if we fill in lines of constant z-value. Mathematically, they are the curves of the surface which are at a constant distance from the x–y plane; they are the intersecting curves of the surface with planes parallel to the x–y plane at a given z-value (figure 12.8c).

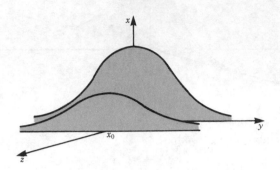

(a)

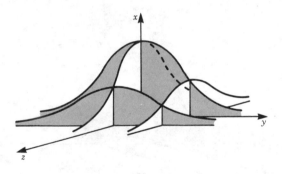

(b)

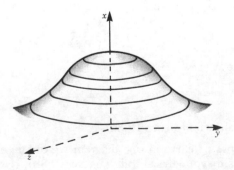

Figure 12.8   (c)

12.2.3 OBTAINING A FUNCTIONAL EXPRESSION FOR A GIVEN SURFACE

In the above discussion we started from a known function and looked for the resulting surface. Now we reverse the process and look for a functional expression for a given surface.

For example, consider a sphere of radius R with the origin of the coordinates at the centre (figure 12.9). Our task is to determine the equation of the spherical surface above the x-y plane.

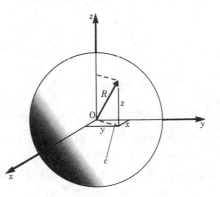

Figure 12.9

Referring to the figure and applying Pythagoras' theorem, we have

$$R^2 = z^2 + c^2, \qquad c^2 = x^2 + y^2$$

Thus we obtain $R^2 = x^2 + y^2 + z^2$

Solving for z gives

$$z_{1,2} = \pm\sqrt{R^2 - x^2 - y^2}$$

The positive root z_1 represents the spherical shell above the x-y plane.

The negative root z_2 represents the spherical shell below the x-y plane.

The domain is $-R \leqslant x \leqslant R$

and $-R \leqslant y \leqslant R$

such that $x^2 + y^2 \leqslant R^2$

Having acquired a pictorial idea of functions of two variables, $z = f(x, y)$, we now give a formal definition.

DEFINITION	$z = f(x, y)$ is called a function of two independent variables if there exists one value of z for each paired value (x, y) within a particular domain.

By plotting points (x, y) and $z = f(x, y)$ in a three-dimensional coordinate system we obtain a graph of the function which represents a surface F within the domain D of the variables (figure 12.10).

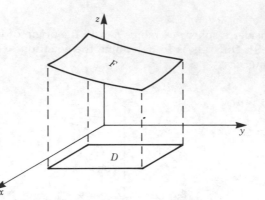

Figure 12.10

It is not possible to represent a function of three variables geometrically since to do so we would need a four-dimensional coordinate system.

However, in engineering and science such relationships play a very important role. For example, we can express the temperature T of the atmosphere as a function of three variables: the latitude x, the longitude y and the altitude (above sea level) z, i.e. $T = T(x, y, z)$.

12.3 PARTIAL DIFFERENTIATION

Remember that the geometrical meaning of the derivative of a function of one variable is the slope of the tangent to the curve $y = f(x)$ (figure 12.11).

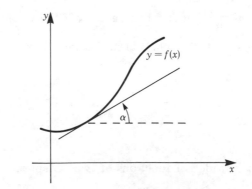

Figure 12.11

In the previous section we considered as an example of a function of two variables the function

$$z(x, y) = \frac{1}{1 + x^2 + y^2} \qquad [1]$$

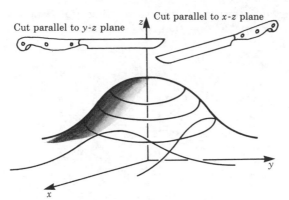

Figure 12.12

It represents a surface in three-dimensional space. By setting one variable at a constant value we obtain an intersecting curve of the surface with a particular plane.

We can slice the surface with planes parallel to the x-z plane (figure 12.13a). If the intersecting plane is at a distance y_0 from the x-z plane, the equation of the resulting curve is obtained by substituting $y = y_0$ in equation 1, i.e. $z(x) = \dfrac{1}{1 + x^2 + y_0^2}$ z is now a function of x only.

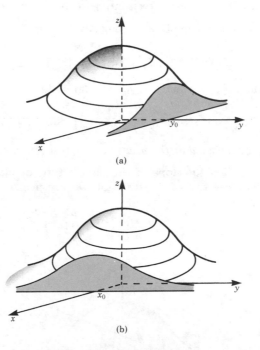

(a)

(b)

Figure 12.13

We can also slice the surface with planes parallel to the y-z plane (figure 12.13b). If the intersecting plane is at a distance x_0 from the y-z plane, the equation of the resulting curve is obtained by substituting $x = x_0$ in equation 1, i.e. $z(y) = \dfrac{1}{1 + x_0^2 + y^2}$. z is now a function of y only.

Let us now consider an intersecting curve of the first type (where y is constant). As it is a function of one variable, $z = f(x)$, we can calculate the slope α at any given point (figure 12.14).

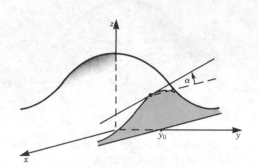

Figure 12.14

In order to distinguish between the ordinary derivative and this new derivative, we use the symbol ∂ instead of d. It indicates that we are differentiating a function of more than one variable with respect to a particular variable only, regarding all other variables as constant. Thus, when differentiating the function $z = f(x, y) = \dfrac{1}{1 + x^2 + y^2}$ where y is kept constant at some fixed value y_0, we obtain

$$\frac{\partial z}{\partial x} = \frac{\partial}{\partial x} f(x, y_0) = \frac{\partial}{\partial x} \left(\frac{1}{1 + x^2 + y_0^2} \right) = -\frac{2x}{(1 + x^2 + y_0^2)^2}$$

Since this holds true for any value $y = y_0$, we can write

$$\frac{\partial z}{\partial x} = \frac{\partial}{\partial x} f(x, y) = \frac{\partial}{\partial x} \left(\frac{1}{1 + x^2 + y^2} \right) = -\frac{2x}{(1 + x^2 + y^2)^2}$$

This operation is called *partial differentiation* with respect to x.

Similarly, we may obtain the slope of the second type of intersecting curves. It is given by the partial derivative of the function with respect to y. x is kept constant at some fixed value (figure 12.15).

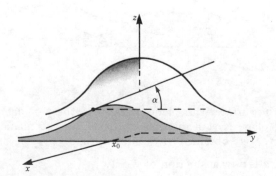

Figure 12.15

Thus we obtain

$$\frac{\partial z}{\partial y} = \frac{\partial}{\partial y} f(x, y) = \frac{\partial}{\partial y} \left(\frac{1}{1 + x^2 + y^2} \right) = -\frac{2y}{(1 + x^2 + y^2)^2}$$

Functions of three variables, $f = f(x, y, z)$, are treated similarly, but it is not possible to present them in geometrical form. There exist, of course, three partial derivatives. Table 12.2 contains a summary of rules and an example of each one.

TABLE 12.2

Partial derivative	Rule	Example: $f(x, y, z) = 2x^3y + z^2$	
Partial derivative with respect to x: $\dfrac{\partial}{\partial x}$	Treat all variables as constants except for x	$\dfrac{\partial f}{\partial x} = 6x^2y$	(12.1a)
Partial derivative with respect to y: $\dfrac{\partial}{\partial y}$	Treat all variables as constants except for y	$\dfrac{\partial f}{\partial y} = 2x^3$	(12.1b)
Partial derivative with respect to z: $\dfrac{\partial}{\partial z}$	Treat all variables as constants except for z	$\dfrac{\partial f}{\partial z} = 2z$	(12.1c)

The partial derivative may be written in another way. Let $f(x, y, z)$ be a function of the three variables x, y and z, then the partial derivatives may be abbreviated as follows:

$$\frac{\partial f}{\partial x} = f_x, \qquad \frac{\partial f}{\partial y} = f_y, \qquad \frac{\partial f}{\partial z} = f_z \qquad (12.2)$$

EXAMPLE $f(x, y, z) = xyz$

then

$$f_x = \frac{\partial f}{\partial x} = yz$$

$$f_y = \frac{\partial f}{\partial y} = xz$$

and

$$f_z = \frac{\partial f}{\partial z} = xy$$

EXAMPLE   Obtain the partial derivatives of the function

$$z = 5x^2 + 2xy - y^2 + 3x - 2y + 3 \text{ at } x = 1, \ y = -2$$

The partial derivatives are

$$\frac{\partial z}{\partial x} = 10x + 2y + 3$$

$$\frac{\partial z}{\partial y} = 2x - 2y - 2$$

Substituting the values $x = 1, \ y = -2$ gives

$$\left(\frac{\partial z}{\partial x}\right)_{\substack{x=1 \\ y=-2}} = 10 \times 1 + 2(-2) + 3 = 9$$

$$\left(\frac{\partial z}{\partial y}\right)_{\substack{x=1 \\ y=-2}} = 2 \times 1 - 2(-2) - 2 = 4$$

Note that we have introduced the expression $\left(\dfrac{\partial z}{\partial x}\right)_{\substack{x=1 \\ y=-2}}$. It means that the partial derivative with respect to x of the function z is to be evaluated at $x = 1$ and $y = -2$.

12.3.1   HIGHER PARTIAL DERIVATIVES

The partial derivatives are themselves functions of the independent variables $x, y, \ldots$, in general. We can, therefore, differentiate them partially again.

EXAMPLE   Let $f(x, y, z) = \dfrac{x}{y} + 2z$.

Evaluate $\dfrac{\partial}{\partial x}\left(\dfrac{\partial f}{\partial y}\right)$ and $\dfrac{\partial}{\partial y}\left(\dfrac{\partial f}{\partial x}\right)$.

The first expression means that we differentiate the function f first with respect to y and then with respect to x.

$$\frac{\partial}{\partial x}\left(\frac{\partial f}{\partial y}\right) = \frac{\partial^2 f}{\partial x \, \partial y}$$

$$= \frac{\partial}{\partial x} f_y = f_{yx}$$

Therefore $$f_{yx} = -\frac{1}{y^2}$$

Similarly

$$\frac{\partial}{\partial y}\left(\frac{\partial f}{\partial x}\right) = \frac{\partial^2 f}{\partial y\, \partial x} = \frac{\partial}{\partial y} f_x = f_{xy}$$

Therefore

$$f_{xy} = -\frac{1}{y^2}$$

The order of partial differentiation is immaterial.

For most functions encountered by engineers and scientists the following holds true:

$$f_{xy} = f_{yx}, \quad \text{etc.} \tag{12.3}$$

EXAMPLE   For the function $u = \dfrac{x^2}{y} \sin z$ show that the following mixed third derivatives are equal:

$$u_{xyz} = u_{zyx}$$

$$u_x = \frac{2x}{y} \sin z \qquad\qquad u_z = \frac{x^2}{y} \cos z$$

$$u_{xy} = -\frac{2x}{y^2} \sin z \qquad\qquad u_{zy} = -\frac{x^2}{y^2} \cos z$$

$$u_{xyz} = -\frac{2x}{y^2} \cos z \qquad\qquad u_{zyx} = -\frac{2x}{y^2} \cos z$$

12.4   TOTAL DIFFERENTIAL

12.4.1   TOTAL DIFFERENTIAL OF FUNCTIONS

A function $z = f(x, y)$ represents a surface in space; on this surface there are lines at the same level, $z = $ constant. If we drop perpendiculars from these lines on to the x-y plane we obtain their projections on this plane. These projections are called *contour lines*. They are extensively used in geographical maps.

Algebraically, contour lines are obtained by setting the function $z = f(x, y) = C$ (where C is a constant). When $f(x, y) = \dfrac{1}{1 + x^2 + y^2}$, we have

$$\frac{1}{1 + x^2 + y^2} = C$$

This is an implicit representation of a curve in the x-y plane.

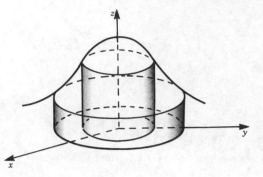

Figure 12.16

In this case we obtain contour lines which are circular, as shown in figure 12.16. This can be proven as follows. Rearranging gives

$$x^2 + y^2 = \frac{1}{C} - 1$$

Remember that the equation of a circle of radius R in the x-y plane is $x^2 + y^2 = R^2$. Hence, in this case, $R = \sqrt{\frac{1}{C} - 1}$. The larger we choose C the smaller is the radius of the circle. (But, of course, C must not exceed 1, and it must be positive.)

Following these preliminary remarks we are now in a position to look for the direction of steepest rise or decrease of the surface at a given point:

$$z = \frac{1}{1 + x^2 + y^2}$$

It is clear from figure 12.17 that the surface decreases most steeply in a radial direction. (Note that, for the sake of clarity, the surface has been redrawn to a different scale.)

Let us look more closely at this figure.

If we travel the same distance d**r** from point A′ in the x-y plane:

(a) in an arbitrary direction d**r**,
(b) perpendicularly to a contour line d**r**$_2$,
(c) along a contour line d**r**$_3$,

then, on the given surface, this corresponds to the paths $\overrightarrow{AC}$, $\overrightarrow{AB}$, $\overrightarrow{AD}$ respectively.

The path $\overrightarrow{AD}$ is along a contour line. Hence d$z_3 = 0$ and the function does not change at all.

In contrast, the function z changes most rapidly along the path $\overrightarrow{AB}$ which is in a direction perpendicular to the contour lines.

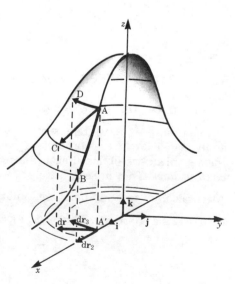

Figure 12.17

We are interested in finding how much the function $z = f(x, y)$ changes when we travel a distance dr in an arbitrary direction $\mathbf{dr} = (dx, dy)$. The total displacement, in vector notation, is

$$\mathbf{dr} = dx\,\mathbf{i} + dy\,\mathbf{j}$$

In Chapter 5 we saw that for functions of one variable the differential is an approximation for the change of the function for a given Δx, i.e. $\Delta y \approx \dfrac{df}{dx}\,\Delta x$. In the same way the total differential is an approximation for the change in the function for small changes in x and y. The change of $f(x, y)$ is obtained in two steps:
(1) by proceeding in the x-direction through a distance dx, with y remaining constant;
(2) by proceeding in the y-direction through a distance dy, with x remaining constant.

Let us be more explicit.

Step 1 For the change in z in the x-direction, with y remaining constant, we have:

$$dz_{(x)} = \frac{\partial}{\partial x} f(x, y)\,dx$$

Step 2 For the change in z in the y-direction, with x remaining constant, we have:

$$dz_{(y)} = \frac{\partial}{\partial y} f(x, y)\,dy$$

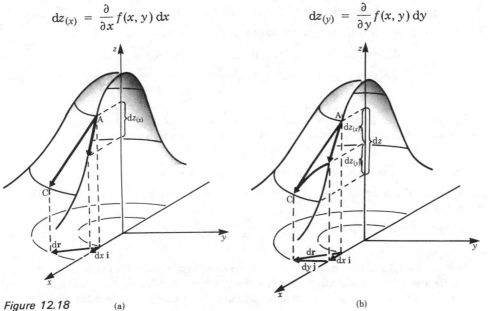

Figure 12.18   (a)   (b)

The total change in z is the sum of these partial changes. Thus

$$dz = dz_{(x)} + dz_{(y)} = \frac{\partial z}{\partial x}\,dx + \frac{\partial z}{\partial y}\,dy$$

DEFINITION   The *total differential* of a function $z = f(x, y)$ is given by

$$dz = \frac{\partial f}{\partial x}\,dx + \frac{\partial f}{\partial y}\,dy \qquad (12.4)$$

The total differential is an approximation of the true change Δz in the function z as we proceed from a point (x, y) a short distance in the direction $\mathbf{dr} = (dx, dy)$

$$\Delta z \approx dz$$

EXAMPLE   The function $z = x^2 + y^2$ has a total differential

$$dz = 2x \, dx + 2y \, dy$$

EXAMPLE   The total differential of the function $f(x, y) = \dfrac{1}{1 + x^2 + y^2}$ is

$$df(x, y) = -\frac{2x}{(1 + x^2 + y^2)^2} \, dx - \frac{2y}{(1 + x^2 + y^2)^2} \, dy$$

EXAMPLE   Calculate the values of Δz (the true change in z) and dz (the approximate change in z) if $z = 5x^2 + 3y$ at the point $(2, 3)$ with $dx = 0.1$ and $dy = 0.05$.

The true change in z is given by

$$\Delta z = f(x + \Delta x, y + \Delta y) - f(x, y)$$

Hence with $\Delta x = dx = 0.1$, $\Delta y = dy = 0.05$ we have

$$\Delta z = 5(2.1)^2 + 3(3.05) - (5 \times 2^2 + 3 \times 3) = 2.20$$

The approximate change is given by the total differential

$$dz = 10x \, dx + 3 \, dy = 10 \times 2(0.1) + 3(0.05) = 2.15$$

The difference between the true value Δz and the total differential dz is small. Thus

$$\Delta z \approx dz$$

In practice, if dx and dy are small then the approximation $\Delta z \approx dz$ is acceptable and commonly used in many of the problems engineers and scientists encounter.

Extension to Functions of Three Independent Variables $f(x,y,z)$

In the case of a function $f(x, y, z)$ of three independent variables the total differential is given by

$$df = \frac{\partial f}{\partial x} \, dx + \frac{\partial f}{\partial y} \, dy + \frac{\partial f}{\partial z} \, dz \tag{12.5}$$

As before, the total differential is a measure of the change in the function $u = f(x, y, z)$. If we proceed by a small displacement along $d\mathbf{r} = (dx, dy, dz)$, as shown in figure 12.19, the function changes by an amount equal to the total differential.

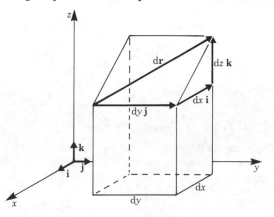

Figure 12.19

EXAMPLE   The volume of a parallelepiped is given by $V = f(x, y, z) = xyz$, where x, y and z are the lengths of the three sides. The total differential $\mathrm{d}V$ is

$$\mathrm{d}V = yz\,\mathrm{d}x + xz\,\mathrm{d}y + xy\,\mathrm{d}z$$

12.4.2 APPLICATION: SMALL TOLERANCES

We know for a function of a single variable $y = f(x)$ that if x is subject to an increment or decrement Δx, then the change in y is approximately given by $\Delta y \approx f'(x)\,\Delta x$.

In the preceeding section this has been generalised to several variables. For the sake of concreteness, let us concern ourselves with the tolerances of finished products due to the tolerances of their components. Thus the increment or decrement is determined by the tolerances δ of these components. (In this section we will use the symbol δ instead of Δ.)

If we have a function of several variables, such as $u = u(x, y, z)$, then the total tolerance δu due to individual tolerances δx, δy, δz is

$$\delta u \approx \frac{\partial u}{\partial x}\delta x + \frac{\partial u}{\partial y}\delta y + \frac{\partial u}{\partial z}\delta z$$

It is assumed that the tolerances δx, δy and δz are small. In practice, this is the case in most situations.

Since u is a linear function in δx, δy and δz, it follows that the total tolerance is obtained by adding the effects due to each one separately.

For example, consider a dimension of a link in a mechanism. It would be specified by its length x and a manufacturing tolerance imposed on it of $\pm \delta x$. When the part has been made and a check on its dimension is carried out we would hope to find that its length will lie in the interval $x - \delta x \leqslant x \leqslant x + \delta x$ as a result of the manufacturing process.

If a device consists of a number of parts, it will be affected by the tolerances imposed on those parts. If u, the output, is a function of n parts of lengths x_i, $i = 1, 2, \ldots, n$, and tolerances δx_i, then the tolerance in the output will be

$$\delta u \approx \sum_{i=1}^{n} \frac{\partial u}{\partial x_i}\delta x_i$$

The individual tolerance δx_i may have either sign and usually it does not attain its maximum value. But, if we assume the worst, we must add the effects of all maximum individual tolerances in order to obtain the maximum possible tolerance.

The maximum tolerance is then given, approximately, by

$$\delta u \approx \pm \sum_{i=1}^{n} \left| \frac{\partial u}{\partial x_i}\delta x_i \right|$$

EXAMPLE   Figure 12.20 shows diagramatically a link pivoted at O and connected to another link at A (not shown). The position of the link at A (its output) will have an influence on the position of the link to which it is connected. This position will then depend on the tolerances $\pm \delta l$ on its length and $\pm \delta\theta$ on its angle relative to some datum, i.e. the x-axis in this instance. If $l = 95.00\,\mathrm{mm}$ and $\delta l = \pm 0.10\,\mathrm{mm}$, $\theta = 35.00°$ and $\delta\theta = \pm 0.25°$, calculate the maximum tolerance in y and compare the result with its true maximum value.

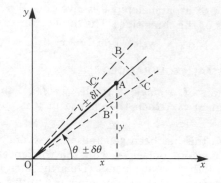

Figure 12.20

As a result of the manufacturing process, we know that $l - \delta l \leqslant l \leqslant l + \delta l$ and $\theta - \delta\theta \leqslant \theta \leqslant \theta + \delta\theta$. It follows, therefore, that A will lie somewhere inside the boundaries BCB′C′.

Now let us calculate this maximum tolerance using the total differential approach.

$$y = f(l, \theta) = l \sin \theta$$

Hence

$$\delta y = \frac{\partial f}{\partial l} \delta l + \frac{\partial f}{\partial \theta} \delta\theta$$

$$\frac{\partial f}{\partial l} = \sin\theta, \qquad \frac{\partial f}{\partial \theta} = l \cos\theta$$

Substituting in the above equation gives

$$\delta y = \sin\theta \, (\pm\delta l) + l \cos\theta \, (\pm\delta\theta)$$

$$= \sin 35° \, (\pm 0.1) + 95 \cos 35° \left(\pm \frac{0.25}{57.3} \right)$$

$$= \pm \left((\sin 35°) \, 0.1 + (95 \cos 35°) \, \frac{0.25}{57.3} \right)$$

$$= \pm 0.3969 \approx \pm 0.40 \text{ mm} \quad \text{to 2 d.p.}$$

(Note that the factor $\dfrac{1}{57.3} = \dfrac{2\pi}{360}$ is necessary to convert the $\delta\theta$ value to radians.)

Now let us calculate the true maximum and minimum value of y. Considering the y position, its maximum value will correspond to B and its minimum value to B′. It is easy to calculate these values:

$$y_{\max} = (l + \delta l) \sin(\theta + \delta\theta)$$

$$= 95.1 \sin 35.25 = 54.8865 = 54.89 \text{ mm} \quad \text{to 2 d.p.}$$

and
$$y_{\min} = 94.9 \sin 34.75 = 54.0927 = 54.09 \text{ mm} \quad \text{to 2 d.p.}$$

The exact or nominal value of y, ignoring tolerances, is

$$y = l \sin\theta = 95 \sin 35 = 54.4898 = 54.49 \text{ mm}$$

Hence the maximum tolerances in y are

$$y_{max} - y = \delta y = 54.89 - 54.49 = 0.4 \, \text{mm}$$

and $$y_{min} - y = \delta y = 54.49 - 54.09 = 0.4 \, \text{mm}$$

i.e. $$\delta y = \pm 0.4 \, \text{mm}$$

Both methods give the same result.

This was a very simple case in which we could visualise easily where the output position of the links was likely to be, but in practice the problems encountered are much more involved and visualisation can be almost impossible.

An example of this can be found in precision mechanisms where the output depends on the accuracy of a number of links, cams and gears. In such cases, the influence of tolerances can only be calculated using the total differential approach.

12.4.3 GRADIENT

In section 12.4.1 the total differential of a function $z = f(x, y)$ was defined as

$$dz = \frac{\partial f}{\partial x} dx + \frac{\partial f}{\partial y} dy$$

It is possible to regard the total differential as a scalar product of two vectors:

first vector: $\mathbf{dr} = dx \, \mathbf{i} + dy \, \mathbf{j}$ (called the *path element*)

second vector: $\operatorname{grad} f = \dfrac{\partial f}{\partial x} \mathbf{i} + \dfrac{\partial f}{\partial y} \mathbf{j}$ (called the *gradient of f*)

It is easy to verify by inserting these vectors

$$dz = \mathbf{dr} \; \operatorname{grad} f$$

> **DEFINITION**   The vector $\left(\dfrac{\partial f}{\partial x}, \dfrac{\partial f}{\partial y} \right)$ is called the gradient of a function $z = f(x, y)$.
>
> $$\operatorname{grad} f(x, y) = \left(\frac{\partial f}{\partial x}, \frac{\partial f}{\partial y} \right) \tag{12.6}$$

The gradient has two properties:

(1) The gradient is a vector normal to the contour lines. Thus it points in the direction of the greatest change in z.

(2) The absolute value of the gradient is proportional to the change in z per unit of length in its direction.

In order to explain these properties, we will consider the scalar product

$$\operatorname{grad} f \; \mathbf{dr} = dz$$

If $\mathbf{dr}$ coincides with a contour line we obtain $dz = 0$, since a contour line is the projection of a line of constant z-value on to the x–y plane. Thus, in this case

$$\operatorname{grad} f \, \mathbf{dr} = 0$$

We know from Chapter 4 that the scalar product of two non-zero vectors vanishes if and only if they are perpendicular to each other. Thus it follows that **grad** f is perpendicular to the contour lines, i.e. **grad** f is a vector normal to the contour lines.

Let us illustrate this with the example used earlier:

$$f(x, y) = \frac{1}{1 + x^2 + y^2}$$

Remember that the contour lines are circles.

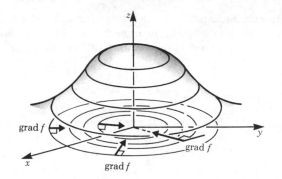

Figure 12.21

The gradient of the given function is

$$\mathbf{grad}\, f = \left(\frac{-2x}{(1 + x^2 + y^2)^2}, \; \frac{-2y}{(1 + x^2 + y^2)^2} \right)$$

This vector always points in the direction of the origin of the coordinate system. Thus it is perpendicular to circles around the origin (figure 12.21).

The absolute value of the gradient is a measure of the change of z. This follows directly from the equation

$$dz = \mathbf{dr}\; \mathbf{grad}\, f$$

For the particular case when **dr** is normal to a contour line, the equation can be rearranged:

$$\frac{dz}{|\mathbf{dr}|} = |\mathbf{grad}\, f|$$

Thus the change in dz per unit of length in the direction of the gradient is given by the absolute value of the gradient.

The concept of a gradient can be extended to functions of more than two variables. In the case of a function of three independent variables, $f(x, y, z)$, constant function values are represented by surfaces in three-dimensional space. The gradient is normal to these surfaces:

$$\mathbf{grad}\, f(x, y, z) = \left(\frac{\partial f}{\partial x}, \frac{\partial f}{\partial y}, \frac{\partial f}{\partial z} \right)$$

Its magnitude gives the change in the value of the function in the direction of the gradient.

12.5 **TOTAL DERIVATIVE**

12.5.1 **EXPLICIT FUNCTIONS**

Up to now we have assumed that x and y are independent variables. It may happen that x and y are both functions of one independent variable t. If this is the case, $z = f(x, y)$ is, in fact, a function of the single independent variable t. z will, therefore, have a derivative with respect to t.

Let $x = g(t)$ and $y = h(t)$; it is assumed that the functions can be differentiated. If t is given a small increment Δt, then x, y and z will have corresponding increments Δx, Δy and Δz. The change or increment in the function z is given by

$$\Delta z \approx \frac{\partial f}{\partial x} \Delta x + \frac{\partial f}{\partial y} \Delta y$$

Dividing by Δt we have

$$\frac{\Delta z}{\Delta t} \approx \frac{\partial f}{\partial x} \frac{\Delta x}{\Delta t} + \frac{\partial f}{\partial y} \frac{\Delta y}{\Delta t}$$

We can now proceed to the limit $\Delta t \to 0$. The result is stated below.

$\dfrac{dz}{dt}$ is called the *total derivative* or *total differential coefficient*.

$$\frac{dz}{dt} = \frac{\partial f}{\partial x} \frac{dx}{dt} + \frac{\partial f}{\partial y} \frac{dy}{dt}$$

Similarly if $u = u(x, y, z)$ where x, y and z are functions of t, we obtain the total derivative of u with respect to t:

$$\frac{du}{dt} = \frac{\partial u}{\partial x} \frac{dx}{dt} + \frac{\partial u}{\partial y} \frac{dy}{dt} + \frac{\partial u}{\partial z} \frac{dz}{dt} \qquad (12.7)$$

Note that we write $\dfrac{dz}{dt}$, not $\dfrac{\partial z}{\partial t}$, since z is a function of a single variable.

This concept can obviously be generalised for any number of variables.

EXAMPLE   Let $z = f(x, y) = x + y$ and $x = \dfrac{e^t}{2}$, $y = \dfrac{e^{-t}}{2}$

Obtain the derivative $\dfrac{dz}{dt}$.

Firstly we need the partial derivatives:

$$\frac{\partial z}{\partial x} = 1, \qquad \frac{\partial z}{\partial y} = 1$$

Secondly we need the derivatives of x and y with respect to t:

$$\frac{dx}{dt} = \frac{e^t}{2} \qquad \frac{dy}{dt} = -\frac{e^{-t}}{2}$$

Hence the total derivative is

$$\frac{dz}{dt} = \frac{e^t}{2} - \frac{e^{-t}}{2}$$

Note that z is familiar: $z(t) = \cosh t$, and the derivative with respect to t is $\frac{dz}{dt} = \sinh t$

12.5.2 IMPLICIT FUNCTIONS

Remember that the function $f(x, y) = C$ is an implicit function y of one single variable x in the x-y plane (contour line). We wish to obtain the derivative $\frac{d}{d}$ without explicitly solving for y. In Chapter 5, section 5.9.1, we showed how this could be done. Using the concept of the total derivative, we will now obtain general expression.

The geometrical meaning of a total derivative of the function $z = f(x, y)$ is that gives the total change in z.

The total derivative with respect to x is

$$\frac{dz}{dx} = \frac{\partial f}{\partial x}\frac{dx}{dx} + \frac{\partial f}{\partial y}\frac{dy}{dx}$$

But since z is constant for a contour line (i.e. $dz = 0$), and $\frac{dx}{dx} = 1$, we obtain

$$0 = \frac{\partial f}{\partial x} + \frac{\partial f}{\partial y}\frac{dy}{dx}$$

Solving for $\frac{dy}{dx}$ gives

$$\frac{dy}{dx} = -\frac{\partial f/\partial x}{\partial f/\partial y}$$ (12.

EXAMPLE   If $x^3 - y^3 + 4xy = 0$,

calculate the value of the derivative $\frac{dy}{dx}$ at $x = 2$, $y = -2$.

Let $f = x^3 - y^3 + 4xy$

Then $\frac{\partial f}{\partial x} = 3x^2 + 4y$ and $\frac{\partial f}{\partial y} = -3y^2 + 4x$

Therefore $\frac{dy}{dx} = -\frac{3x^2 + 4y}{-3y^2 + 4x}$

Hence the value of the derivative at $x = 2, y = -2$ is

$$\frac{dy}{dx} = -\frac{3 \times 2^2 + 4(-2)}{-3(-2)^2 + 4 \times 2} = -\frac{4}{-4} = 1$$

The equation $f(x, y, z) = 0$ can be considered as defining z as an implicit function of two variables x and y. We are interested in finding expressions for the part: derivatives of z, i.e. $\frac{\partial z}{\partial x}$ and $\frac{\partial z}{\partial y}$.

We first concentrate our attention on $\dfrac{\partial z}{\partial x}$.

We know that

$$0 = f(x, y, z)$$

The total derivative of this expression with respect to x gives

$$0 = \frac{\partial f}{\partial x} + \frac{\partial f}{\partial y}\frac{\partial y}{\partial x} + \frac{\partial f}{\partial z}\frac{\partial z}{\partial x}$$

Remember that forming the partial derivative with respect to x implies regarding y as constant. Thus $\dfrac{\partial y}{\partial x} = 0$, and solving for $\dfrac{\partial z}{\partial x}$ gives

$$\frac{\partial z}{\partial x} = -\frac{\partial f/\partial x}{\partial f/\partial z}$$

Similarly, we obtain

$$\frac{\partial z}{\partial y} = -\frac{\partial f/\partial y}{\partial f/\partial z}$$

EXAMPLE   Given $\dfrac{x^2}{25} + \dfrac{y^2}{15} + \dfrac{z^2}{9} = 1$, calculate the partial derivatives of z.

$$\frac{\partial f}{\partial x} = \frac{2x}{25}, \qquad \frac{\partial f}{\partial y} = \frac{2y}{15}, \qquad \frac{\partial f}{\partial z} = \frac{2z}{9}$$

Substituting in the equations for the partial derivatives we find

$$\frac{\partial z}{\partial x} = -\frac{2x/25}{2z/9} = -\frac{9x}{25z}, \qquad \frac{\partial z}{\partial y} = -\frac{2y/15}{2z/9} = -\frac{9y}{15z}$$

12.6 MAXIMA AND MINIMA OF FUNCTIONS OF TWO OR MORE VARIABLES

In Chapter 5 we derived the necessary conditions for a function of a single variable to have a maximum or a minimum. We now consider the conditions for maximum and minimum values in the case of functions of several independent variables.

A function of two variables, $z = f(x, y)$ (see figure 12.22), is said to have a maximum at the point (x_0, y_0) if

$$\Delta f = f(x_0 + h, y_0 + k) - f(x_0, y_0) < 0$$

for all sufficiently small values of h and k, positive or negative.

The function will have a minimum if

$$\Delta f = f(x_0 + h, y_0 + k) - f(x_0, y_0) > 0$$

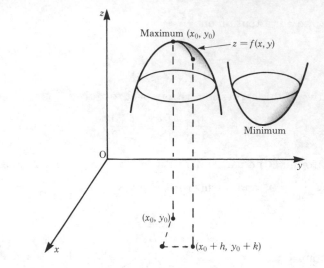

Figure 12.22

From a geometrical standpoint, this means that when the point (x_0, y_0, z_0) on the surface $z = f(x, y)$ is higher than any other point in its neighbourhood then (x_0, y_0, z_0) is a maximum, but if the point is lower than any other neighbouring point on the surface it is a minimum.

At a maximum or a minimum, the tangent plane to the surface is parallel to the x–y plane. This condition will be satisfied if

$$f_x(x_0, y_0) = 0 \qquad \text{and} \qquad f_y(x_0, y_0) = 0 \qquad (12.9)$$

The condition is necessary but not sufficient, as the following considerations show.

Consider, for example, two adjacent hills (figure 12.23). If we go from the top of one hill to the other one (path 1) there is a minimum. If we go through the passage between the hills (path 2) there is a maximum. Even though the condition $f_x = f_y = 0$ is satisfied at P, P is a saddle point and not an extreme point.

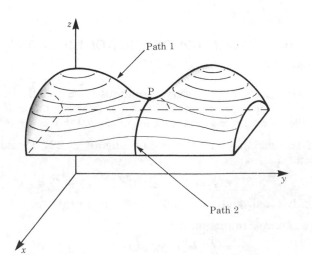

Figure 12.23

We will now state a sufficient condition. The following comments do not constitute a complete mathematical proof: they are only included as a hint for diligent readers. The expansion of the function with respect to h and k is

$$\Delta f = f(x_0+h, y_0+k) - f(x_0, y_0)$$

$$= (hf_x + kf_y)_{x_0 y_0} + \frac{1}{2!}(h^2 f_{xx} + 2hk f_{xy} + k^2 f_{yy})_{x_0 y_0} + \ldots$$

The first term of the expansion is zero at a maximum or a minimum, since $f_x = f_y = 0$. For a maximum the expression must be negative, independently of h and k. For a minimum it must be positive, independently of h and k. Thus the sign of the second term determines whether there is a maximum, a minimum or a saddle point.

Taking k^2 outside the brackets we have

$$\Delta f = \frac{k^2}{2!}\left[\left(\frac{h}{k}\right)^2 f_{xx} + 2\frac{h}{k}f_{xy} + f_{yy}\right]_{x_0 y_0} + \ldots$$

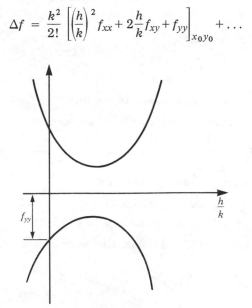

Figure 12.24

The expression in the brackets is a quadratic function in $\frac{h}{k}$.

From Chapter 1, section 1.4, we know that the graph of a quadratic function — a parabola — lies entirely above or below the horizontal axis only if this function has no real roots (figure 12.24). This means that the radicand must be negative in this case. Therefore, the supplementary criterion for the extreme value we were seeking is

$$(f_{xy})^2 - f_{xx}f_{yy} < 0$$

Furthermore, there will be a maximum if

$$f_{xx}(x_0, y_0) < 0 \quad \text{and} \quad f_{yy}(x_0, y_0) < 0 \qquad (12.10a)$$

and there will be a minimum if

$$f_{xx}(x_0, y_0) > 0 \quad \text{and} \quad f_{yy}(x_0, y_0) > 0 \qquad (12.10b)$$

This can be memorised by recalling the corresponding criteria for a function of one variable.

There remains one other case to consider:

$$[f_{xy}(x_0, y_0)]^2 - f_{xx}(x_0, y_0) f_{yy}(x_0, y_0) = 0$$

However, we shall not enlarge on this case; generally speaking, one needs further information to decide whether or not the point (x_0, y_0) corresponds to an extreme value.

Let us recapitulate the results obtained.

The necessary condition for the existence of an extreme value at $x_0 y_0$ is

$$f_x(x_0, y_0) = f_y(x_0, y_0) = 0$$

This condition becomes sufficient if it is supplemented by the following condition:

$$[f_{xy}(x_0, y_0)]^2 - f_{xx}(x_0, y_0) f_{yy}(x_0, y_0) < 0$$

Maximum if $f_{xx} < 0$ and $f_{yy} < 0$.

Minimum if $f_{xx} > 0$ and $f_{yy} > 0$.

EXAMPLE   Calculate the extreme values of the function $f = 6xy - x^3 - y^3$.

(1) Apply the necessary condition for a horizontal tangent plane.

$$f_x = 6y - 3x^2 = 0, \qquad f_y = 6x - 3y^2 = 0$$

Solving the two equations in x and y, we find that

(a) $x = 0$, $y = 0$ is one solution, i.e. the point $(0, 0)$

(b) $x = 2$, $y = 2$ is the other solution, i.e. the point $(2, 2)$.

(2) Apply the sufficient condition for a maximum or a minimum.

$$f_{xx} = -6x, \qquad f_{xy} = 6, \quad f_{yy} = -6y$$

The condition is

$$f_{x^2 y} - f_{xx} f_{yy} = 36 - 36xy < 0$$

We must check whether this condition is satisfied at the two points under consideration.

Inserting $x = 0$ and $y = 0$ gives $36 - 0 > 0$

Therefore there is no extreme value at the point $(0, 0)$.

Inserting $x = 2$ and $y = 2$ gives

$$36 - 36 \times 2 \times 2 < 0$$

In this case the sufficient condition is fulfilled: there is a minimum or a maximum at the point $(2, 2)$.

(3) Check the type of the extreme value.

Since $f_{xx} = -6x = -12$ and $f_{yy} = -6y = -12$ there is a maximum at the point $(2, 2)$. Its value is $f_{max} = 8$.

EXAMPLE   A container of $10 \, \text{m}^3$ capacity with an open top is to be made from thin sheet metal. Calculate what the dimensions of the sides and the height must be for a minimum amount of metal to be used. What is the saving compared with a container of equal sides?

If V is the volume, A the surface area of metal and x, y and z the lengths of the container, we have

$$V = xyz, \qquad A = 2xz + 2zy + xy \qquad \text{(where } z \text{ is the height)}$$

We need to eliminate one of the variables, z say; thus $z = \dfrac{V}{xy}$. Substituting in the equation for the surface area A, we obtain

$$A = \frac{2V}{y} + \frac{2V}{x} + xy$$

i.e. $A = f(x, y)$ is a function of two independent variables.

(1) Apply the necessary condition for an extreme value; calculate the partial derivatives and set them equal to zero:

$$\frac{\partial A}{\partial y} = -\frac{2V}{y^2} + x = 0, \qquad \frac{\partial A}{\partial x} = -\frac{2V}{x^2} + y = 0$$

Solving the two equations gives

$$x = y = \sqrt[3]{2V} = \sqrt[3]{20} = 2.714 \, \text{m}$$

Obviously, this makes sense only for a minimum of the area. But let us use the formal procedure to verify this common-sense judgement.

(2) Apply the sufficient condition for an extreme value:

$$\frac{\partial^2 A}{\partial y^2} = \frac{4V}{y^3}, \qquad \frac{\partial^2 A}{\partial x \partial y} = 1, \qquad \frac{\partial^2 A}{\partial x^2} = \frac{4V}{x^3}$$

With $x = y = \sqrt[3]{2V}$ it follows that

$$(A_{xy})^2 - A_{xx}A_{yy} = -3 < 0$$

(3) A_{xx} and A_{yy} are positive; hence we have a minimum.

The numerical result is given by

$$z = \frac{V}{xy} = \sqrt[3]{\frac{V}{4}} = \sqrt[3]{2.5} \approx 1.357 \, \text{m}$$

The amount of metal required $= 4.2 \times 714 \times 1.357 + 2.714^2$. Hence $A = 16.58 \, \text{m}^2$. If we made $x = y = z = a$, say, then $a = \sqrt[3]{10} = 2.15 \, \text{m}$, and the amount of metal would be $A_1 = 5a^2 = 5.2 \times 15^2 = 23.11 \, \text{m}^2$. The saving is $A_1 - A = 6.53 \, \text{m}^2$.

EXAMPLE   A trough, $12 \, \text{m}$ long, is to be made out of a steel sheet $1.65 \, \text{m}$ wide by bending it into the shape ABCD, as shown in figure 12.25. Calculate the lengths x of the sides and the angle θ if the sectional area is to be a maximum.

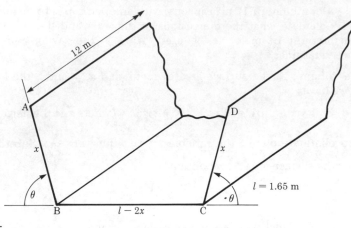

Figure 12.25

We first observe that the length of the trough is not relevant, as we are concerned only with the cross section. The relevant variables are the length x and the angle θ. Let us denote the cross section by A:

$$A = \tfrac{1}{2}(AD + BC) \times (\text{vertical depth}) = \tfrac{1}{2}(l - 2x + 2x \cos \theta + l - 2x)x \sin \theta$$
$$= lx \sin \theta - 2x^2 \sin \theta + x^2 \sin \theta \, \cos \theta$$

$A = f(x, \theta)$ is a function of two independent variables.

(1) Calculate the partial derivatives and equate them to zero:

$$\frac{\partial A}{\partial x} = l \sin \theta - 4x \sin \theta + 2x \sin \theta \, \cos \theta = 0$$

$$\frac{\partial A}{\partial \theta} = lx \cos \theta - 2x^2 \cos \theta + x^2(\cos^2 \theta - \sin^2 \theta) = 0$$

Solve the two equations in x and θ:

(a) By inspection, we find a trivial solution $x = 0$, $\sin \theta = 0$; but this has no physical meaning.

(b) Assuming $x \neq 0$ and $\sin \theta \neq 0$, we divide the first equation by $\sin \theta$ and the second equation by x, so that

$$l = 4x - 2x \cos \theta$$
$$l \cos \theta = 2x \cos \theta - x(2 \cos^2 \theta - 1)$$

By combining these equations we obtain

$$\frac{l}{x} = 4 - 2 \cos \theta = 2 - \frac{2 \cos^2 \theta - 1}{\cos \theta}$$

We solve for $\cos \theta$ and x:

$2 \cos \theta = 1$, i.e. $\cos \theta = \dfrac{1}{2}$ or $\theta = \dfrac{\pi}{3}$ (60°), and $\dfrac{l}{x} = 3$, i.e. $x = \dfrac{l}{3} = 0.55 \, \text{m}$.

Steps 2 and 3, which are necessary to prove that this is in fact a maximum, are left to the reader.

12.7   APPLICATIONS: WAVE FUNCTION AND WAVE EQUATION

12.7.1   WAVE FUNCTION

Let us consider the function of two variables:

$$z = f(x, y) = \sin(x - y)$$

It can be represented by a surface in space.

To gain a proper understanding of this function, we first draw the intersecting curves with planes parallel to the x–z plane for the values

$$y = 0, \qquad y = \frac{\pi}{2}, \qquad y = \pi, \qquad y = \frac{3\pi}{2}, \qquad y = 2\pi \qquad \text{(see figure below)}$$

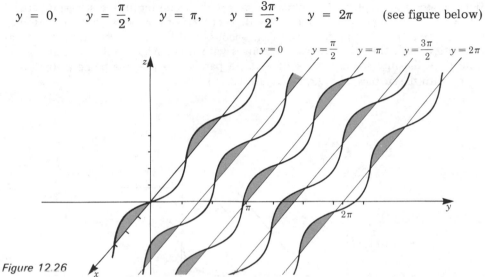

Figure 12.26

In the x-direction, the intersection curves are sine functions with a period of 2π.

We may see already that the surface is a series of parallel hills and valleys. The direction of the hills and valleys is an angle of $45°$ to the x- and the y-axes.

We now draw the intersecting curves with planes parallel to the y–z plane (see figure 12.27).

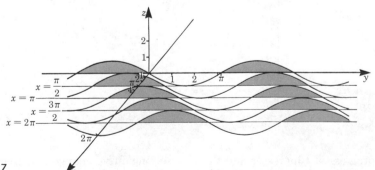

Figure 12.27

We obtain sine functions with the same period. Again we see that the surface is a series of parallel hills and valleys.

The maxima and minima lie on parallel lines making an angle $\dfrac{\pi}{4}$ with the x- and y-axes. These are given by

$$x - y \;=\; (2n + \tfrac{1}{2})\pi \qquad \text{for maxima}$$

$$x - y \;=\; (2n - \tfrac{1}{2})\pi \qquad \text{for minima, where } n \text{ is an integer}$$

Now let us examine the special case when one variable, say y, represents time. If we consider the value of time without its dimension, we have

$$f(x, t) \;=\; \sin(x - t)$$

If we sketch the graph of this function we get the same picture, a series of parallel hills and valleys. The only difference is that $f(x, t)$ of course is now a function of position and time. In practice, we often encounter this type of function $f(x, t)$ as a function of x, the value of which changes with time. What is observed at a given time t_0 is the intersecting curve of a plane parallel to the x–z plane at the point $t = t_0$ with the surface $z = f(x, t)$ (figure 12.28).

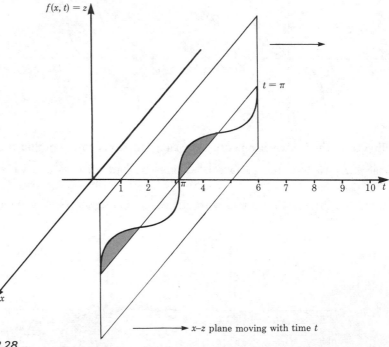

Figure 12.28

As time progresses, t increases and the intersecting curve changes. In this case, this results in a sine function travelling in the x-direction. We obtain a function which behaves like a wave on a cable. This type of function is called a *wave function*. The period in both the x-direction and the t-direction is 2π in our example.

The length of one period in the x-direction is called the *wavelength*. If we want to describe a wave function with an arbitrary wavelength λ, we must set

$$f(x, t) \;=\; \sin\left(2\pi\frac{x}{\lambda} - t\right)$$

The period, usually denoted by T, is the time interval of one oscillation at a given point x_0. The inverse value is called the *frequency* ν, i.e. the given function $\sin(x - t)$ has the frequency $\nu = \dfrac{1}{T}$.

If we wish to describe an arbitrary frequency $\nu = \dfrac{1}{T}$ we must write

$$f(x, t) \;=\; \sin\left(2\pi\frac{x}{\lambda} - \frac{2\pi}{T}t\right)$$

We have now obtained a general formula for the one-dimensional wave function. It must be noted that now we can reintroduce the dimensions of the physical quantities position and time. The argument of the sine function remains dimensionless.

The value of $2\pi\nu$ is also referred to as the *circular frequency* ω. From another aspect, it can be observed that during the time $T = \dfrac{1}{\nu} = \dfrac{2\pi}{\omega}$ the wave travels one wavelength λ.

The velocity v of the wave is called the *phase velocity*:

$$v \;=\; \frac{\lambda}{T}, \qquad \text{i.e.} \qquad v \;=\; \lambda\nu$$

Using the circular frequency, the phase velocity can be expressed

$$v \;=\; \frac{\omega}{2\pi}\lambda$$

Particular positions of the wave such as maxima, minima or zeros travel in the x-direction with this velocity.

Generally, the one-dimensional harmonic wave function is written in two equivalent forms:

$$f(x, t) \;=\; A \sin\left(\frac{2\pi}{\lambda}x - \omega t + \phi_0\right)$$

$$=\; A \sin\left(\frac{2\pi}{\lambda}(x - vt) + \phi_0\right)$$

The argument of the wave function is called the *phase*. ϕ_0 is the phase at $t = 0$ and $x = 0$. A is the amplitude. It may be the physical displacement of a point, an electrical quantity, an air pressure (sound waves), a distortion etc. A may be a scalar or A may be a vector, like a displacement $\mathbf{r}$ or an electrical field vector $\mathbf{E}$.

Figure 12.29 shows the usual graphical representation of the wave function. Figure 12.29a gives the wave for a fixed time t_0; figure 12.29b gives the oscillation of a fixed point x_0.

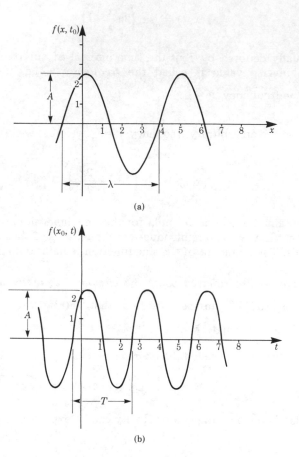

(a)

(b)

Figure 12.29

A harmonic wave propagating in the opposite direction is given by

$$f(x, t) = A \sin\left(\frac{2\pi}{\lambda}x + \omega t + \phi_0\right)$$

$$\text{or } f(x, t) = A \sin\left(\frac{2\pi}{\lambda}(x + vt) + \phi_0\right)$$

Spherical Waves

In physics we frequently encounter waves which travel in all directions from the origin of a point source. The wavefronts of such waves are concentric spheres with the source as centre. The separation of two adjacent wavefronts having the same phase is equal to one wavelength. Electromagnetic and acoustic waves are often spherical waves.

In the case of spherical waves, we have to take into account the fact that their amplitudes decrease with the distance from the source; for instance, this is the case for sound waves.

In the case of an acoustic wave, the amplitude of the air pressure p is given by the following function:

$$p = \left(\frac{p_0}{r}\right) \sin\left(\frac{2\pi}{\lambda} r - \omega t + \phi_0\right)$$

where p is the pressure difference compared with the air pressure of the air at rest and r is the distance from the centre of a harmonic sound source.

12.7.2 WAVE EQUATION

Wave functions are solutions of differential equations of the form

$$\frac{\partial^2 f(x, t)}{\partial x^2} = \frac{1}{c^2} \frac{\partial^2 f(x, t)}{\partial t^2}$$

The RHS is the second derivative with respect to time, and the LHS is the second derivative with respect to displacement.

Equations in which differential coefficients appear are called differential equations (Chapter 10). In the above case, since the differential coefficients are partial ones, the equation is referred to as a *partial differential equation*.

The equation is known as the one-dimensional *wave equation* with a velocity of propagation c.

Waves on the surface of a liquid, or on a stretched membrane, lead to a wave equation being two-dimensional for the function $f(x, y, t)$:

$$\frac{\partial^2 f}{\partial x^2} + \frac{\partial^2 f}{\partial y^2} = \frac{1}{c^2} \frac{\partial^2 f}{\partial t^2}$$

In the case of sound and electromagnetic waves we find the wave equation to be three-dimensional for the function $f(x, y, z, t)$:

$$\frac{\partial^2 f}{\partial x^2} + \frac{\partial^2 f}{\partial y^2} + \frac{\partial^2 f}{\partial z^2} = \frac{1}{c^2} \frac{\partial^2 f}{\partial t^2}$$

Thus the wave equation occurs in many different fields of engineering and science. J. Maxwell showed the relationship between electric and magnetic fields by means of the wave equation, and this started the search for electromagnetic waves. H. Hertz proved their existence experimentally in 1888.

The general solution of partial differential equations is a subtle problem in mathematics. There is no general method like, for instance, the exponential solution for ordinary differential equations (see Chapter 10).

Starting with the general solution of ordinary differential equations, we were able to obtain particular solutions from the statement of the boundary conditions. Partial differential equations have no general solutions, only particular ones; consequently, the boundary conditions have a marked influence on their solutions.

The wave equation has a great number of solutions. Which one is chosen depends on the boundary conditions of the problem.

We will concern ourselves with the one-dimensional case and show that any function of the following form is a solution of the one-dimensional wave equation. Thus u can be any function which is differentiable twice with respect to x and t:

$$f(x, t) = u(x - ct)$$

D'Alembert's solution

D'Alembert noticed that x and ct have the dimensions of length and he therefore introduced two new independent variables:

$$p = x - ct$$

$$q = x + ct$$

He then reduced the wave equation to a form which can readily be integrated. It can be shown that by these substitutions the wave equation becomes

$$\frac{\partial^2 f}{\partial p\, \partial q} = 0$$

or

$$\frac{\partial}{\partial p} \left(\frac{\partial f}{\partial q} \right) = 0$$

Integrating once, we obtain

$$\frac{\partial f}{\partial q} = Q(q), \text{ an arbitrary function of } q \text{ only.}$$

Integrating once again gives $f = \int Q(q)\, dq + P(p)$

where $P(p)$ is an arbitrary function of p only.

If we let $\int Q(q)\, dq = G(q)$ the solution becomes

$$f = P(p) + G(q)$$

Substituting for p and q, we obtain

$$f = P(x - ct) + G(x + ct)$$

This is d'Alembert's solution of the wave equation. P and G are arbitrary functions.

This result implies that, physically, the wave equation is satisfied for waves travelling in opposite directions. It gives rise to *stationary waves*. A stationary wave can be produced by the superposition of two harmonic waves of equal frequency travelling in opposite directions. It can also be produced by a progressive wave being reflected at a boundary, provided that the conditions are suitable. Examples are the vibrations in a pipe and the vibrations of a string fixed at each end.

In this book we will deal no further with partial differential equations. Further examination belongs to more advanced texts.

EXERCISES

1. Construct a table of values for the function $f(x, y) = x^2y + 6$ where $x = -2$, $-1, 0, 1$ and $y = -2, -1, 0, 1, 2$.

2. What surfaces are represented by the following functions? Sketch them!

    (a) $z = -x - 2y + 2$ (b) $z = x^2 + y^2$ (c) $z = \sqrt{1 - \dfrac{x^2}{4} - \dfrac{y^2}{9}}$

12.3 Partial Differentiation

3. Obtain the partial derivatives of

    (a) $f(x, y) = \sin x + \cos y$ (b) $f(x, y) = x^2\sqrt{1 - y^2}$
    (c) $f(x, y) = e^{-(x^2 + y^2)}$ (d) $f(x, y, z) = xyz + xy + z$
    (e) $f(x, y, z) = e^x \ln y + z^4$ (f) $f(x, y) = e^{\sin x} + e^{\cos(x + y)}$

4. Determine the slope of the tangent in the x- and y-directions to the surface $z = x^2 + y^2$ at the point $P = (0, 1)$.

5. Determine the partial derivatives f_{xx}, f_{xy}, f_{yx} and f_{yy} of the function
    $$z = R^2 - x^2 - y^2$$

6. Show that the function $z = e^{(x/y)^2}$ satisfies the relation $xf_x + yf_y = 0$.

12.4 Total Differentiation

7. Determine the total differential of the functions

    (a) $z = \sqrt{1 - x^2 - y^2}$ (b) $z = x^2 + y^2$

    (c) $f(x, y, z) = \dfrac{1}{(x^2 + y^2 + z^2)}$

8. A container in the form of an inverted right circular cone has a radius of 1.75 m and a height of 4 m. The radius is subject to a tolerance of 50 mm and the height to a tolerance of 75 mm.

    (a) Calculate the total percentage tolerance in the volume.

    (b) What is the total percentage tolerance in the surface area of the container?

9. Find the contour lines and calculate the gradient for the following functions:

    (a) $f(x, y) = -x - 2y + 2$ (b) $f(x, y) = \sqrt{1 - \dfrac{x^2}{4} - \dfrac{y^2}{9}}$

    (c) $f(x, y) = \dfrac{10}{\sqrt{x^2 + y^2}}$

10. Find the surfaces of constant functional values and calculate the gradient.

    (a) $f(x, y, z) = x + y - 3z$

    (b) $f(x, y, z) = x^2 + y^2$

    (c) $f(x, y, z) = (x^2 + y^2 + z^2)^{3/2}$

12.5   Total Derivative

11.   Obtain $\dfrac{du}{dt}$ when

(a) $u = x^2 - 3xy + 2y^2$ and $x = \cos t,\ y = \sin t$

(b) $u = x + 4\sqrt{xy} - 3y,\ x = t^3,\ y = \dfrac{1}{t}$

12.   (a) $u = x^2 + y^2,\ y = ax + b$ (b) $x^3 - y^3 + 4xy = 0$

Obtain $\dfrac{\partial u}{\partial x}$ Obtain $\dfrac{dy}{dx}$ at $x = 2,\ y = -2$

(c) $xy + \sin y = 2$

Obtain $\dfrac{dy}{dx}$ at $x = 4,\ y = \dfrac{\pi}{2}$

12.6   Maxima and Minima

13.   Examine the following functions for maxima and minima:

(a) $f(x, y) = x^3 - 3xy + y^3$

(b) $f(x, y) = 12x + 6y - x^2 + xy - y^2$

(c) A conduit along a wall is to have a cross-section (as shown in figure 12.30) made by bending a sheet of metal of width 0.75 m and length 5.5 m along the line ABCD. Calculate h, l and θ for maximum cross-sectional area.

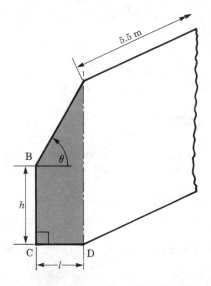

Figure 12.30

12.7   Wave Function and Wave Equation

14.   Two cables considered as being infinitely long are excited at the left-hand end with an amplitude A and a frequency f. Write down the wave function for

cable (a) $A = 0.5\,\text{m},\quad f = 5\,\text{Hz},\quad \lambda = 1.2\,\text{m}$

cable (b) $A = 0.2\,\text{m},\quad f = 0.8\,\text{Hz},\quad \lambda = 4.0\,\text{m}$

15.   Verify that the function $f(x, t) = e^{-(vt - x)^2}$ satisfies the wave equation

$$\frac{\partial^2 f}{\partial t^2} = v^2 \frac{\partial^2 f}{\partial x^2}$$

CHAPTER 13

Multiple integrals; coordinate systems

13.1 MULTIPLE INTEGRALS

Let us develop the problem by a simple example.

A solid cube, as shown in figure 13.1, has a volume V. If the density ρ is constant throughout the entire volume then the mass is given by

$$M = \rho V$$

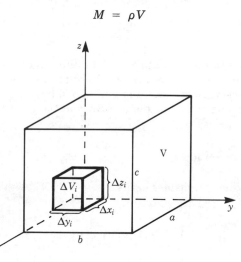

Figure 13.1

There are cases, however, in which the density ρ is not constant throughout the volume. The density of the Earth is greater near the centre than at the surface. The density of the atmosphere is at a maximum at the surface of the Earth: it decreases exponentially with the altitude.

Let us assume that the variation in the density is determined empirically and exists in a three-dimensional table of values or in the form of an equation as a function of position, i.e. $\rho = \rho(x, y, z)$.

To obtain an approximation for the mass when the density varies we proceed as follows. The volume is divided into N cells, the volume of the ith cell being

$$\Delta V_i = \Delta x_i \Delta y_i \Delta z_i$$

If the density at the point $P_i(x_i, y_i, z_i)$ is ρ the mass of the cell is

$$\Delta M_i \approx \rho(x_i, y_i, z_i)\Delta x_i \Delta y_i \Delta z_i$$

The mass of the entire cube is obtained, approximately, by adding up the masses of all the cells. Hence

$$M \approx \sum_{i=1}^{N} \Delta M_i = \sum_{i=1}^{N} \rho(x_i, y_i, z_i)\Delta x_i \Delta y_i \Delta z_i$$

Now let the size of the cells be taken smaller and smaller, so that N tends to infinity. In this way we get closer and closer to the exact value and can write

$$M = \lim_{N \to \infty} \sum_{i=1}^{N} \rho(x_i, y_i, z_i)\Delta x_i \Delta y_i \Delta z_i$$

When we were dealing with a function of only one variable, such a limit was called an integral. We now extend this concept to the above sum. In the limiting process, the differences Δx_i, Δy_i and Δz_i become differentials dx, dy, dz; and to express our limiting process we use three integrals, one for each variable, and write

$$M = \iiint\limits_{V} \rho(x, y, z)\, dx\, dy\, dz$$

In words, we describe this as the integral of the function ρ over the volume V. Such an integral is also referred to as a *multiple integral*; the special case of three variables is called a *triple integral* or a *space integral*. To solve for M we have to carry out three integrations, taking each variable in turn and paying attention to the limits of integration.

There are two cases to consider:

(a) *Multiple integrals with constant limits*
   All limits of integration are constant.

   EXAMPLE $\displaystyle\int_{z=0}^{10} \int_{y=-\pi}^{\pi} \int_{x=3}^{4} \rho(x, y, z)\, dx\, dy\, dz$

(b) *Multiple integrals with variable limits*
   Not all the limits of integration are constant.

   EXAMPLE $\displaystyle\int_{z=0}^{1} \int_{y=-\pi}^{x^2} \int_{x=3}^{y} \rho(x, y, z)\, dx\, dy\, dz$

The analytical evaluation of multiple integrals is discussed in the following sections.

Many multiple integrals can be solved analytically. There are, however, cases which lead to very complex expressions or which cannot be solved at all. In such cases, the values of multiple integrals can be computed approximately by means of sums which are sufficiently exact for practical purposes if the subdivision is fine enough.

13.2 MULTIPLE INTEGRALS WITH CONSTANT LIMITS

The actual execution of a multiple integration is particularly easy if all the limits of integration are constant. It is thus reduced to the repeated integration of simple, definite integrals. In our example of a solid cube, the computation of the mass of the cube is obtained by integrating throughout the entire volume (cf. figure 13.1), i.e.

along the x-axis from O to a
along the y-axis from O to b
along the z-axis from O to c

The triple integral sign denotes the following operations:

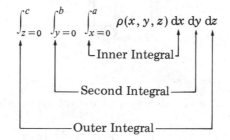

Step 1: Obtain the inner integral. y and z are regarded as constants; the only variable is x.

Step 2: The result of the first integration is a function of the variables y and z. We now solve the second integral assuming z to be constant by integrating with respect to y.

Step 3: Finally, we are left with a function of z alone and the outer integral is obtained.

Note that in the case of constant limits the order of integration can be changed, provided the integrand is continuous.

Multiple integrals are referred to as simple, double, triple, quadruple etc., depending on how many integrations are to be performed.

EXAMPLE   Calculate the mass of the rectangular prism shown in figure 13.2 of base a, b and height h.
The density decreases exponentially with height according to the relationship

$$\rho = \rho_0 \, e^{-\alpha z}$$

This example is of practical interest in the calculation of the mass of a rectangular column of air above the Earth's surface. Due to gravity, the density decreases exponentially with increasing altitude.

ρ_0 is the density at $z = 0$

The constant α has the form $\alpha = \dfrac{\rho_0}{P_0} g$, where g = acceleration due to gravity and P_0 = barometric pressure at $z = 0$.

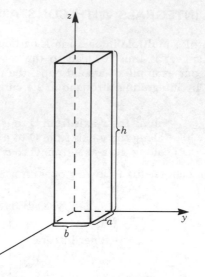

Figure 13.2

The mass of the column of air is calculated by the following multiple integral:

$$M = \int_0^h \int_0^b \int_0^a \rho_0 e^{-\alpha z} \, dx \, dy \, dz$$

Evaluation of the inner integral gives

$$M = \int_0^h \int_0^b \rho_0 e^{-\alpha z} [x]_0^a \, dy \, dz = a \int_0^h \int_0^b \rho_0 e^{-\alpha z} \, dy \, dz$$

Evaluation of the second integral gives

$$M = a \int_0^h \rho_0 e^{-\alpha z} [y]_0^b \, dz = ab \int_0^h \rho_0 e^{-\alpha z} \, dz$$

Evaluation of the outer integral gives

$$M = ab\rho_0 \left[\frac{1}{-\alpha} e^{-\alpha z} \right]_0^h = \frac{ab\rho_0}{\alpha} (1 - e^{-\alpha h})$$

Figure 13.3 shows that the mass of the column of air approaches a limiting value M_∞.

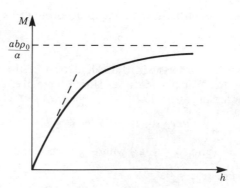

Figure 13.3

13.2.1   DECOMPOSITION OF A MULTIPLE INTEGRAL INTO A PRODUCT OF INTEGRALS

There are cases in which the integrand can be expressed as the product of functions, each in terms of a single variable.

$$f(x, y, z) \; = \; g(x) \, h(y) \, m(z)$$

Hence the multiple integral is the product of simple integrals:

$$\int_{z=0}^{1} \int_{y=0}^{2} \int_{x=0}^{1} f(x, y, z) \, dx \, dy \, dz \; = \; \int_{0}^{1} g(x) \, dx \; \int_{0}^{2} h(y) \, dy \; \int_{0}^{1} m(z) \, dz$$

EXAMPLE   Evaluate $I = \displaystyle\int_{y=0}^{\pi} \int_{x=0}^{\pi/4} \sin x \, \cos y \, dx \, dy$

The integrand is the product of two independent functions. Hence

$$I \; = \; \int_{0}^{\pi/4} \sin x \, dx \; \int_{0}^{\pi} \cos y \, dy \; = \; [-\cos x]_{0}^{\pi/4} \, [\sin y]_{0}^{\pi} \; = \; 0$$

13.3 MULTIPLE INTEGRALS WITH VARIABLE LIMITS

Multiple integrals with constant limits of integration are a special case. Generally, the limits of integration are variable. Now we will consider the general case of variable limits.

We will demonstrate the procedure with an example of the calculation of the area shown shaded in figure 13.4.

The area is obtained by summing all the small areas or meshes, such as $\Delta A = \Delta x \, \Delta y$, within the boundaries so that

$$A \; \approx \; \sum_{i=1}^{N} \Delta x_i \Delta y_i$$

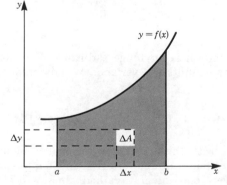

Figure 13.4

By letting $N \to \infty$ we obtain a double integral:

$$A = \iint dA = \iint dx\, dy$$

The problem now is how to regard the boundaries of the area.

Let us consider a small strip (figure 13.5) of width dx corresponding to a summation in the y direction.

The limits of this integral with respect to y are

$$\text{lower limit} \quad y = 0$$
$$\text{upper limit,} \quad y = f(x)$$

In this case the upper limit is a function of x. We now insert this into the formula and obtain

$$A = \iint_{y=0}^{f(x)} dx\, dy$$

The limits of the variable x are constant:

$$\text{lower limit,} \quad x = a$$
$$\text{upper limit,} \quad x = b$$

Inserting these limits into our double integral gives

$$A = \int_{x=a}^{b} \int_{y=0}^{f(x)} dx\, dy$$

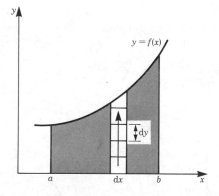

Figure 13.5

Here the order of integration is no longer arbitrary. We must integrate first with respect to variables whose limits are variable. We integrate first with respect to y obtaining

$$A = \int_{a}^{b} [f(x) - 0]\, dx = \int_{a}^{b} f(x)\, dx$$

We are already familiar with this result. We see that if we solve the area problem systematically we first obtain a double integral. In Chapters 6 and 7, when dealing with areas under curves, one integration had already been performed, without mentioning it, by considering a strip of height $f(x)$ and width dx.

EXAMPLE   Calculate the area A bounded by the curves shown in figure 13.6.

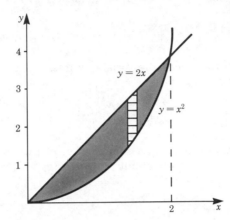

Figure 13.6

The area A has the following boundaries:

$$y \;=\; x^2 \qquad \text{for the lower one,}$$

$$y \;=\; 2x \qquad \text{for the upper one.}$$

If we integrate in the direction of y from $y = x^2$ to $y = 2x$, we obtain the area of the strip of width dx. The required area A is then obtained by integrating along x from $x = 0$ to $x = 2$, since both curves intersect at $x = 0$ and $x = 2$. Hence

$$A \;=\; \int_{x=0}^{2} \int_{y=x^2}^{2x} dy \; dx$$

Now we have first to integrate with respect to y, since its limits are variable:

$$A \;=\; \int_{0}^{2} (2x - x^2)\, dx \;=\; \left[x^2 - \frac{1}{3}x^3 \right]_{0}^{2} \;=\; \frac{4}{3}$$

We can generalise the process as follows.

Multiple integrals with variable limits are evaluated step by step, evaluating the integrals with variable limits first, up to the last integral, whose limits must be constant.

Thus at least one integral must have constant limits.

You can proceed as follows.

Find a variable which does not appear in any limit of the integrals. Now integrate with respect to this variable. Repeat this procedure until all integrals are dealt with. Following this procedure, integrals with variable limits are dealt with first.

Multiple integrals often occur in practical problems.

EXAMPLE   Obtain the position $(\bar{x}, \bar{y})$ of the centre of mass for the area A in the previous example (figure 13.6). A problem of this type in mechanics has already been discussed in Chapter 7.

First let us find the component $\bar{y}$ of the position of the centre of mass. To do this we take the first moment about the x-axis of an elemental area $dx\,dy$ and then sum for the whole area. Hence

$$A\bar{y} \;=\; \int_{x=0}^{2} \int_{y=x^2}^{2x} y\,dy\,dx \qquad (A \;=\; \text{area})$$

Let us evaluate this integral. Since the variable y does not appear in any limit, we integrate with respect to y first:

$$A\bar{y} \;=\; \int_0^2 \left[\frac{y^2}{2}\right]_{x^2}^{2x} dx \;=\; \int_0^2 \left(2x^2 - \frac{x^4}{2}\right) dx \;=\; \left[\frac{2}{3}x^3 - \frac{x^5}{10}\right]_0^2 \;=\; 2.133$$

Second, we find the component $\bar{x}$ in a similar way. In this case, consider the first moment of the elemental areas $dx\,dy$ about the y-axis.

$$A\bar{x} \;=\; \int_{x=0}^{2} \int_{y=x^2}^{2x} x\,dy\,dx$$

and $\qquad A\bar{x} \;=\; \int_0^2 [xy]_{x^2}^{2x}\,dx \;=\; \int_0^2 (2x^2 - x^3)\,dx \;=\; \left[\frac{2x^3}{3} - \frac{x^4}{4}\right]_0^2 \;=\; 1.333$

Since $A = \dfrac{4}{3}$, it follows that $\bar{x} = 1$, $\bar{y} = 1.6$.

EXAMPLE   Determine the area of a circle with radius R.

The area is given by a multiple integral,

$$A \;=\; \iint dx\,dy$$

The problem is how to take the boundaries of the circle into account. Let us look at figure 13.7. We wish to sum the elemental areas $dx\,dy$ in the y-direction, indicated by the small strip. This means that we integrate with respect to y first. The limits of this integration are given by the boundaries of this small strip and these in turn are given by the familiar equation of a circle

$$x^2 + y^2 \;=\; R^2 \qquad \text{or} \qquad y \;=\; \pm\sqrt{R^2 - x^2}$$

Lower limit, $\quad y_0 \;=\; -\sqrt{R^2 - x^2}$

Upper limit, $\quad y_1 \;=\; +\sqrt{R^2 - x^2}$

Thus we obtain

$$A \;=\; \iint_{y_0 = \sqrt{R^2 - x^2}}^{y_1 = \sqrt{R^2 - x^2}} dy\,dx$$

$$A \;=\; \int [y]_{y_0}^{y_1}\,dx$$

$$A \;=\; \int 2\sqrt{R^2 - x^2}\,dx$$

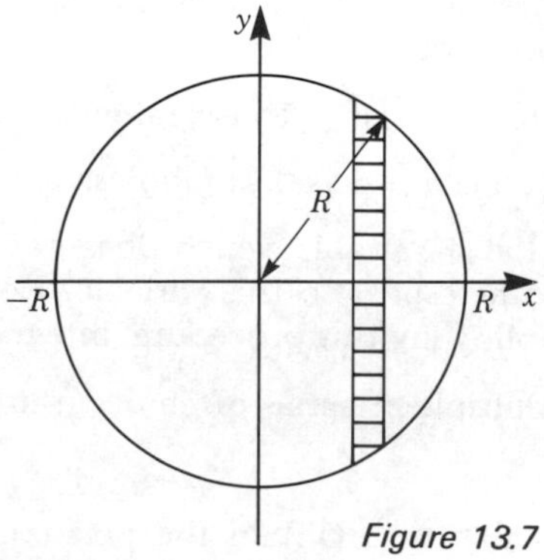

Figure 13.7

The expression $2\sqrt{R^2 - x^2}\,dx$ is the area of the small strip.

The remaining integral with respect to x sums together the elemental strips from $-R$ to $+R$. Thus the limits are

$$\text{lower limit,} \quad x_0 = -R$$
$$\text{upper limit,} \quad x_1 = +R$$

$$A = \int_{x_0=-R}^{+R} 2\sqrt{R^2-x^2}\ dx$$

Using the table of integrals in the appendix to Chapter 6, we obtain

$$A = 2\left[\frac{x}{2}\sqrt{R^2-x^2} + \frac{R^2}{2}\sin^{-1}\left(\frac{x}{R}\right)\right]_{-R}^{R}$$

$$A = R^2\pi$$

We will see in the next section that this result can be obtained much more easily by using polar coordinates.

13.4 COORDINATE SYSTEMS

The evaluation of volumes, masses, moments of inertia, load distributions and many other physical quantities leads to multiple integrals. The integrals are not always of a simple type with constant limits of integration. However, in many cases we can obtain simpler types if we replace the variables x, y, z by other more appropriate ones. This implies that we should select our coordinate system carefully according to the particular symmetry of the problem. For circular symmetries we choose polar coordinates or cylindrical ones. For radial symmetries spherical coordinates are advisable. In the following discussion we examine polar coordinates, cylindrical coordinates and spherical coordinates and relate them to cartesian coordinates.

13.4.1 POLAR COORDINATES

A point P in the x-y plane can be represented by the position vector $\mathbf{r}$, as shown in figure 13.8. In cartesian coordinates, the position vector is given by its x-y components. The same position vector can be defined by two other quantities:

the length of r,

the angle ϕ with respect to the x-axis, or any other fixed direction.

These two quantities are called *polar coordinates*.

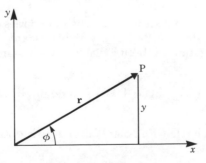

Figure 13.8

Polar coordinates can be obtained from cartesian coordinates and vice versa. The equations of transfer can be derived from figure 13.8. We obtain cartesian coordinates from polar coordinates by

$$x = r \cos \phi$$
$$y = r \sin \phi$$

We obtain polar coordinates from cartesian coordinates by

$$r = \sqrt{x^2 + y^2}$$
$$\tan \phi = \frac{y}{x}$$

An *elemental area* in cartesian coordinates is given by

$$\mathrm{d}A = \mathrm{d}x\,\mathrm{d}y \qquad \text{(see figure 13.9a)}$$

In polar coordinates, an elemental area is given by

$$\mathrm{d}A = r\,\mathrm{d}\phi\,\mathrm{d}r \qquad \text{(see figure 13.9b)}$$

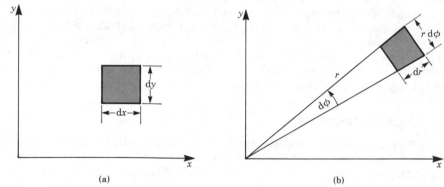

(a)   (b)

Figure 13.9

EXAMPLE   Compute the area of a circle of radius R.

We have to sum the elemental areas within the boundaries of the circle. The variable ϕ extends from $\phi = 0$ to $\phi = 2\pi$. The variable r extends from $r = 0$ to $r = R$. Thus the limits of integration for both variables are constant.

$$A = \int \mathrm{d}A = \int_{r=0}^{R} \int_{\phi=0}^{2\pi} r\,\mathrm{d}\phi\,\mathrm{d}r = \pi R^2$$

Note that the area of a circle with polar coordinates is obtained far more easily than with cartesian coordinates.

EXAMPLE   Compute the area within the spiral $r = a\phi$, $a > 0$ for one rotation of the radius vector (see figure 13.10).

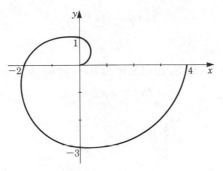

Figure 13.10

Consider the elemental area $dA = r\, d\phi\, dr$. The total area A is given by the integral

$$A = \iint r\, dr\, d\phi$$

The variable ϕ extends from $\phi = 0$ to $\phi = 2\pi$, since we are considering one rotation. The variable r extends from $r = 0$ to $r = a\phi$. Thus the limits of r are variable. Inserting the limits and solving the integral gives

$$A = \int_{\phi=0}^{2\pi} \int_{r=0}^{a\phi} r\, dr\, d\phi = \int_0^{2\pi} \left[\frac{r^2}{2}\right]_0^{a\phi} d\phi = \int_0^{2\pi} \frac{a^2}{2} \phi^2 d\phi = \left[\frac{a^2}{6} \phi^3\right]_0^{2\pi} = \frac{4}{3} a^2 \pi^3$$

13.4.2   CYLINDRICAL COORDINATES

Cylindrical coordinates are polar coordinates for a point in three-dimensional space obtained by the addition of the coordinate z to specify its height, as shown in figure 13.11.

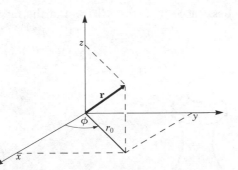

Figure 13.11

The equations of transformation between cylindrical and cartesian coordinates are

$$x = r_0 \cos \phi$$
$$y = r_0 \sin \phi$$
$$z = z$$

or, in the reverse direction,

$$r_0 = \sqrt{x^2 + y^2}$$

$$\tan\phi = \frac{y}{x}$$

$$z = z$$

The *elemental volume* dV in figure 13.12 is then given by

$$dV = r_0\,d\phi\,dr\,dz$$

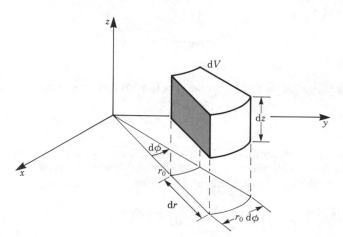

Figure 13.12

Cylindrical coordinates can facilitate calculation in the case of either of the following symmetries.

Axial Symmetry

For cylindrical coordinates we only need the functional relationship between r_0 and z, since these are independent of the angle ϕ.

Examples are the chess piece in figure 13.13, and the magnetic field round a coil (figure 13.14).

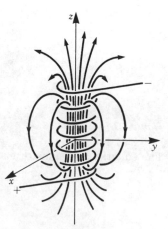

Figure 13.13   *Figure 13.14*

Cylindrical Symmetry

In cylindrical coordinates the function which describes the quantity under consideration depends only on the distance r_0 from the z-axis: it is independent of both z and ϕ.

An example is the magnetic field $\mathbf{H} = \mathbf{H}(r)$ surrounding a straight conductor carrying an electric current (figure 13.15). It possesses cylindrical symmetry.

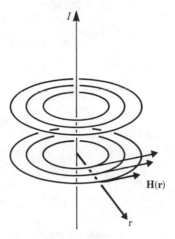

Figure 13.15

13.4.3 SPHERICAL COORDINATES

Spherical coordinates are particularly useful in problems where radial symmetry exists. Furthermore, these coordinates are used in geography to fix a point on the Earth's surface; it is assumed that the surface of the Earth is spherical. Spherical coordinates are also called *spatial polar coordinates*.

To fix the position of a point in these coordinates, we need three quantities:

   r, the position of the radius vector,

   θ, the angle between the radius vector and the z-axis, known as the
       polar angle,

   ϕ, the angle which the projection of the radius vector in the x–y plane
       makes with the x-axis, known as the *meridian*.

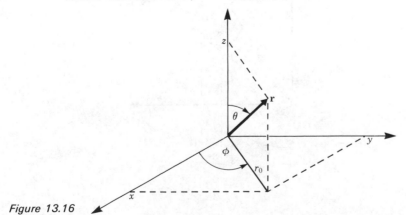

Figure 13.16

To determine the equations of transformation between cartesian and spherical coordinates, we start with the projection of the position vector $\mathbf{r}$ upon the x-y plane. The projection of the position vector upon the x-y plane has the length $r\sin\theta$. The relationships are then easily shown (see figure 13.16) to be

$$x = r\sin\theta\cos\phi$$

$$y = r\sin\theta\sin\phi$$

$$z = r\cos\theta$$

The equations in the reverse direction are

$$r = \sqrt{x^2+y^2+z^2}$$

$$\cos\theta = \frac{z}{\sqrt{x^2+y^2+z^2}}$$

$$\tan\phi = \frac{y}{x}$$

The *elemental volume* is given by

$$dV = r^2\sin\theta\; d\theta\; d\phi\; dr$$

It is a little more difficult to determine. Let us find it by taking one step at a time.

dV in the direction of the radius vector has a thickness dr and a base area dA' (figure 13.17), so that

$$dV = dA\; dr$$

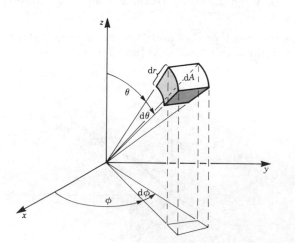

Figure 13.17

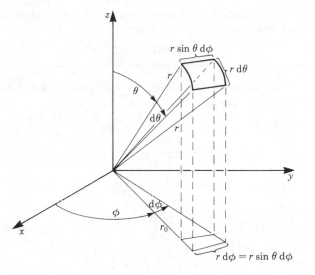

Figure 13.18

From figure 13.18, we see that

$$dA = (r \sin \theta \, d\phi)(r \, d\theta) = r^2 \sin \theta \, d\theta \, d\phi$$

Hence it follows that

$$dV = r^2 \sin \theta \, d\theta \, d\phi \, dr$$

EXAMPLE   Obtain the volume of a sphere of radius R.

In order to obtain V we integrate over the three spherical coordinates.

The volume V of the sphere is

$$V = \int_{\phi=0}^{2\pi} \int_{\theta=0}^{\pi} \int_{r=0}^{R} r^2 \sin \theta \, dr \, d\theta \, d\phi$$

$$= \int_0^{2\pi} \int_0^{\pi} \frac{R^3}{3} \sin \theta \, d\theta \, d\phi$$

$$= \int_0^{2\pi} \frac{R^3}{3} \left[-\cos \theta \right]_0^{\pi} d\phi = \int_0^{2\pi} \frac{2R^3}{3} \, d\phi = \frac{4}{3} \pi R^3$$

Spherical Symmetry

Examples of spherical symmetry include the gravitational field of the Earth, the electric field of a point charge and the sound intensity of a point source.

In spherical coordinates, the describing function depends only on the distance r from the origin and not on the angles θ and ϕ.

$$f = f(r)$$

Table 13.1 shows the important characteristics of cylindrical and spherical coordinates and their relationship with cartesian coordinates.

TABLE 13.1

Coordinates	Equations of transformation		Elemental volume	Suitable for
Cartesian	x y z		$dV = dx\,dy\,dz$	
Cylindrical	$x = r\cos\phi$ $y = r\sin\phi$ $z = z$	$r = \sqrt{x^2 + y^2}$ $\tan\phi = \dfrac{y}{x}$ $z = z$	$dV = r\,d\phi\,dr\,dz$	axial symmetry; cylindrical symmetry
Spherical	$x = r\sin\theta\cos\phi$ $y = r\sin\theta\sin\phi$ $z = r\cos\theta$	$r = \sqrt{x^2 + y^2 + z^2}$ $\cos\theta = \dfrac{z}{\sqrt{x^2 + y^2 + z^2}}$ $\tan\phi = \dfrac{y}{x}$	$dV = r^2\sin\theta\,d\theta\,d\phi\,dr$	spherical symmetry

13.5 APPLICATION: MOMENTS OF INERTIA OF A SOLID

In Chapter 7, section 7.4.4 we dealt to some extent with moments of inertia. Here we will show the calculation of moments of inertia using the concept of multiple integrals. The moment of inertia plays a major role in the dynamics of rotary motions.

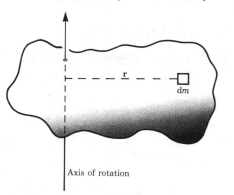

Figure 13.19

The energy of rotation of a mass element dm rotating about the axis of rotation with the constant angular velocity ω (see figure 13.19) is

$$dE_{\text{rot}} = \tfrac{1}{2}\omega^2 r^2\,dm$$

r denotes the perpendicular distance from the axis of rotation. Thus the energy of rotation of the total body is

$$E_{\text{rot}} = \tfrac{1}{2}\omega^2 \int_{\text{body}} r^2\,dm$$

The following quantity is called the *moment of inertia*:

$$I = \int_{\text{body}} r^2\,dm$$

EXAMPLE   Calculate the moment of inertia of the cylinder shown in figure 13.20. The axis of rotation is the axis of the cylinder. The density is constant throughout the body. This problem can best be solved using cylindrical coordinates.

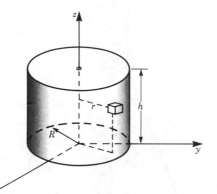

Figure 13.20

Consider a mass element dm inside the body at a distance r from the axis of rotation. The moment of inertia of the mass element is

$$dI = r^2\, dm$$

The mass element can be expressed in terms of its volume and density so that

$$dm = \rho\, dV$$

The moment of inertia of the whole cylinder is

$$I = \int_V r^2 \rho\, dV = \rho \int r^2\, dV$$

The elemental volume dV in cylindrical coordinates is

$$dV = r\, d\phi\, dr\, dz$$

Thus

$$I = \rho \int_{z=0}^{h} \int_{r=0}^{R} \int_{\phi=0}^{2\pi} r^3\, d\phi\, dr\, dz$$

$$I = \rho\, \frac{R^4 \pi h}{2} = \frac{R^2}{2} M \qquad (M = \text{total mass})$$

EXAMPLE   Now we will consider an example in cartesian coordinates. The solid shown in figure 13.21 has a square base OABC, vertical faces and a sloping top O'A'B'C'. We want to calculate the moment of inertia about the z-axis.

OA = 25 mm, AA' = 50 mm, CC' = 75 mm, OO' = 100 mm. The density is uniform throughout the body $(\rho = 7800 \text{ kg/m}^3)$.

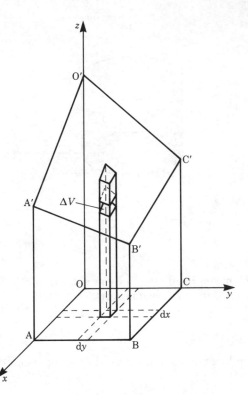

Figure 13.21

We must first find the equation of the plane O'A'B'C' to get the upper limit of the variable z.

The general equation of a plane is given by $ax + by + cz + d = 0$. The constants a, b, c and d are determined by inserting the given values of the four points O' A'B'C' in the general equation. For example,

O' = (0, 0, 100),   so that $c \times 100 + d = 0$ and $d = -100c;$

A' = (25, 0, 50),   so that $a \times 25 + c \times 50 + d = 0$ and $d = -25a - 50c.$

There are four such linear equations giving $a = 25$, $b = 1$, $c = 1$, $d = -100$. The required equation is $25x + y + z - 100 = 0$.

$$I_z = \int_x \int_y \int_z \rho(x^2 + y^2) \, dx \, dy \, dz$$

$$= \int_{x=0}^{25} \int_{y=0}^{25} \int_{z=0}^{100-25x-y} \rho(x^2 + y^2) \, dz \, dy \, dx = 13.875 \times 10^6 \rho$$

$$= 1.079 \times 10^{-4} \text{ kg m}^2$$

To conclude this example, let us also calculate the radius of gyration (see Chapter 7, section 7.0):

$$M = \text{mass} = \iiint \rho \, dx \, dy \, dz$$

$$= \rho \int_{x=0}^{25} \int_{y=0}^{25} \int_{z=0}^{100-25x-y} dz \, dy \, dz, \qquad \text{since } \rho \text{ is constant}$$

$$= \int_{0}^{25} \int_{0}^{25} (100 - 25x - y) \, dy \, dx = 39062.5\rho = 0.305 \, \text{kg}$$

The radius of gyration with respect to the z-axis is

$$k_z = \sqrt{\frac{I}{M}} = \sqrt{\frac{1.079 \times 10^{-4}}{0.305}} = 1.88 \times 10^{-2} \, \text{m} \qquad \text{or} \qquad 18.8 \, \text{mm}$$

APPENDIX

APPLICATIONS OF DOUBLE INTEGRALS

Field of application		General	Expression Cartesian coordinates $y = f(x)$	Polar coordinates $r = g(\phi)$
Area A	A	$\int_A dA$	$= \int_{x_1}^{x_2} \int_{y_1}^{y_2} dy \, dx$	$= \int_{\phi_1}^{\phi_2} \int_{r_1}^{r_2} r \, dr \, d\phi$
First moment or static moment	M_x	$= \int_A y \, dA$	$= \int_{x_1}^{x_2} \int_{y_1}^{y_2} y \, dy \, dx$	$= \int_{\phi_1}^{\phi_2} \int_{r_1}^{r_2} r^2 \sin\phi \, dr \, d\phi$
	M_y	$= \int_A x \, dA$	$= \int_{x_1}^{x_2} \int_{y_1}^{y_2} x \, dy \, dx$	$= \int_{\phi_1}^{\phi_2} \int_{r_1}^{r_2} r^2 \cos\phi \, dr \, d\phi$
Centroid	$\bar{x}$	$= \dfrac{M_x}{A}$		
	$\bar{y}$	$= \dfrac{M_y}{A}$		
Moment of inertia	I_x	$= \int_A y^2 \, dA$	$= \int_{x_1}^{x_2} \int_{y_1}^{y_2} y^2 \, dy \, dx$	$= \int_{\phi_1}^{\phi_2} \int_{r_1}^{r_2} r^3 \sin^2\phi \, dr \, d\phi$
	I_y	$= \int_A x^2 \, dA$	$= \int_{x_1}^{x_2} \int_{y_1}^{y_2} x^2 \, dy \, dx$	$= \int_{\phi_1}^{\phi_2} \int_{r_1}^{r_2} r^3 \cos^2\phi \, dr \, d\phi$
Polar moment of inertia	I_0	$= \int_A r^2 \, dA$	$= \int_{x_1}^{x_2} \int_{y_1}^{y_2} (x^2 + y^2) \, dy \, dx$	$= \int_{\phi_1}^{\phi_2} \int_{r_1}^{r_2} r^3 \, dr \, d\phi$

EXERCISES

13.2   Multiple Integrals with Constant Limits

1. Evaluate the following multiple integrals:

(a) $\displaystyle\int_{y=0}^{b}\int_{x=0}^{a} dx\, dy$

(b) $\displaystyle\int_{y=0}^{2}\int_{x=0}^{1} x^2 \, dx\, dy$

(c) $\displaystyle\int_{x=0}^{\pi}\int_{y=0}^{\pi} \sin x \, \sin y \, dx \, dy$

(d) $\displaystyle\int_{n=1}^{2}\int_{v=2}^{4} n(1+v)\, dn\, dv$

(e) $\displaystyle\int_{x=-1/2}^{1/2}\int_{y=-1}^{1}\int_{z=0}^{2} dx\, dy\, dz$

(f) $\displaystyle\int_{x=0}^{1}\int_{y=y_0}^{y_1}\int_{z=z_0}^{z_1} e^{az} \, dx\, dy\, dz$

13.3   Multiple Integrals with Variable Limits

2. Evaluate the integrals

(a) $\displaystyle\int_{x=0}^{2}\int_{y=x-1}^{3x} x^2 \, dx\, dy$

(b) $\displaystyle\int_{x=0}^{1}\int_{y=0}^{2x}\int_{z=0}^{x+y} dx\, dy\, dz$

Pay particular attention to the order of integration!

(c) Using a double integral, obtain the area of an ellipse and the position of the centre of mass of the half ellipse $(x \geqslant 0)$.

The equation of an ellipse is $\dfrac{x^2}{a^2} + \dfrac{y^2}{b^2} = 1$

13.4   Coordinate Systems

3. (a) A point has cartesian coordinates $P = (3,3)$. What are its polar coordinates?

(b) Give the equation of a circle of radius R in cartesian coordinates and polar coordinates.

(c) Obtain the equation of the spiral shown in figure 13.22 in polar coordinates.

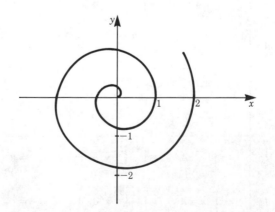

Figure 13.22

(d) Evaluate $\displaystyle\int_{\theta=0}^{\pi/4}\int_{r=0}^{a} r^2 \cos\theta \, dr\, d\theta$.

4. (a) Compute the volume of the hollow cylinder shown in figure 13.23 using cylindrical coordinates.

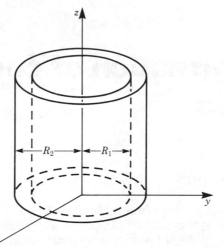

Figure 13.23

(b) Evaluate the volume of a cone of radius R and height h. Obtain the moment of inertia of the cone about its centre axis. The density ρ is constant.

5. Calculate the moment of inertia of a sphere of radius R and of constant density ρ about an axis through its centre, using spherical coordinates.

CHAPTER 14

Transformation of coordinates; matrices

14.1 INTRODUCTION

One important aspect in the solution of scientific and engineering problems is the choice of coordinate systems. The right choice may considerably reduce the degree of difficulty and the length of the necessary computations.

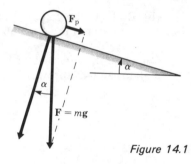

Consider, for example, the motion of a spherical particle down an inclined plane, as shown in figure 14.1.

Figure 14.1

The force of gravity, $\mathbf{F} = m\mathbf{g}$, directed vertically down can be resolved into two components, one parallel to the inclined plane and the other perpendicular to the plane, as shown in the figure.

The component parallel to the inclined plane is

$$\mathbf{F}_p = m\mathbf{g}\sin\alpha$$

and the component perpendicular to the plane is $m\mathbf{g}\cos\alpha$.

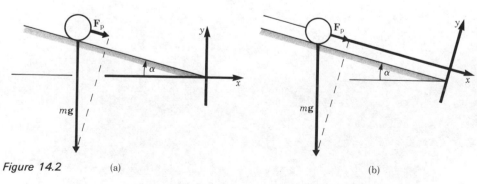

Figure 14.2   (a)   (b)

To analyse the motion of the particle we must define a frame of reference, i.e. a suitable coordinate system. Two obvious possibilities exist:

(a) the x-axis horizontal (figure 14.2a); or

(b) the x-axis parallel to the inclined plane (figure 14.2b).

Of course the actual motion of the particle is totally independent of the choice of a coordinate system. But it is important to note that a judicious choice of coordinates can simplify the calculations, as we will now demonstrate.

First let us consider case (a). As the particle rolls down the incline, it has motion in both the x and y directions. In order to determine the motion, we need to divide the force $\mathbf{F}_p$ into its components in the x-direction and the y-direction (figure 14.3a):

$$\mathbf{F}_{px} = \mathbf{F}_p \cos\alpha = mg \sin\alpha \cos\alpha$$

$$\mathbf{F}_{py} = \mathbf{F}_p \sin\alpha = -mg \sin\alpha \sin\alpha$$

According to Newton's second law of motion,

$$m\ddot{x} = mg \sin\alpha \cos\alpha$$

$$m\ddot{y} = -mg \sin^2\alpha$$

The negative sign takes care of the fact that the direction of the force and the chosen positive direction of the y-coordinate are opposite to each other.

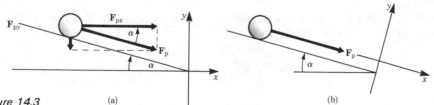

Figure 14.3   (a)   (b)

Now let us consider case (b). The motion is restricted to the x-direction (figure 14.3b). Hence

$$m\ddot{x} = mg \sin\alpha$$

$$m\ddot{y} = 0$$

These equations are obviously much simpler than those for case (a).

This example shows the importance of the choice of coordinates; in fact in some cases a problem can only be solved through the choice of appropriate coordinates. Therefore, before commencing the solution of a problem we should spend a little time selecting the most appropriate system of coordinates.

In very complex problems it sometimes happens that during a calculation it becomes apparent that a different choice of coordinates would have been more sensible. In such cases, we can either start again from scratch, or, if a lot of work has already been done, transform the old coordinates into new ones.

In this chapter we consider the second alternative, i.e. the transformation of a rectangular x-y-z coordinate system into a different rectangular x'-y'-z' coordinate system, as illustrated in figure 14.4.

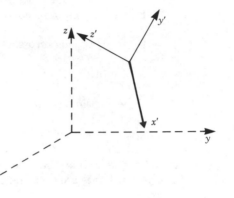

Figure 14.4

The following two transformations are particularly important.

Translation

The origin of the coordinates is shifted by a vector $\mathbf{r}_0$ in such a way that the old and the new axes are parallel (figure 14.5a).

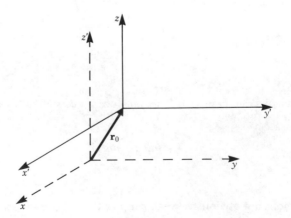

Figure 14.5a

Rotation

The new system of coordinates is rotated by an angle ϕ relative to the old system. (Figure 14.5b shows, as an example, a rotation about the x-axis through an angle ϕ.)

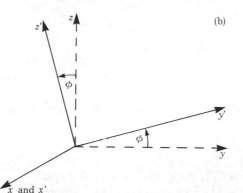

(b)

Figure 14.5b

A more general transformation of a rectangular coordinate system into a different rectangular coordinate system is composed of a translation and a rotation. In this book we will not consider inversion. For this topic and for general transformations of coordinate systems the reader should consult advanced mathematical texts.

14.2 PARALLEL SHIFT OF COORDINATES: TRANSLATION

Figure 14.6 shows a point P whose position is defined by the vector $\mathbf{r} = (x, y, z)$ in an x-y-z coordinate system.

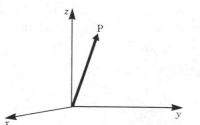

Figure 14.6

We now shift the origin O of the coordinate system to a new origin O' by the vector $\mathbf{r}_0 = (x_0, y_0, z_0)$, as shown in figure 14.7, and denote the new set of coordinates by the axes x', y' and z'.

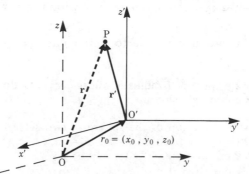

$$r_0 = (x_0, y_0, z_0)$$

Figure 14.7

What are the coordinates of P in the new system of coordinates? The vector $\mathbf{r}$ in the x-y-z system corresponds to the vector $\mathbf{r}'$ in the x'-y'-z' system of coordinates. From figure 14.7 we see that

$$\mathbf{r} = \mathbf{r}_0 + \mathbf{r}'$$

or

$$\mathbf{r}' = \mathbf{r} - \mathbf{r}_0$$

This is the required vectorial transformation when the axes remain parallel.

When expressed in terms of the coordinates, we obtain the transformation rule for a shift.

Transformation rule   If an x-y-z coordinate system is shifted by a vector $\mathbf{r}_0 = (x_0, y_0, z_0)$, the coordinates of a point in the shifted $x'-y'-z'$ system are given by

$$
\begin{aligned}
x' &= x - x_0 & \qquad x &= x' + x_0 \\
y' &= y - y_0 & \quad \text{or} \quad y &= y' + y_0 \\
z' &= z - z_0 & \qquad z &= z' + z_0
\end{aligned}
\qquad (14.1)
$$

EXAMPLE   Consider a position vector $\mathbf{r} = (5, 2, 3)$ of a point P. Now shift the coordinate system by the vector $\mathbf{r}_0 = (2, -3, 7)$. Calculate the position vector in the new system.

According to the transformation rule we have $r' = r - r_0$

$$x' = 5 - 2 = 3$$

$$y' = 2 - (-3) = 5$$

$$z' = 3 - 7 = -4$$

Hence P in the x'-y'-z' system is given by the position vector $\mathbf{r}' = (3, 5, -4)$.

To make clear how useful it can be to shift coordinates, let us consider another case.

Figure 14.8a shows a sphere of radius R whose centre O' does not coincide with the origin of an x-y-z coordinate system.

Let us investigate the equation of the sphere in two sets of coordinates. The centre of the sphere is fixed by the position vector $\mathbf{r}_0 = (x_0, y_0, z_0)$. The position vector for an arbitrary point P on the sphere (figure 14.8b) is

$$\mathbf{r} = \mathbf{r}_0 + \mathbf{R} \qquad \text{or} \qquad \mathbf{R} = \mathbf{r} - \mathbf{r}_0$$

Taking the scalar product, we get the equation of the sphere:

$$\mathbf{R} \cdot \mathbf{R} = R^2 = (x - x_0)^2 + (y - y_0)^2 + (z - z_0)^2$$

Figure 14.8

Now we will consider an x'-y'-z' coordinate system which is obtained by shifting the old system by the vector $\mathbf{r}_0$.

The new origin of the coordinate system coincides with the centre of the sphere (see figure 14.9). The equation of the sphere in the x'-y'-z' coordinate system is well known to be

$$R^2 = x'^2 + y'^2 + z'^2$$

This equation is obtained by applying the transformation rule to the previous equation.

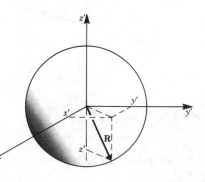

Figure 14.9

Hence the equation of a sphere and other equations as well can often be made simpler by shifting the origin of the coordinate system.

14.3 ROTATION

14.3.1 ROTATION IN A PLANE

Consider the position vector $\mathbf{r} = x\mathbf{i} + y\mathbf{j}$ in an x-y system of coordinates. We can now rotate this system through an angle ϕ into a new position, as shown in figure 14.10. The new coordinate axes are denoted by x' and y' and the unit vectors by $\mathbf{i}'$ and $\mathbf{j}'$, respectively.

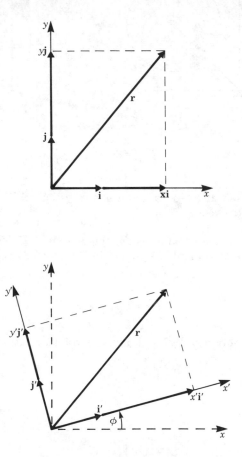

Figure 14.10

In the x'-y' coordinate system the vector $\mathbf{r}$ is given by

$$\mathbf{r} = x'\mathbf{i}' + y'\mathbf{j}'$$

The problem now is to find the relationship between the original coordinates (x, y) and the new coordinates (x', y').

We start with the components (x, y) of $\mathbf{r}$ in the original system. These are separated into components in the direction of the new axes; we need to find these components. Finally, we will collect corresponding terms.

From figure 14.11, we have

$$x\mathbf{i} = x\cos\phi\mathbf{i}' - x\sin\phi\mathbf{j}' \qquad \text{for the } x \text{ component}$$

$$y\mathbf{j} = y\sin\phi\mathbf{i}' + y\cos\phi\mathbf{j}' \qquad \text{for the } y \text{ component}$$

In the original system the vector $\mathbf{r}$ was given by

$$\mathbf{r} = x\mathbf{i} + y\mathbf{j}$$

In the new system, the vector $\mathbf{r}$ is obtained by using the relationships for $x\mathbf{i}$ and $y\mathbf{j}$. Thus

$$\mathbf{r} = x\cos\phi\mathbf{i}' - x\sin\phi\mathbf{j}' + y\sin\phi\mathbf{i}' + y\cos\phi\mathbf{j}'$$

or $$\mathbf{r} = (x\cos\phi + y\sin\phi)\mathbf{i}' + (-x\sin\phi + y\cos\phi)\mathbf{j}'$$

The expressions in brackets are the components x' and y' in the new coordinate system:

$$x' = x\cos\phi + y\sin\phi$$

$$y' = -x\sin\phi + y\cos\phi$$

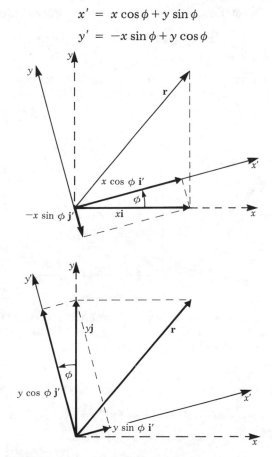

Figure 14.11

By the reverse argument, the vector $\mathbf{r} = (x, y)$ is obtained from $\mathbf{r} = (x', y')$ by replacing ϕ by $-\phi$. We get

$$x = x'\cos\phi - y'\sin\phi$$

$$y = x'\sin\phi + y'\cos\phi$$

Rule   If a vector $\mathbf{r} = (x, y)$ is transformed into the vector $\mathbf{r} = (x', y')$ when a two-dimensional system of coordinates is rotated through an angle ϕ, the transformation equations are

$$x' = x\cos\phi + y\sin\phi$$
$$y' = -x\sin\phi + y\cos\phi$$
$$x = x'\cos\phi - y'\sin\phi \qquad (14.2)$$
$$y = x'\sin\phi + y'\cos\phi$$

EXAMPLE   Given the position vector $\mathbf{r} = (2, 2)$ of a point P in an x-y system, what is the position vector of this point when the coordinate system is rotated through an angle of 45°?

$\mathbf{r}$ is transformed by the rotation according to equation 14.2 as follows:

$$x'\mathbf{i}' = (2\cos 45° + 2\sin 45°)\mathbf{i}' = 2\sqrt{2}\mathbf{i}'$$
$$y'\mathbf{j}' = (-2\sin 45° + 2\cos 45°)\mathbf{j}' = 0 \times \mathbf{j}'$$

Hence $\mathbf{r}$ is given in the rotated system by $\mathbf{r} = (2\sqrt{2}, 0)$ as shown in figure 14.12. It is obvious that in the new system the y'-component vanishes since the x'-axis coincides with $\mathbf{r}$.

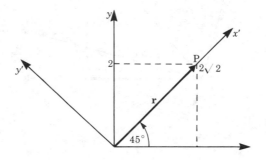

Figure 14.12

EXAMPLE   Given the hyperbola $x^2 - y^2 = 1$ in an x-y coordinate system, what is the equation of this same hyperbola in an x'-y' coordinate system after the coordinate system has been rotated through an angle of $-45°$?

From equation 14.2, we have

$$x = x'\cos(-45°) - y'\sin(-45°) = \tfrac{1}{2}\sqrt{2}(x' + y')$$
$$y = x'\sin(-45°) + y'\cos(-45°) = \tfrac{1}{2}\sqrt{2}(y' - x')$$

Substituting these into the equation $x^2 - y^2 = 1$ gives

$$[\tfrac{1}{2}\sqrt{2}(x' + y')]^2 - [\tfrac{1}{2}\sqrt{2}(y' - x')]^2 = 1$$

or
$$\tfrac{1}{2}(x'^2 + 2x'y' + y'^2) - \tfrac{1}{2}(y'^2 - 2x'y' + x'^2) = 1$$

Hence
$$x'y' = \tfrac{1}{2} \qquad \text{or} \qquad y' = \frac{1}{2x'}$$

This is the required equation of the hyperbola in the x'-y' coordinate system.

Let us consider the reverse problem.

Given $xy = \frac{1}{2}$, we rotate the system of coordinates through an angle of $+45°$. Then the equation of the hyperbola in the new x'-y' system of coordinates is

$$(x'\cos 45° - y'\cos 45°)(x'\sin 45° + y'\cos 45°) = \frac{1}{2}$$

Hence
$$x'^2 - y'^2 = 1$$

14.3.2 SUCCESSIVE ROTATIONS

We will now derive transformation equations for the case when an x-y coordinate system is rotated through an angle ϕ into an x'-y' system and then taken through a further rotation ψ into an x''-y'' system. We require the formula describing the transformation from the x-y system into the x''-y'' system.

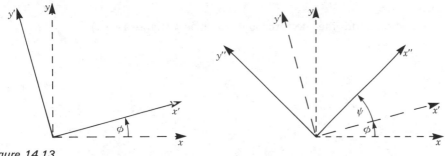

Figure 14.13

We note from figure 14.13 that the two successive rotations, ϕ and ψ, are equal to a single rotation, $\phi + \psi$. We will show analytically that this assumption is justified.

From equation 14.2 we have

(a) the rotation of the x-y system into the x'-y' system through an angle ϕ gives

$$x' = x\cos\phi + y\sin\phi$$
$$y' = -x\sin\phi + y\cos\phi$$

(b) the rotation of the x'-y' system into the x''-y'' system through an angle ψ gives

$$x'' = x'\cos\psi + y'\sin\psi$$
$$y'' = -x'\sin\psi + y'\cos\psi$$

Substituting the expressions for x' and y' in the last two equations gives

$$x'' = (x\cos\phi + y\sin\phi)\cos\psi + (-x\sin\phi + y\cos\phi)\sin\psi$$
$$y'' = -(x\cos\phi + y\sin\phi)\sin\psi + (-x\sin\phi + y\cos\phi)\cos\psi$$

Expanding, rearranging and using the addition theorems of trigonometry (Chapter 1, equation 1.10) gives the rule for successive rotations.

Transformation rule for successive rotations in the x–y plane:
$$x'' = x \cos(\phi + \psi) + y \sin(\phi + \psi)$$
$$y'' = -x \sin(\phi + \psi) + y \cos(\phi + \psi)$$

Reverse transformation: (14.3)
$$x = x'' \cos(\phi + \psi) - y'' \sin(\phi + \psi)$$
$$y = x'' \sin(\phi + \psi) + y'' \cos(\phi + \psi)$$

The vector $\mathbf{r} = (x, y)$ is transformed into the vector $\mathbf{r} = (x'', y'')$ if the coordinate system is rotated successively through the angles ϕ and ψ.

Thus our assumption was correct and the rule is established.

14.3.3 ROTATIONS IN THREE-DIMENSIONAL SPACE

In this section, we will restrict ourselves to rotations about one of the coordinate axes.

Case 1: Rotation about the z-axis through an angle ϕ (figure 14.14).

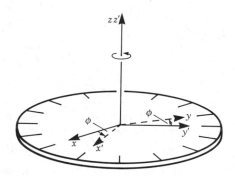

Figure 14.14

The x-axis is rotated to the x'-axis and the y-axis to the y'-axis, the z-axis remaining unchanged, i.e. the z- and z'-axes coincide.

It follows that the z-component of a vector $\mathbf{r} = (x, y, z)$ is unchanged for a rotation ϕ about the z-axis, i.e. $z' = z$.

The transformations of the x- and y-components are those of a rotation through an angle ϕ in a plane. The transformations from the x–y plane to the x'–y' plane are thus the same as in equation 14.2 in section 14.3.1.
$$x' = x \cos\phi + y \sin\phi$$
$$y' = -x \sin\phi + y \cos\phi$$
$$z' = z$$

Case 2: Now consider a rotation about the x-axis, as shown in figure 14.15.

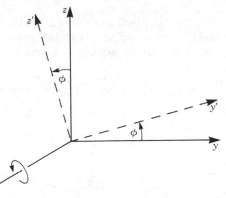

Figure 14.15

In this case we see that $y \to y'$, $z \to z'$ and $x' = x$; the transformation takes place in the y–z plane.

From equation 14.2 we find

$$y' = y \cos\phi + z \sin\phi$$

$$z' = -y \sin\phi + z \cos\phi$$

$$x' = x$$

Similarly, for rotations about the y-axis the transformation would take place in the x–z plane and $y' = y$.

Rule For a rotation about the z-axis of a three-dimensional x–y–z system through an angle ϕ, the transformation equations are

$$x' = x \cos\phi + y \sin\phi$$
$$y' = -x \sin\phi + y \cos\phi \qquad (14.4)$$
$$z' = z$$

For a rotation about the x-axis, they are

$$x' = x$$
$$y' = y \cos\phi + z \sin\phi \qquad (14.5)$$
$$z' = -y \sin\phi + z \cos\phi$$

For a rotation about the y-axis, they are

$$x' = x \cos\phi + z \sin\phi$$
$$y' = y \qquad (14.6)$$
$$z' = -x \sin\phi + z \cos\phi$$

We have obtained transformation equations for rotations about the x-, y- or z-axis only. Successive rotations can be described as a single rotation about some axis. Conversely, any rotation about any given axis can be described as a succession of rotations about the axes of the coordinate system. Details of this can be found in more advanced texts on algebra.

14.4 MATRIX ALGEBRA

Matrix algebra is a powerful tool in linear algebra. It also has the advantage of being very concise. The transformation equations in the preceding sections can be more clearly arranged by introducing the concept of matrix operations.

> **DEFINITION**   A rectangular array or set of real numbers is called a *real matrix*.
>
> $$A = \begin{pmatrix} a_{11} & a_{12} & \cdots & a_{1n} \\ a_{21} & a_{22} & \cdots & a_{2n} \\ \vdots & \vdots & & \vdots \\ a_{m1} & a_{m2} & \cdots & a_{mn} \end{pmatrix}$$

The horizontal lines of numbers are referred to as *rows of the matrix*.

$$\begin{pmatrix} \cdot & \cdot & \cdots & \cdot \\ \cdot & a_{22} & \cdots & \cdot \\ \vdots & \vdots & & \vdots \\ \cdot & \cdot & \cdots & \cdot \end{pmatrix}$$ This example shows the second row.

The vertical lines of numbers are referred to as *columns of the matrix*.

$$\begin{pmatrix} \cdot & \cdot & \cdots & \cdot \\ \cdot & a_{22} & \cdots & \cdot \\ \vdots & \vdots & & \vdots \\ \cdot & \cdot & \cdots & \cdot \end{pmatrix}$$ This example shows the second column.

Please note that in this book we are dealing only with real matrices (as opposed to complex matrices). Also, all vectors considered are real vectors (as opposed to complex vectors).

A matrix with m rows and n columns is said to be an $m \times n$ matrix or an (m, n) matrix. When $m \neq n$ the matrix is rectangular, and when $m = n$ it is square. An $m \times n$ matrix is said to be of the *order* $m \times n$.

Matrices are often denoted by a bold upper-case letter or by an upper-case letter underlined, e.g. $\underline{A}$ in a manuscript.

EXAMPLE

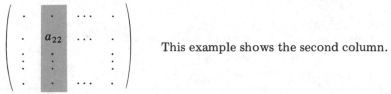

$$A = \begin{pmatrix} a_{11} & a_{12} \\ a_{21} & a_{22} \\ a_{31} & a_{32} \end{pmatrix} \quad \text{is a } 3 \times 2 \text{ matrix, or a matrix of order } 3 \times 2 = 6$$

This matrix can also be written thus:

$$A = (a_{ik}) \quad \text{with} \quad i = 1, 2, 3$$
$$k = 1, 2$$

The numbers a_{ik} are called the *elements of the matrix*; the first subscript, i, refers to the row and the second, k, to the column.

In the case of square matrices, the elements a_{ii} are found on a diagonal which is called the *leading diagonal*.

Matrices with one row or one column only are referred to as *vectors*.

Column and row vectors are denoted by bold lower-case letters.

For example a *row vector* is given by

$$\mathbf{a} = (a_{ik})_{(1,n)} = (a_{11} \quad a_{12} \quad \dots \quad a_{1n}) = (a_1 \quad a_2 \quad \dots \quad a_n)$$

For example a *column vector* is given by

$$\mathbf{a} = (a_{ik})_{(n,1)} = \begin{pmatrix} a_{11} \\ a_{21} \\ \cdot \\ \cdot \\ \cdot \\ a_{n1} \end{pmatrix} = \begin{pmatrix} a_1 \\ a_2 \\ \cdot \\ \cdot \\ \cdot \\ a_n \end{pmatrix}$$

14.4.1 ADDITION AND SUBTRACTION OF MATRICES

DEFINITION   The sum (or difference) of two matrices $\mathbf{A}$ and $\mathbf{B}$ of the same order $m \times n$ is another matrix $\mathbf{C}$ of the order $m \times n$ whose elements c_{ik} are the sum (or difference) $a_{ik} \pm b_{ik}$ of the corresponding elements of matrices $\mathbf{A}$ and $\mathbf{B}$.

$$\mathbf{C} = \mathbf{A} \pm \mathbf{B} = \begin{pmatrix} a_{11} \pm b_{11} & a_{12} \pm b_{12} & \dots & a_{1n} \pm b_{1n} \\ a_{21} \pm b_{21} & a_{22} \pm b_{22} & \dots & a_{2n} \pm b_{2n} \\ \cdot & \cdot & & \cdot \\ \cdot & \cdot & & \cdot \\ a_{m1} \pm b_{m1} & \dots & \dots & a_{mn} \pm b_{mn} \end{pmatrix} \qquad (14.7)$$

EXAMPLE   Given

$$\mathbf{A} = \begin{pmatrix} 3 & -1 & 4 & 10 \\ 1 & 3 & 3 & -2 \\ -7 & 1 & 5 & 3 \end{pmatrix} \text{ and } \mathbf{B} = \begin{pmatrix} -2 & 5 & -8 & 0 \\ -4 & 1 & -3 & 0 \\ 1 & -3 & 0 & -1 \end{pmatrix}$$

obtain $\mathbf{A} + \mathbf{B}$.

We simply add the elements with the same subscript, i.e. in the same position. Hence

$$\mathbf{C} = \mathbf{A} + \mathbf{B} = \begin{pmatrix} 1 & 4 & -4 & 10 \\ -3 & 4 & 0 & -2 \\ -6 & -2 & 5 & 2 \end{pmatrix}$$

14.4.2 MULTIPLICATION OF A MATRIX BY A SCALAR

DEFINITION   A matrix $\mathbf{A}$ multiplied by a scalar quantity k is a new matrix whose elements are multiplied by k.   (14.8)

EXAMPLE   If

$$A = \begin{pmatrix} a_{11} & a_{12} & a_{13} \\ a_{21} & a_{22} & a_{23} \end{pmatrix}$$

then

$$kA = \begin{pmatrix} ka_{11} & ka_{12} & ka_{13} \\ ka_{21} & ka_{22} & ka_{23} \end{pmatrix}$$

EXAMPLE   If $A = \begin{pmatrix} 5 & -7 & -1 \\ 3 & 2 & 2 \end{pmatrix}$ and $k = 2.5$

then

$$kA = \begin{pmatrix} 12.5 & -17.5 & -2.5 \\ 7.5 & 5 & 5 \end{pmatrix}$$

14.4.3   PRODUCT OF A MATRIX AND A VECTOR

We illustrate the product of a matrix and a vector by considering the 2×2 matrix

$$A = \begin{pmatrix} a_{11} & a_{12} \\ a_{21} & a_{22} \end{pmatrix}$$

and the vector $r = (x, y)$ or $r = \begin{pmatrix} x \\ y \end{pmatrix}$

We can therefore state the following definition.

DEFINITION   The product Ar of a matrix A and a vector r is a new vector r' whose components are given by

$$r' = Ar = \begin{pmatrix} a_{11} & a_{12} \\ a_{21} & a_{22} \end{pmatrix}\begin{pmatrix} x \\ y \end{pmatrix} = \begin{pmatrix} a_{11}x + a_{12}y \\ a_{21}x + a_{22}y \end{pmatrix} \qquad (14.9)$$

The components x' and y' of the vector r' are obtained by forming the product of rows and columns as for the scalar product. Hence, if $r' = \begin{pmatrix} x' \\ y' \end{pmatrix}$, then

$$x' = a_{11}x + a_{12}y \qquad \text{and} \qquad y' = a_{21}x + a_{22}y$$

EXAMPLE   Obtain Ar if $A = \begin{pmatrix} 1 & -3 \\ 6 & 4 \end{pmatrix}$ and $r = \begin{pmatrix} x \\ y \end{pmatrix}$

$$Ar = r' = \begin{pmatrix} x' \\ y' \end{pmatrix} = \begin{pmatrix} 1 & -3 \\ 6 & 4 \end{pmatrix}\begin{pmatrix} x \\ y \end{pmatrix} = \begin{pmatrix} x - 3y \\ 6x + 4y \end{pmatrix}$$

If we have a 3×3 matrix and a three-dimensional vector $\begin{pmatrix} x \\ y \\ z \end{pmatrix}$ then

$$\begin{pmatrix} x' \\ y' \\ z' \end{pmatrix} = \begin{pmatrix} a_{11} & a_{12} & a_{13} \\ a_{21} & a_{22} & a_{23} \\ a_{31} & a_{32} & a_{33} \end{pmatrix}\begin{pmatrix} x \\ y \\ z \end{pmatrix} = \begin{pmatrix} a_{11}x + a_{12}y + a_{13}z \\ a_{21}x + a_{22}y + a_{23}z \\ a_{31}x + a_{32}y + a_{33}z \end{pmatrix}$$

EXAMPLE   Obtain $r' = Ar$ if $r = (x, y, z)$ and

$$A = \begin{pmatrix} 1 & 0 & 3 \\ 4 & -2 & 0 \\ 0 & 0 & 5 \end{pmatrix}$$

The solution is

$$r' = \begin{pmatrix} x' \\ y' \\ z' \end{pmatrix} = \begin{pmatrix} 1 & 0 & 3 \\ 4 & -2 & 0 \\ 0 & 0 & 5 \end{pmatrix} \begin{pmatrix} x \\ y \\ z \end{pmatrix} = \begin{pmatrix} x + 3z \\ 4x - 2y \\ 5z \end{pmatrix}$$

14.4.4   MULTIPLICATION OF TWO MATRICES

The product AB of two matrices A and B is defined as follows, provided that the number of columns in A is the same as the number of rows in B.

> **DEFINITION**   The product AB of a matrix A of order $(m \times n)$ and a matrix B of order $(n \times p)$ is a matrix C of order $(m \times p)$. The coefficients of the matrix C are denoted by $c_{ik}(i = 1, 2, \ldots m; \ k = 1, 2, \ldots p)$.
>
> They are obtained by multiplying the ith row of matrix A by the kth column of B, both being considered as vectors and forming the 'scalar product' as follows:
>
> $$c_{ik} = \sum_{\nu=1}^{n} a_{i\nu} b_{\nu k} = a_{i1} b_{1k} + a_{i2} b_{2k} + \ldots + a_{in} b_{nk} \qquad (14.10)$$

The following diagram indicates the way the coefficients of the matrix $C = AB$ are generated. It shows how the coefficient c_{22} is calculated.

2nd
column

$$\begin{pmatrix} b_{11} & b_{12} \\ b_{21} & b_{22} \\ b_{31} & b_{32} \end{pmatrix}$$

2nd row $\begin{pmatrix} a_{11} & a_{12} & a_{13} \\ a_{21} & a_{22} & a_{23} \end{pmatrix} \begin{pmatrix} c_{11} & c_{12} \\ c_{21} & c_{22} \end{pmatrix}$

This shows what is meant by saying that we form the scalar product of the ith row (2nd in this example) of matrix A by the kth column (2nd in this example) of matrix B.

EXAMPLE   Obtain the product of

$$A = \begin{pmatrix} 5 & 2 \\ 0 & 1 \end{pmatrix} \quad \text{and} \quad B = \begin{pmatrix} -3 & 7 \\ 1 & -1 \end{pmatrix}$$

The solution is

$$AB = \begin{pmatrix} 5 & 2 \\ 0 & 1 \end{pmatrix}\begin{pmatrix} -3 & 7 \\ 1 & -1 \end{pmatrix} = \begin{pmatrix} -15+2 & 35-2 \\ 0+1 & 0-1 \end{pmatrix}$$

$$= \begin{pmatrix} -13 & 33 \\ 1 & -1 \end{pmatrix}$$

EXAMPLE   Find the product of

$$A = \begin{pmatrix} 1 & 0 & 1 \\ 2 & -7 & 8 \\ 0 & 1 & -4 \\ 6 & 2 & 1 \end{pmatrix} \quad \text{and} \quad B = \begin{pmatrix} 2 & 0 \\ -3 & -1 \\ 4 & 5 \end{pmatrix}$$

A is a 4×3 matrix and B a 3×2 matrix. Hence the product will be a 4×2 matrix.

The solution is

$$AB = \begin{pmatrix} 2+0+4 & 0+0+5 \\ 4+21+32 & 0+7+40 \\ 0-3-16 & 0-1-20 \\ 12-6+4 & 0-2+5 \end{pmatrix} = \begin{pmatrix} 6 & 5 \\ 57 & 47 \\ -19 & -21 \\ 10 & 3 \end{pmatrix}$$

Note that the product of two matrices A and B is, in general, not commutative, i.e. $AB \neq BA$.

14.5   ROTATIONS EXPRESSED IN MATRIX FORM

14.5.1   ROTATION IN TWO-DIMENSIONAL SPACE

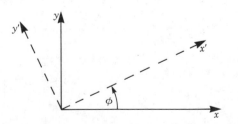

Figure 14.16

The transformation equations for a rotation through an angle ϕ (figure 14.16) were obtained in section 14.3.1 (equation 14.2). They are

$$x' = x \cos\phi + y \sin\phi$$
$$y' = -x \sin\phi + y \cos\phi$$

These equations can now be expressed in matrix form to give the new vector as a product of a rotation matrix and the original vector:

$$\begin{pmatrix} x' \\ y' \end{pmatrix} = \begin{pmatrix} \cos\phi & \sin\phi \\ -\sin\phi & \cos\phi \end{pmatrix} \begin{pmatrix} x \\ y \end{pmatrix}$$

EXAMPLE   Let us rotate an x-y coordinate system through an angle $\phi = \dfrac{\pi}{2}$ so that the x-axis moves into the y-axis and the y-axis into the negative x-axis. Calculate the rotation matrix.

Substituting the value of $\dfrac{\pi}{2}$ into the general rotation matrix we get

$$\begin{pmatrix} \cos\dfrac{\pi}{2} & \sin\dfrac{\pi}{2} \\ -\sin\dfrac{\pi}{2} & \cos\dfrac{\pi}{2} \end{pmatrix} = \begin{pmatrix} 0 & 1 \\ -1 & 0 \end{pmatrix}$$

The transformation of coordinates is obtained:

$$\begin{pmatrix} x' \\ y' \end{pmatrix} = \begin{pmatrix} 0 & 1 \\ -1 & 0 \end{pmatrix}\begin{pmatrix} x \\ y \end{pmatrix} = \begin{pmatrix} y \\ -x \end{pmatrix}$$

We now determine the matrix for successive rotations ϕ and ψ.

We first rotate through an angle ϕ. Thus x goes to x' and y to y'. Second, we rotate through an angle ψ. Thus x' goes to x'' and y' to y''. The equations obtained in section 14.3.2 look in matrix form as follows:

$$\begin{pmatrix} x' \\ y' \end{pmatrix} = \begin{pmatrix} \cos\phi & \sin\phi \\ -\sin\phi & \cos\phi \end{pmatrix} \begin{pmatrix} x \\ y \end{pmatrix} \qquad [1]$$

$$\begin{pmatrix} x'' \\ y'' \end{pmatrix} = \begin{pmatrix} \cos\psi & \sin\psi \\ -\sin\psi & \cos\psi \end{pmatrix} \begin{pmatrix} x' \\ y' \end{pmatrix} \qquad [2]$$

Substituting equation 1 into equation 2 gives

$$\begin{pmatrix} x'' \\ y'' \end{pmatrix} = \begin{pmatrix} \cos\psi & \sin\psi \\ -\sin\psi & \cos\psi \end{pmatrix}\begin{pmatrix} \cos\phi & \sin\phi \\ -\sin\phi & \cos\phi \end{pmatrix} \begin{pmatrix} x \\ y \end{pmatrix}$$

Multiplying out these matrices gives

$$\begin{pmatrix} x'' \\ y'' \end{pmatrix} = \begin{pmatrix} \cos\psi\cos\phi - \sin\psi\sin\phi & \cos\psi\sin\phi + \sin\psi\cos\phi \\ -\sin\psi\cos\phi - \cos\psi\sin\phi & -\sin\psi\sin\phi + \cos\psi\cos\phi \end{pmatrix} \begin{pmatrix} x \\ y \end{pmatrix}$$

By applying the addition theorems in trigonometry, the transformation matrix finally becomes

$$\begin{pmatrix} \cos(\phi+\psi) & \sin(\phi+\psi) \\ -\sin(\phi+\psi) & \cos(\phi+\psi) \end{pmatrix}$$

14.5.2 SPECIAL ROTATION IN THREE-DIMENSIONAL SPACE

In section 14.3.3 we derived the transformation equations for a rotation about the z-axis (figure 14.17) through an angle ϕ.

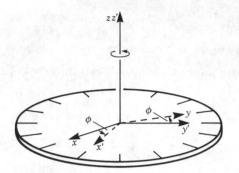

Figure 14.17

These transformation equations are

$$x' = x \cos \phi + y \sin \phi$$
$$y' = -x \sin \phi + y \cos \phi$$
$$z' = z$$

We can express these equations in matrix form thus:

$$\begin{pmatrix} x' \\ y' \\ z' \end{pmatrix} = \begin{pmatrix} \cos \phi & \sin \phi & 0 \\ -\sin \phi & \cos \phi & 0 \\ 0 & 0 & 1 \end{pmatrix} \begin{pmatrix} x \\ y \\ z \end{pmatrix}$$

The transformation matrix for rotation about the y-axis through an angle ψ is

$$\begin{pmatrix} \cos \psi & 0 & \sin \psi \\ 0 & 1 & 0 \\ -\sin \psi & 0 & \cos \psi \end{pmatrix}$$

You should verify this for yourself.

14.6 SPECIAL MATRICES

In this section we introduce the definitions of some important special matrices. Some assertions will be made without proof.

Unit Matrix

A *unit matrix* is a quadratic matrix of, for example, the following form:

$$\mathbf{I} = \begin{pmatrix} 1 & 0 & 0 \\ 0 & 1 & 0 \\ 0 & 0 & 1 \end{pmatrix}$$

All elements on the leading diagonal are unity and all other elements are zero.

If a vector $\mathbf{r}$ or a matrix $\mathbf{A}$ is multiplied by a unit matrix $\mathbf{I}$, the vector or the matrix remains unchanged.

$$\mathbf{Ir} = \mathbf{r}$$

$$\mathbf{IA} = \mathbf{A}$$

These relationships are easy to verify.

Diagonal Matrices

A *diagonal matrix* is a quadratic matrix whose elements are all zero except those on the leading diagonal.

$$\mathbf{D} = \begin{vmatrix} a_{11} & 0 & 0 \\ 0 & a_{22} & 0 \\ 0 & 0 & a_{33} \end{vmatrix}$$

A unit matrix is thus a special diagonal matrix.

Null Matrix

A *null matrix* is one whose elements are all zero; it is denoted by $\mathbf{0}$. We should note that if $\mathbf{AB} = \mathbf{0}$, it does not necessarily follow that $\mathbf{A} = \mathbf{0}$ or $\mathbf{B} = \mathbf{0}$.

Transposed Matrix

If we interchange the rows and columns of a matrix $\mathbf{A}$ of order $m \times n$ we obtain a new matrix of order $n \times m$. This matrix is called the *transposed matrix* or the *transpose* of the original matrix and it is denoted by $\mathbf{A}^T$ or $\tilde{\mathbf{A}}$.

If $\quad\quad \mathbf{A} = \begin{vmatrix} a_{11} & a_{12} \\ a_{21} & a_{22} \\ a_{31} & a_{32} \end{vmatrix} \quad$ then $\quad \mathbf{A}^T = \begin{pmatrix} a_{11} & a_{21} & a_{31} \\ a_{12} & a_{22} & a_{32} \end{pmatrix} \quad\quad$ (14.11)

Note that $\mathbf{A}^T$ is the mirror image of $\mathbf{A}$.

EXAMPLE

If $\quad\quad \mathbf{A} = \begin{vmatrix} 2 & 0 & 0 \\ 2 & 1 & -6 \\ 6 & 0 & -1 \end{vmatrix} \quad$ then $\quad \mathbf{A}^T = \begin{vmatrix} 2 & 2 & 6 \\ 0 & 1 & 0 \\ 0 & -6 & -1 \end{vmatrix}$

Note that the first row becomes the first column, the second row becomes the second column, etc.

You should verify the following assertions for yourself.

(1) The transpose of the transposed matrix gives the original matrix $\mathbf{A}$.

$$(\mathbf{A}^T)^T = \mathbf{A}$$

(2) An important relationship is

$$(\mathbf{AB})^T = \mathbf{B}^T\mathbf{A}^T$$

and generally $\quad\quad\quad (\mathbf{ABC} \ldots \mathbf{Z})^T = \mathbf{Z}^T \ldots \mathbf{B}^T\mathbf{A}^T$

Orthogonal Matrices

A square matrix **A** which satisfies the following identity is called an *orthogonal matrix*:

$$\mathbf{A}\mathbf{A}^{\mathrm{T}} = \mathbf{I} \qquad \text{(orthogonality)} \qquad (14.12)$$

This relationship is equivalent to

$$\mathbf{A}^{\mathrm{T}}\mathbf{A} = \mathbf{I}$$

It can be interpreted in terms of rows and columns of the matrix **A** as follows. The nth column of **A** is the nth row of $\mathbf{A}^{\mathrm{T}}$. Now consider the equation $\mathbf{A}\mathbf{A}^{\mathrm{T}} = \mathbf{I}$. If we think of rows and columns as vectors and compute their scalar product, then we observe for an orthogonal matrix **A** that

(1) the scalar product of a column by itself is 1;

(2) the scalar product of a column by a different column is always zero.

The following assertions are equivalent:

(3) the scalar product of a row by itself is 1;

(4) the scalar product of a row by a different row is always zero.

For example, the matrices describing rotations are always orthogonal matrices.

The name 'orthogonal' is derived from the fact that if an orthogonal matrix **A** is applied to two vectors **r** and **s**, their scalar product remains unaffected, i.e.

$$\mathbf{r}\cdot\mathbf{s} = (\mathbf{Ar})\cdot(\mathbf{As})$$

This implies that the lengths and angles of the vectors are preserved, and, in particular, a system of orthogonal coordinate axes is transformed into another orthogonal system.

Singular Matrix

A matrix whose determinant is zero is called a *singular matrix* (for determinants see Chapter 15).

Symmetric Matrices and Skew-Symmetric (or Antisymmetric) Matrices

For square matrices two new properties may be relevant. A square matrix is called *symmetric* if for all i and j $a_{ij} = a_{ji}$. This means it equals its transpose.

$$\mathbf{A} = \mathbf{A}^{\mathrm{T}} \qquad \text{(symmetry)} \qquad (14.13)$$

A square matrix is called *skew-symmetric* or *antisymmetric* if all $a_{ij} = -a_{ji}$. This means it equals the negative of its transpose. Note that for antisymmetric matrices all elements on the leading diagonal are zero.

$$A = -A^T \qquad \text{(skew-symmetry)} \qquad (14.14)$$

It is useful to note that any square matrix can be expressed as the sum of a symmetric and a skew-symmetric matrix.

PROOF $\qquad A = \frac{1}{2}(A + A^T) + \frac{1}{2}(A - A^T)$

Observe that the first term is a symmetric matrix and the second term is a skew-symmetric matrix.

EXAMPLE   Write as the sum of a skew-symmetric and a symmetric matrix

$$A = \begin{vmatrix} 798 & 29 & 26 \\ 1 & 8 & 27 \\ 74 & 69 & 88 \end{vmatrix}$$

The solution is

$$A = \begin{vmatrix} 0 & 14 & -24 \\ -14 & 0 & -21 \\ 24 & 21 & 0 \end{vmatrix} + \begin{vmatrix} 798 & 15 & 50 \\ 15 & 8 & 48 \\ 50 & 48 & 88 \end{vmatrix}$$

14.7 INVERSE MATRIX

If a square matrix A when multiplied by another matrix B results in the unit matrix, then the matrix B is called the *inverse matrix* of A; it is denoted by A^{-1}. Not all matrices possess an inverse, and the criterion for a matrix to have an inverse is that its determinant must be different from zero, i.e. it must not be singular. Determinants are dealt with in Chapter 15. If an inverse exists it is unique.

The following equations must hold true:

$$\begin{aligned} AA^{-1} &= I, & \text{(post multiplication by } A^{-1}) \\ A^{-1}A &= I, & \text{(pre multiplication by } A^{-1}) \end{aligned} \qquad (14.15)$$

We will not give details here of how A^{-1} is calculated. This will be done in Chapter 15, section 15.2.3. For the time being we will give only the following example for an inverse matrix.

If $\qquad A = \begin{vmatrix} 2 & 0 & 0 \\ 2 & 1 & -6 \\ 6 & 0 & -1 \end{vmatrix} \qquad$ then $\qquad A^{-1} = \begin{vmatrix} \frac{1}{2} & 0 & 0 \\ 17 & 1 & -6 \\ 3 & 0 & -1 \end{vmatrix}$

As an exercise, you should verify for yourself that

$$AA^{-1} = A^{-1}A = I$$

Returning to the concept of orthogonal matrices introduced earlier, we can now state the following criterion. A square matrix A is orthogonal if its inverse is equal to its transpose, i.e. if $A^{-1} = A^T$.

EXERCISES

14.2   Parallel Shift of Coordinates

1. The vertex of the paraboloid shown in figure 14.18 is at a distance 2 from the origin of the coordinates. The equation is

$$z = 2 + x^2 + y^2$$

What is the transformation which will shift the parapoloid so that its vertex coincides with the origin O?

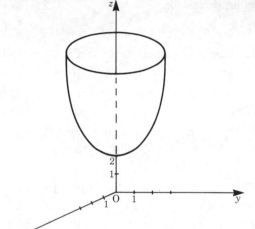

Figure 14.18

2. The equation of a certain straight line is $y = -3x + 5$. What will its equation be in a new $x'-y'$ coordinate system due to a shift of the origin of $(-2, 3)$?

14.3   Rotation

3. A two-dimensional system of coordinates is rotated through an angle of $\dfrac{\pi}{3}$. The transformation matrix is

$$\begin{pmatrix} \dfrac{1}{2} & \dfrac{\sqrt{3}}{2} \\ -\dfrac{\sqrt{3}}{2} & \dfrac{1}{2} \end{pmatrix}$$

What is the new vector $\mathbf{r}'$ if $\mathbf{r} = (2, 4)$?

4. Given the equation $y = -\dfrac{x}{\sqrt{3}} + 2$, if the system of coordinates is rotated through an angle of $60°$ obtain an expression for the equation in the rotated system.

5. A three-dimensional system of coordinates is rotated about the z-axis through an angle of $30°$. Obtain the transformed vector $\mathbf{r}'$ if $\mathbf{r} = (3, 3, 3)$.

14.4   Matrix Algebra

6. Given the two matrices $\mathbf{A}$ and $\mathbf{B}$ where

$$\mathbf{A} = \begin{vmatrix} 1 & 3 \\ 2 & 5 \\ 0 & 7 \end{vmatrix} \quad \text{and} \quad \mathbf{B} = \begin{vmatrix} 2 & 0 \\ -1 & 3 \\ -1 & 2 \end{vmatrix}$$

evaluate (a) $\mathbf{A} + \mathbf{B}$, (b) $\mathbf{A} - \mathbf{B}$

7. Let

$$A = \begin{pmatrix} 2 & 7 \\ 3 & 0 \\ 9 & -1 \end{pmatrix} \quad \text{and} \quad B = \begin{pmatrix} 9 & 3 & 0 \\ 1 & -2 & 4 \end{pmatrix}$$

(a) Evaluate the matrix 6A.

(b) Show that the expression $AB \neq BA$.

8. Given

$$A = \begin{pmatrix} 1 & 2 \\ 7 & 3 \\ 5 & 9 \end{pmatrix} \quad \text{and} \quad B = \begin{pmatrix} -1 & 0 \\ 2 & 3 \\ -1 & -1 \end{pmatrix}$$

evaluate **AB**

9. Evaluate the product $Ar = r'$ if

$$A = \begin{pmatrix} 1 & -2 \\ 5 & 7 \end{pmatrix}, \quad r = \begin{pmatrix} x \\ y \end{pmatrix}$$

10. Given

$$A = \begin{pmatrix} 1 & 2 \\ 4 & -3 \\ 3 & 0 \end{pmatrix}$$

evaluate (a) A^T, (b) $(A^T)^T$

11. How many independent entries are there in a skew-symmetric 3×3 matrix?

12. Decompose into a symmetric and a skew-symmetric matrix:

$$\begin{pmatrix} 54 & 1 & 1 \\ 0 & 26 & 20 \\ 8 & 84 & 9 \end{pmatrix}$$

14.7 Inverse Matrix

13. If

$$A = \begin{pmatrix} 1 & 0 & 3 \\ 2 & -3 & 1 \\ 1 & 2 & 2 \end{pmatrix} \quad \text{and} \quad A^{-1} = \frac{1}{13} \begin{pmatrix} -8 & 6 & 9 \\ -3 & -1 & 5 \\ 7 & -2 & -3 \end{pmatrix}$$

show that $AA^{-1} = A^{-1}A = I$

CHAPTER 15

Sets of linear equations; determinants

15.1 INTRODUCTION

In this chapter we will investigate the solution of sets of linear algebraic equations. First, we show a method which will be used in most practical cases. This is the Gaussian method of elimination and its refinements. The basic idea is quite clear and elementary. Notation in matrix form will prove to be helpful. In Chapter 17 the numerical calculation will be discussed further.

Second, the concept of determinants and a second method of solution, Cramer's rule, will be developed. This concept is of theoretical importance, e.g. a determinant shows whether a set of simultaneous equations is uniquely solvable.

15.2 SETS OF LINEAR EQUATIONS

15.2.1 GAUSSIAN ELIMINATION: SUCCESSIVE ELIMINATION OF VARIABLES

Our problem is to solve a set of linear algebraic equations. For the time being we will assume that a unique solution exists and that the number of equations equals the number of variables.

Consider a set of three equations:

$$a_{11}x_1 + a_{12}x_2 + a_{13}x_3 = b_1$$
$$a_{21}x_1 + a_{22}x_2 + a_{23}x_3 = b_2$$
$$a_{31}x_1 + a_{32}x_2 + a_{33}x_3 = b_3$$

The basic idea of the Gaussian elimination method is the transformation of this set of equations into a staggered set:

$$a'_{11}x_1 + a'_{12}x_2 + a'_{13}x_3 = b'_1$$
$$a'_{22}x_2 + a'_{23}x_3 = b'_2$$
$$a'_{33}x_3 = b'_3$$

All coefficients a'_{ij} below the diagonal are zero. The solution in this case is straightforward. The last equation is solved for x_3. Now, the second can be solved by inserting the value of x_3. This procedure can be repeated for the uppermost equation.

The question is how to transform the given set of equations into a staggered set. This can be achieved by the method of successive elimination of variables. The following steps are necessary:

(1) We have to eliminate x_1 in all but the first equation. This can be done by subtracting $\dfrac{a_{21}}{a_{11}}$ times the first equation from the second equation and $\dfrac{a_{31}}{a_{11}}$ times the first equation from the third equation.

(2) We have to eliminate x_2 in all but the second equation. This can be done by subtracting $\dfrac{a_{32}}{a_{22}}$ times the second equation from the third equation.

(3) Determination of the variables. Starting with the last equation in the set and proceeding upwards, we obtain first x_3, then x_2, and finally x_1.

This procedure is called the *Gaussian method of elimination*. It can be extended to sets of any number of linear equations.

EXAMPLE   We can solve the following set of equations according to the procedure given:

$$6x_1 - 12x_2 + 6x_3 = 6 \qquad [1]$$
$$3x_1 - 5x_2 + 5x_3 = 13 \qquad [2]$$
$$2x_1 - 6x_2 + 0 = -10 \qquad [3]$$

(1) Elimination of x_1. We multiply equation 1 by $\frac{3}{6}$ and subtract it from equation 2. Then we multiply equation 1 by $\frac{2}{6}$ and subtract it from equation 3. The result is

$$6x_1 - 12x_2 + 6x_3 = 6 \qquad [1]$$
$$x_2 + 2x_3 = 10 \qquad [2']$$
$$-2x_2 - 2x_3 = -12 \qquad [3']$$

(2) Elimination of x_2. We multiply equation 2' by 2 and add it to equation 3'. The result is

$$6x_1 - 12x_2 + 6x_3 = 6 \qquad [1]$$
$$x_2 + 2x_3 = 10 \qquad [2']$$
$$2x_3 = 8 \qquad [3'']$$

(3) Determination of the variables x_1, x_2, x_3. Starting with the last equation in the set, we obtain

$$x_3 = \tfrac{8}{2} = 4$$

Now equation 2' can be solved for x_2 by inserting the value of x_3. Thus

$$x_2 = 2$$

This procedure is repeated for equation 1 giving

$$x_1 = 1$$

15.2.2 GAUSS–JORDAN ELIMINATION

Let us consider whether a set of n linear equations with n variables can be transformed by successive elimination of the variables into the form

$$x_1 + 0 + 0 + \ldots + 0 = C_1$$
$$0 + x_2 + 0 + \ldots + 0 = C_2$$
$$0 + 0 + x_3 + \ldots + 0 = C_3$$
$$\vdots \quad \vdots \quad \vdots \qquad \vdots \quad \vdots$$
$$0 + 0 + 0 + \ldots + x_n = C_n$$

The transformed set of equations gives the solution for all variables directly. The transformation is achieved by the following method, which is basically an extension of the Gaussian elimination method.

At each step, the elimination of x_j has to be carried out not only for the coefficients below the diagonal, but also for the coefficients above the diagonal. In addition, the equation is divided by the coefficient a_{jj}.

This method is called *Gauss-Jordan elimination*.

We show the procedure by using the previous example.

This is the set

$$6x_1 - 12x_2 + 6x_3 = 6$$
$$3x_1 - 5x_2 + 5x_3 = 13$$
$$2x_1 - 6x_2 + 0 = -10$$

To facilitate the numerical calculation, we will begin each step by dividing the respective equation by a_{jj}.

(1) We divide the first equation by $a_{11} = 6$ and eliminate x_1 in the other two equations.

        Second equation: we subtract $3 \times$ first equation
        Third equation: we subtract $2 \times$ first equation

This gives

$$x_1 - 2x_2 + x_3 = 1 \qquad [1]$$
$$0 + x_2 + 2x_3 = 10 \qquad [2]$$
$$0 - 2x_2 - 2x_3 = -12 \qquad [3]$$

(2) We eliminate x_2 above and below the diagonal.

        Third equation: we add $2 \times$ second equation
        First equation: we add $2 \times$ second equation

This gives

$$x_1 + 0 + 5x_3 = 21 \qquad [1']$$
$$0 + x_2 + 2x_3 = 10 \qquad [2']$$
$$0 + 0 + 2x_3 = 8 \qquad [3']$$

(3) We divide the third equation by a_{33} and eliminate x_3 in the two equations above it.

> Second equation: we subtract $2 \times$ third equation
> First equation: we subtract $5 \times$ third equation

This gives

$$x_1 + 0 + 0 \ = \ 1 \qquad\qquad [1'']$$
$$0 + x_2 + 0 \ = \ 2 \qquad\qquad [2'']$$
$$0 + 0 + x_3 \ = \ 4 \qquad\qquad [3'']$$

This results in the final form which shows the solution.

15.2.3 MATRIX NOTATION OF SETS OF EQUATIONS AND DETERMINATION OF THE INVERSE MATRIX

Let us consider the following set of linear algebraic equations:

$$a_{11}x_1 + a_{12}x_2 + a_{13}x_3 \ = \ b_1$$
$$a_{21}x_1 + a_{22}x_2 + a_{23}x_3 \ = \ b_2$$
$$a_{31}x_1 + a_{32}x_2 + a_{33}x_3 \ = \ b_3$$

This set of equations can formally be written as a matrix equation. Let $\mathbf{A}$ be a matrix, whose elements are the coefficients a_{ij}. It is called a *matrix of coefficients*.

$$\mathbf{A} \ = \ \begin{vmatrix} a_{11} & a_{12} & a_{13} \\ a_{21} & a_{22} & a_{23} \\ a_{31} & a_{32} & a_{33} \end{vmatrix}$$

$\mathbf{x}$ and $\mathbf{b}$ are column vectors:

$$\mathbf{x} \ = \ \begin{vmatrix} x_1 \\ x_2 \\ x_3 \end{vmatrix} \qquad \mathbf{b} \ = \ \begin{vmatrix} b_1 \\ b_2 \\ b_3 \end{vmatrix}$$

The set of equations can now be written

$$\mathbf{Ax} \ = \ \mathbf{b}$$

In Chapter 14 we discussed the rules for the multiplication of matrices, the concepts of the inverse matrix $\mathbf{A}^{-1}$ and of the unit matrix $\mathbf{I}$. You will remember that $\mathbf{A}^{-1}\mathbf{A} = \mathbf{I}$.

Let us consider a matrix equation representing a set of linear algebraic equations:

$$\mathbf{Ax} \ = \ \mathbf{b}$$

We will now multiply both sides of this matrix equation from the left (premultiplication) by the inverse of $\mathbf{A}$.

$$\mathbf{A}^{-1}\mathbf{Ax} \ = \ \mathbf{A}^{-1}\mathbf{b}$$

Since $\mathbf{A}^{-1}\mathbf{A} = \mathbf{I}$, we obtain

$$\mathbf{Ix} \ = \ \mathbf{A}^{-1}\mathbf{b}$$

This equation is in fact the solution of the set of linear equations in matrix notation. But at present we do not have the inverse A^{-1} of the matrix of coefficients A to perform this multiplication. On the other hand, we do know a method for solving a set of linear equations, e.g. the Gauss–Jordan elimination. We want to find if a relationship exists between the solution of a set of equations and the determination of A^{-1}.

Without giving the proof we can state the answer. We transform the matrix of coefficients A by the Gauss–Jordan elimination into a unit matrix I. If we apply all operations simultaneously to a unit matrix I, the latter will be transformed into the inverse A^{-1}.

Thus we do not, in practice, gain a new method for solving a set of linear equations, but a method for calculating the inverse of a given matrix.

Consequently, if we form the inverse of the matrix of coefficients A and premultiply matrix b by it, then we obtain as a column vector the solution of x.

An $n \times m$ matrix can formally be augmented by another $n \times o$ matrix B thus forming an *augmented $n \times (m + o)$ matrix* denoted $A|B$. For example, $A|I$ is an augmented matrix whose first part consists of A and whose second part consists of I.

> **Rule**   Calculation of the *inverse* A^{-1} of a matrix A.
>
> Augment A by a unit matrix I. Execute the Gauss–Jordan elimination to transform the first part A of the augmented matrix into a unit matrix. Then the second part I will be transformed into A^{-1}.

As an example, we will show the calculation of the inverse matrix of A cited in section 14.7. Consider

$$A = \begin{pmatrix} 2 & 0 & 0 \\ 2 & 1 & -6 \\ 6 & 0 & -1 \end{pmatrix}$$

We extend A by I and get the augmented matrix $A|I$:

$$A|I = \begin{pmatrix} 2 & 0 & 0 & 1 & 0 & 0 \\ 2 & 1 & -6 & 0 & 1 & 0 \\ 6 & 0 & -1 & 0 & 0 & 1 \end{pmatrix}$$

Now we carry out the Gauss–Jordan elimination to transform its first part A into the unit matrix, following the steps described in the previous section.

(1) Division of the first row by $a_{11} = 2$ and elimination of the elements of the first column below the diagonal results in

$$\begin{pmatrix} 1 & 0 & 0 & \frac{1}{2} & 0 & 0 \\ 0 & 1 & -6 & -1 & 1 & 0 \\ 0 & 0 & -1 & -3 & 0 & 1 \end{pmatrix}$$

(2) The elements of the second column above and below the diagonal are already zero, so nothing has to be done for this step.

(3) Division of the third row by $a_{33} = -1$ and elimination of the element above in the third column results in

$$\left(\begin{array}{ccc|ccc} 1 & 0 & 0 & \frac{1}{2} & 0 & 0 \\ 0 & 1 & 0 & 17 & 1 & -6 \\ 0 & 0 & 1 & 3 & 0 & -1 \end{array}\right)$$

The second part of the augmented matrix represents A^{-1}:

$$A^{-1} = \begin{pmatrix} \frac{1}{2} & 0 & 0 \\ 17 & 1 & -6 \\ 3 & 0 & -1 \end{pmatrix}$$

Further, we make use of the matrix notation to facilitate the writing while transforming the system of equations.

Each row of the matrix equation $Ax = b$ represents a linear algebraic equation.

Suppose we multiply row i by a factor. Then all terms $a_{ij}x_j (j = 1 \ldots n)$ and b_i have to be multiplied by this factor. This is carried out by multiplying all elements in row i of the matrix of coefficients and b_i by this factor.

Suppose we add row i to row j. Then we have a new row whose coefficients are $(a_{i1} + a_{j1})$, $(a_{i2} + a_{j2})$, $\ldots$, $(a_{in} + a_{jn})$ and the value of b'_j is then $(b_i + b_j)$.

This equals the addition of corresponding elements of the coefficient matrix A of row i to row j and of b_i to b_j. It can be generalised for the addition of multiples of an equation and for subtraction of multiples of equations.

Thus the Gaussian elimination method and the Gauss–Jordan elimination can be carried out by performing the transformations with the elements of the matrix of coefficients and with the corresponding elements of b. This can be done using matrix notation if we augment the matrix of coefficients A with the column vector b and transform this augmented matrix $A|b$ according to the Gaussian or Gauss–Jordan elimination. Then the first part A will be transformed into a unit matrix and the column b will be transformed into the column vector of solutions. This is more concise and reduces the chance of making errors.

15.2.4   EXISTENCE OF SOLUTIONS

Number of Variables and Equations

We know that from one equation we can only determine one unknown variable. If we have one equation and two variables, one of the variables can only be expressed in terms of the other.

In order to determine n variables we need n equations. These equations must be linearly independent. An equation is linearly dependent if it can be expressed as a sum of multiples of the other equations.

If we have n variables and m linearly independent equations $(m < n)$, only m variables can be determined and $n - m$ variables can be freely chosen. Let us explain: in a system of m equations, $(n - m)$ variables can be shifted to the RHS, m variables remain at the LHS. The Gauss–Jordan elimination can now be carried out, giving a solution for m variables. But this solution contains the $n - m$ variables previously shifted to the RHS. Thus these are the freely chosen parameters.

If $m > n$, the system is overdetermined. It is solvable only if $m - n$ equations are linearly dependent.

Existence of a Solution

Let us consider a set of n linear equations containing n variables. If at any stage in the elimination procedure the coefficient a_{jj} of a variable x_j happens to be zero, the equation has to be changed for an equation whose coefficient of x_j below the diagonal is $\neq 0$. If all coefficients of x_j below the diagonal are zero too, the set has *no unique solution* or *no solution at all*. In this case, we proceed to the next variable and continue the elimination procedure.

The set of equations has no *unique solution* if on the RHS of row j the value of b_j is zero. This happens when the equation is linearly dependent on the other ones. The value of this variable is not determined and it is freely chosen. It should be added that if this happens r times we will have r variables freely chosen.

This can be understood if we note that a row of zeros reduces the number of equations. In this case, the number of variables n exceeds the number of remaining equations $(m = n - r)$ and, as has been stated above, $n - m = r$ parameters are freely chosen.

The set has *no solution at all* if on the RHS of row j the value of b_j is not zero. In this case we have the equation $0 = b_j$, which is impossible. Thus the set of equations contains contradictions and has no solution at all.

Solution of a Homogenous Set of Linear Equations

Consider a set of n linear equations with n variables. If all constants b_j on the RHS are zero, we have what is referred to as a set of *homogeneous linear equations*. There is a trivial solution with

$$x_j = 0, \qquad j = 1, \ldots, n$$

A non-trivial solution may also exist. In this case, there must be at least one equation linearly dependent on the others. Consequently, the solution is not unique and contains at least one parameter freely chosen.

EXAMPLE   Given a set of linear equations

$$\begin{pmatrix} 4 & -8 & 0 & -4 \\ 1 & 1 & 3 & 5 \\ 2 & -2 & 2 & 4 \\ -3 & 7 & 1 & 7 \end{pmatrix} x = \begin{pmatrix} -12 \\ 12 \\ 8 \\ 18 \end{pmatrix}$$

Augmented matrix $\mathbf{A} | \mathbf{b}$

$$\mathbf{A} | \mathbf{b} = \left(\begin{array}{cccc|c} 4 & -8 & 0 & 4 & -12 \\ 1 & 1 & 3 & 5 & 12 \\ 2 & -2 & 2 & 4 & 8 \\ -3 & 7 & 1 & 7 & 18 \end{array} \right)$$

We use matrix notation and carry out the transformations with the augmented matrix $\mathbf{A} | \mathbf{b}$.

(1) Division of the first row by a_{11} and then elimination of the coefficients in the first column: Subtraction of row 1 from row 2, subtraction of row 1 multiplied by 2 from row 3, and addition of row 1 multiplied by 4 to row 4, gives

$$\begin{pmatrix} 1 & -2 & 0 & -1 & -3 \\ 0 & 3 & 3 & 6 & 15 \\ 0 & 2 & 2 & 6 & 14 \\ 0 & 1 & 1 & 4 & 9 \end{pmatrix}$$

(2) Division of the second row by a_{22} and then elimination of the coefficients in the second column: Addition of row 2 multiplied by 2 to row 1, subtraction of row 2 multiplied by 2 from row 3, and subtraction of row 2 from row 4 gives

$$\begin{pmatrix} 1 & 0 & 2 & 3 & 7 \\ 0 & 1 & 1 & 2 & 5 \\ 0 & 0 & 0 & 2 & 4 \\ 0 & 0 & 0 & 2 & 4 \end{pmatrix}$$

(3) In the third column a_{33} and all coefficients below the diagonal happen to be zero. Thus we proceed to the fourth column. We divide the fourth row by a_{44} and eliminate the coefficients above. We obtain

$$\begin{pmatrix} 1 & 0 & 2 & 0 & 1 \\ 0 & 1 & 1 & 0 & 1 \\ 0 & 0 & 0 & 0 & 0 \\ 0 & 0 & 0 & 1 & 2 \end{pmatrix}$$

In the third row all elements are zero. Thus the set has no unique solution. The value of x_3 can be freely chosen and hence the values of x_1 and x_2 depend on this choice.

$$x_1 = 1 - 2x_3$$
$$x_2 = 1 - x_3$$
$$x_4 = 2$$

EXAMPLE   Solve the following set of homogeneous linear equations.

$$\begin{pmatrix} 1 & 4 & -1 \\ 4 & 16 & -4 \\ 2 & -3 & 1 \end{pmatrix} x = 0$$

Augmented matrix $A|b$

$$\begin{pmatrix} 1 & 4 & -1 & 0 \\ 4 & 16 & -4 & 0 \\ 2 & -3 & 1 & 0 \end{pmatrix}$$

(1) Eliminating the coefficients in the first column gives

$$\begin{pmatrix} 1 & 4 & -1 & 0 \\ 0 & 0 & 0 & 0 \\ 0 & -11 & 3 & 0 \end{pmatrix}$$

We see that the set has a non-trivial solution, since one row consists of zeros and is thus linearly dependent.

(2) Since $a_{22} = 0$, we interchange row 2 and row 3. Dividing the new diagonal element and eliminating the coefficient above the diagonal in the second column gives

$$\left(\begin{array}{ccc|c} 1 & 0 & \frac{1}{11} & 0 \\ 0 & 1 & -\frac{3}{11} & 0 \\ 0 & 0 & 0 & 0 \end{array}\right)$$

We are left with two equations for three variables. We write it down explicitly, shift the third variable to the RHS and obtain the solution:

$$x_1 = -\frac{1}{11}x_3$$

$$x_2 = \frac{3}{11}x_3$$

The variable x_3 is freely chosen. Thus the solution is not unique: it contains one free parameter.

15.3 DETERMINANTS

15.3.1 PRELIMINARY REMARKS ON DETERMINANTS

In this section on determinants we explain the concept and its properties. We give as an application the method of solving sets of linear equations known as Cramer's rule.

We will introduce the concept of a determinant by means of an example.

Consider two linear equations with two unknowns, x_1 and x_2:

$$\begin{array}{l} a_{11}x_1 + a_{12}x_2 = b_1 \\ a_{21}x_1 + a_{22}x_2 = b_2 \end{array} \quad \text{or} \quad \begin{pmatrix} a_{11} & a_{12} \\ a_{21} & a_{22} \end{pmatrix}\begin{pmatrix} x_1 \\ x_2 \end{pmatrix} = \begin{pmatrix} b_1 \\ b_2 \end{pmatrix}$$

These equations, when solved, give

$$x_1 = \frac{b_1 a_{22} - b_2 a_{12}}{a_{11}a_{22} - a_{12}a_{21}}$$

$$x_2 = \frac{b_2 a_{11} - b_1 a_{21}}{a_{11}a_{22} - a_{12}a_{21}}$$

The solutions exist, provided that the denominators are not equal to zero. We notice that these denominators are the same for x_1 and x_2. It is customary to express them as follows:

$$\begin{vmatrix} a_{11} & a_{12} \\ a_{21} & a_{22} \end{vmatrix} = a_{11}a_{22} - a_{12}a_{21}$$

This expression is called the *determinant* of the matrix **A**. If this determinant is different from zero, then unique solutions exist for x_1 and x_2.

The determinant is a prescription to assign a numerical value to a square matrix.

For example, we can speak of the determinant of the 2×2 matrix **A**. There are several notations used in the literature:

$$\det\begin{pmatrix} a_{11} & a_{12} \\ a_{21} & a_{22} \end{pmatrix} = \det \mathbf{A} = \begin{vmatrix} a_{11} & a_{12} \\ a_{21} & a_{22} \end{vmatrix} = \Delta = a_{11}a_{22} - a_{12}a_{21}$$

The given formula applies only for the determinant of a 2×2 matrix. But the evaluation of the determinant of a $n \times n$ matrix can be reduced successively to the evaluation of determinants of 2×2 matrices.

The solution of a set of two linear equations for x_1 and x_2 can be expressed in terms of determinants:

$$x_1 = \frac{\begin{vmatrix} b_1 & a_{12} \\ b_2 & a_{22} \end{vmatrix}}{\det A} \qquad x_2 = \frac{\begin{vmatrix} a_{11} & b_1 \\ a_{21} & b_2 \end{vmatrix}}{\det A}$$

This is Cramer's rule for two linear equations which will be dealt with generally in section 15.3.4.

15.3.2 DEFINITION AND PROPERTIES OF AN n-ROW DETERMINANT

Generally speaking, the determinant of a square matrix of order n (n rows and n columns) is referred to as an n-order determinant. Although the determinant is a prescription to assign one numerical value to a given square matrix consisting of n^2 elements, it is usual, before the numerical evaluation, to refer to elements, rows and columns of the determinant in the notation below. Nevertheless, it is essential to distinguish between a matrix, which is an array of numbers, and its determinant, which is a number.

$$\det \begin{pmatrix} a_{11} & a_{12} & \cdots & a_{1k} & \cdots & a_{1n} \\ \vdots & \vdots & & \vdots & & \vdots \\ a_{i1} & a_{i2} & \cdots & a_{ik} & \cdots & a_{in} \\ \vdots & \vdots & & \vdots & & \vdots \\ a_{n1} & a_{n2} & \cdots & a_{nk} & \cdots & a_{nn} \end{pmatrix} = \begin{vmatrix} a_{11} & a_{12} & \cdots & a_{1k} & \cdots & a_{1n} \\ \vdots & \vdots & & \vdots & & \vdots \\ a_{i1} & a_{i2} & \cdots & a_{ik} & \cdots & a_{in} \\ \vdots & \vdots & & \vdots & & \vdots \\ a_{n1} & a_{n2} & \cdots & a_{nk} & \cdots & a_{nn} \end{vmatrix}$$

With each element a_{ik} is associated a *minor* found by omitting row i and column k. The minors are determinants with $n-1$ rows and columns.

The *cofactor A_{ik}* is obtained by multiplying the minor of a_{ik} by $(-1)^{i+k}$.

The procedure for evaluating the cofactor A_{ik} is shown below.

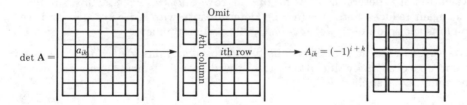

Finally, we define the *expansion of the determinant* by a row (or a column). It is defined by multiplying each element of the row (or column) by its cofactor and summing these products.

EXAMPLE   We expand the given determinant by the first row. First we evaluate the cofactors of the first row:

$$\det A = \begin{vmatrix} 1 & 2 & 3 \\ 3 & 2 & 1 \\ 5 & -3 & 1 \end{vmatrix}$$

Cofactor A_{11}:

$$A_{11} = (-1)^{1+1} \begin{vmatrix} 2 & 1 \\ -3 & 1 \end{vmatrix} = 1 \begin{vmatrix} 2 & 1 \\ -3 & 1 \end{vmatrix} = 2-(-3) = 5$$

Cofactor A_{12}:

$$A_{12} = (-1)^{1+2} \begin{vmatrix} 3 & 1 \\ 5 & 1 \end{vmatrix} = -1 \begin{vmatrix} 3 & 1 \\ 5 & 1 \end{vmatrix} = -1 \times 3 - (-1)5 = 2$$

Cofactor A_{13}:

$$A_{13} = (-1)^{1+3} \begin{vmatrix} 3 & 2 \\ 5 & -3 \end{vmatrix} = 1 \begin{vmatrix} 3 & 2 \\ 5 & -3 \end{vmatrix} = -9-10 = -19$$

Second, we multiply the cofactors by the elements a_{1j} and obtain the sum

$$1 \times 5 + 2 \times 2 + 3(-19) = -48$$

Without giving the proof, we state that the expansion of a determinant by different rows or columns always gives the same value.

Evaluation of determinants

The value of an n-order determinant is defined by the value of its expansion by any row or any column.

Expanding by the ith row gives

$$\det A = a_{i1}A_{i1} + a_{i2}A_{i2} + \ldots + a_{in}A_{in}$$

Expanding by the kth column gives

$$\det A = a_{1k}A_{1k} + a_{2k}A_{2k} + \ldots + a_{nk}A_{nk}$$

The value of the 3×3 determinant in the preceding example is thus given by the expansion which has already been obtained.

The evaluation of a determinant with n rows and n columns is reduced to the evaluation of n determinants with $(n-1)$ rows and $(n-1)$ columns. Applying the rule again reduces it to $n-(n-1)$ determinants with $(n-2)$ rows and $(n-2)$ columns and so on until we are left with 2-row determinants.

As a special case, it should be noted that the determinant of a diagonal matrix is given, up to sign, by the product of the diagonal elements. This follows if the given method is applied.

Hints for the Expansion of Second- and Third-Order Determinants

(a) *Second-order determinants.* The formula for the evaluation of a second-order determinant can be easily remembered with the help of the following scheme:

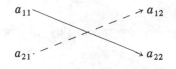

The value is given by the algebraic sum of the products formed by the elements on each of the two diagonals, the product taken downwards being positive, and that taken upwards being negative, i.e. $a_{11}a_{22} - a_{21}a_{12}$.

(b) *Third-order determinant.* We can establish a similar scheme for the expansion known as Sarrus' rule.

Sarrus' rule Repeating the first two columns of the determinant on the right, the expansion may be written down by taking the algebraic sum of the products formed by the elements on each of the six diagonals, as shown below; products taken downwards are positive and products taken upwards are negative.

$$\begin{matrix} a_{11} & a_{12} & a_{13} & a_{11} & a_{12} \\ a_{21} & a_{22} & a_{23} & a_{21} & a_{22} \\ a_{31} & a_{32} & a_{33} & a_{31} & a_{32} \end{matrix}$$

Negative / Positive

EXAMPLE Evaluate the determinant $\det \mathbf{A} = \begin{vmatrix} 2 & 3 & 5 \\ 2 & 1 & -3 \\ 1 & 3 & 4 \end{vmatrix}$

Solution 1 using the Sarrus' rule:

$$\det \mathbf{A} = \begin{vmatrix} 2 & 3 & 5 & 2 & 3 \\ 2 & 1 & -3 & 2 & 1 \\ 1 & 3 & 4 & 1 & 3 \end{vmatrix}$$

$$= 2 \times 1 \times 4 + 3(-3) \times 1 + 5 \times 2 \times 3$$
$$-1 \times 1 \times 5 - 3(-3) \times 2 - 4 \times 2 \times 3$$
$$= 8 - 9 + 30 - 5 + 18 - 24 = 18$$

Solution 2 using cofactors and expanding by the first column:

$$\det \mathbf{A} = \begin{vmatrix} 2 & 3 & 5 \\ 2 & 1 & -3 \\ 1 & 3 & 4 \end{vmatrix}$$

$$= (-1)^{1+1} 2 \begin{vmatrix} 1 & -3 \\ 3 & 4 \end{vmatrix} + (-1)^{2+1} 2 \begin{vmatrix} 3 & 5 \\ 3 & 4 \end{vmatrix} + (-1)^{3+1} 1 \begin{vmatrix} 3 & 5 \\ 1 & -3 \end{vmatrix}$$

$$= 2(4+9) - 2(12-15) + 1(-9-5)$$
$$= 26 + 6 - 14 = 18$$

Properties of Determinants

To evaluate determinants, in practice we frequently make use of the following properties (considering third-order determinants only for simplicity) to simplify the working.

Property 1 The value of the determinant is unaltered if columns and rows are interchanged (transposed).

$$\det \mathbf{A} = \det \mathbf{A}^{\mathrm{T}}$$

Since interchanging rows and columns does not affect the value of the determinant, any property established below for 'rows' also holds for 'columns'. This will not again be mentioned explicitly. Thus

$$\begin{vmatrix} a_{11} & a_{12} & a_{13} \\ a_{21} & a_{22} & a_{23} \\ a_{31} & a_{32} & a_{33} \end{vmatrix} = \begin{vmatrix} a_{11} & a_{21} & a_{31} \\ a_{12} & a_{22} & a_{32} \\ a_{13} & a_{23} & a_{33} \end{vmatrix}$$

Property 2   If two rows of the determinant are interchanged, the absolute value of the determinant is unaltered, but its sign is changed.

$$\begin{vmatrix} a_{11} & a_{12} & a_{13} \\ a_{21} & a_{22} & a_{23} \\ a_{31} & a_{32} & a_{33} \end{vmatrix} = - \begin{vmatrix} a_{21} & a_{22} & a_{23} \\ a_{11} & a_{12} & a_{13} \\ a_{31} & a_{32} & a_{33} \end{vmatrix} \qquad \text{(rows 1 and 2 are interchanged)}$$

Property 3   If all the elements of one row of the determinant are multiplied by a constant k, the new determinant is equal to $k \times$ (value of the original determinant).

$$\det A = \begin{vmatrix} a_{11} & a_{12} & a_{13} \\ ka_{21} & ka_{22} & ka_{23} \\ a_{31} & a_{32} & a_{33} \end{vmatrix} = k \begin{vmatrix} a_{11} & a_{12} & a_{13} \\ a_{21} & a_{22} & a_{23} \\ a_{31} & a_{32} & a_{33} \end{vmatrix}$$

If all the elements of the matrix are multiplied by a constant k the new determinant is equal to $k^n \times$ (value of the original determinant).

Property 4   If two rows of a determinant are identical, the value of the determinant is zero. This applies equally if two rows are proportional to each other.

Property 5   The value of a determinant is not altered by adding to the corresponding elements of any row the multiples of the elements of any other row

$$\begin{vmatrix} a_{11} & a_{12} & a_{13} \\ a_{21} & a_{22} & a_{23} \\ a_{31} & a_{32} & a_{33} \end{vmatrix} = \begin{vmatrix} a_{11}+ka_{21} & a_{12}+ka_{22} & a_{13}+ka_{23} \\ a_{21} & a_{22} & a_{23} \\ a_{31} & a_{32} & a_{33} \end{vmatrix} \qquad \begin{array}{l} k \times \text{(second row)} \\ \text{is added to first} \\ \text{row} \end{array}$$

Property 6   If each element of any row is expressed as the sum of two numbers, the determinant can be expressed as the sum of two determinants whose remaining rows are unaltered.

$$\begin{vmatrix} a_{11}+b_1 & a_{12}+b_2 & a_{13}+b_3 \\ a_{21} & a_{22} & a_{23} \\ a_{31} & a_{32} & a_{33} \end{vmatrix} = \begin{vmatrix} a_{11} & a_{12} & a_{13} \\ a_{21} & a_{22} & a_{23} \\ a_{31} & a_{32} & a_{33} \end{vmatrix} + \begin{vmatrix} b_1 & b_2 & b_3 \\ a_{21} & a_{22} & a_{23} \\ a_{31} & a_{32} & a_{33} \end{vmatrix}$$

Property 7   If the elements of any row are multiplied in order by the cofactors of the corresponding elements of another row, the sum of the products is zero.

$$a_{11}A_{21} + a_{12}A_{22} + a_{13}A_{23} = 0$$

Using properties 2, 3, 5, any determinant can be transformed so that only diagonal elements remain. The product of the diagonal elements is, except for the sign, the value of the determinant. This is equivalent to the Gauss–Jordan elimination. In practice, this method considerably reduces the amount of calculation involved in solving determinants of the fourth order and above. It should be noted that it is sufficient to eliminate the elements below the diagonal (Gaussian elimination), since the elimination of the elements above the diagonal does not affect the diagonal elements.

$$\begin{vmatrix} a_{11} & a_{12} & a_{13} \\ a_{21} & a_{22} & a_{23} \\ a_{31} & a_{32} & a_{33} \end{vmatrix} = \begin{vmatrix} P_1 & 0 & 0 \\ 0 & P_2 & 0 \\ 0 & 0 & P_3 \end{vmatrix} = P_1 P_2 P_3$$

The following example illustrates the application of the above properties before expanding by a row or column.

EXAMPLE   Evaluate the determinant: $\begin{vmatrix} 11 & 3 & 7 \\ 10 & 2 & 6 \\ 5 & 1 & 4 \end{vmatrix}$

Subtraction of row 2 from row 1 (property 5) gives $\begin{vmatrix} 1 & 1 & 1 \\ 10 & 2 & 6 \\ 5 & 1 & 4 \end{vmatrix}$

Subtraction of two times row 3 from row 2 gives $\begin{vmatrix} 1 & 1 & 1 \\ 0 & 0 & -2 \\ 5 & 1 & 4 \end{vmatrix}$

According to property 3, we can write $-2 \begin{vmatrix} 1 & 1 & 1 \\ 0 & 0 & 1 \\ 5 & 1 & 4 \end{vmatrix}$

We interchange row 1 and row 2 (property 2) and evaluate the cofactor A'_{13}, obtaining

$$2 \begin{vmatrix} 0 & 0 & 1 \\ 1 & 1 & 1 \\ 5 & 1 & 4 \end{vmatrix} = 2(-4) = -8$$

The determinant can also be solved by transformation into a diagonal form. In this case we get

$$\begin{vmatrix} 1 & 0 & 0 \\ 0 & -4 & 0 \\ 0 & 0 & +2 \end{vmatrix} = (+2)(-4) \begin{vmatrix} 1 & 0 & 0 \\ 0 & 1 & 0 \\ 0 & 0 & 1 \end{vmatrix} = -8$$

15.3.3   RANK OF A DETERMINANT AND RANK OF A MATRIX

If $\det A \neq 0$, we define the rank r of an n order determinant as $r = n$. If $\det A = 0$, the rank r is less than n. In this case, the rank of the determinant is defined by the order of the largest minor whose determinant does not vanish. Thus its rank r is m if a minor with m rows exists which is not zero, but all minors with more than m rows are zero.

The rank of a square matrix is defined by the rank of its determinant. From a $m \times n$ matrix, submatrices can be formed by deleting some of its rows or columns. The rank of a $m \times n$ matrix is the rank of the square matrix with the highest rank which can be formed.

EXAMPLE   Evaluate the rank of the matrix and of its determinant.

$$\det A = \begin{vmatrix} 1 & 2 & 1 & 2 \\ 2 & 0 & 2 & 0 \\ 1 & 0 & 1 & 0 \\ 2 & 2 & 2 & 2 \end{vmatrix}$$

It is not practical to evaluate the determinant of the matrix by calculating the minors as the calculation involved is rather tedious. We had better try to transform the determinant. If we subtract row 1 and row 3 from row 4, the latter becomes zero. If we subtract half of row 2 from row 3 the latter becomes zero. Hence

$$\det A = \begin{vmatrix} 1 & 2 & 1 & 2 \\ 2 & 0 & 2 & 0 \\ 0 & 0 & 0 & 0 \\ 0 & 0 & 0 & 0 \end{vmatrix}$$

There are only minors of rank two which do not vanish. Thus the rank of the matrix and its determinant is 2. The same result is obtained if we notice that two pairs of columns of the original matrix are equal.

15.3.4   APPLICATIONS OF DETERMINANTS

Cramer's Rule

Cramer's rule is a method for solving sets of linear algebraic equations using determinants. This method is of theoretical interest. In practice, it will only be feasible for sets of two or three equations.

Given a set of equations in matrix notation

$$\begin{pmatrix} a_{11} & \cdots & a_{1n} \\ \vdots & & \vdots \\ a_{n1} & \cdots & a_{nn} \end{pmatrix} \begin{pmatrix} x_1 \\ \vdots \\ x_n \end{pmatrix} = \begin{pmatrix} b_1 \\ \vdots \\ b_n \end{pmatrix}$$

Let $\det A$ be the determinant of the matrix of coefficients A. If $\det A \neq 0$, the system has a unique solution.

Let $\mathbf{A}^{(k)}$ be a matrix which is obtained by replacing in the matrix of coefficients the kth column by the column vector $\mathbf{b}$. The solution is then given by

$$x_k = \frac{\det \mathbf{A}^{(k)}}{\det \mathbf{A}} \qquad (k = 1, 2, 3, \ldots, n)$$

We will refrain from giving the proof. Although it is straightforward it is quite tedious.

Cramer's rule   Given a set of linear algebraic equations $\mathbf{Ax} = \mathbf{b}$, the solution is

$$x_k = \frac{\det \mathbf{A}^{(k)}}{\det \mathbf{A}} \qquad (k = 1, 2, 3, \ldots, n)$$

$\det \mathbf{A}^{(k)}$ is generated from $\det \mathbf{A}$ by replacing the column of coefficients a_{ik} of the variable x_k by the column vector $\mathbf{b}$.

Regarding Cramer's rule, we can draw some conclusions about the existence of a solution which are obvious and plausible and have already been stated in section 15.2.4.

(a) The case of a non-homogeneous set of n linear equations with n unknowns. If $\det \mathbf{A} = 0$, then Cramer's rule cannot be applied. Such a set of equations has either an infinite number of solutions or none at all. In this situation, the concept of the rank of a determinant is of great value.

   (i) If $\det \mathbf{A}$ is of rank $r < n$ and any of the determinants $\det \mathbf{A}^{(k)}$ are of rank greater than r, then no solution exists.
   (ii) If $\det \mathbf{A}$ is of rank $r < n$ and none of the $\det \mathbf{A}^{(k)}$ have a rank greater than r, then there is an infinite number of solutions.

(b) The case of a set of homogeneous linear equations ($\mathbf{b} = 0$).
   (i) This set of linear equations always has the trivial solution $x_1 = x_2 = \ldots = x_n = 0$.
   (ii) A non-trivial solution exists if and only if the rank r of the matrix $\mathbf{A}$ is less than n, i.e. $r < n$.
   (iii) A homogeneous set of equations with m independent equations and n unknowns has a solution which differs from zero if $n > m$. The solution contains $(n - m)$ arbitrary parameters.

EXAMPLE   Consider the following set of non-homogeneous equations:

$$x_1 + x_2 + x_3 = 8$$
$$3x_1 + 2x_2 + x_3 = 49$$
$$5x_1 - 3x_2 + x_3 = 0$$

It can be written in matrix notation thus:

$$\begin{pmatrix} 1 & 1 & 1 \\ 3 & 2 & 1 \\ 5 & -3 & 1 \end{pmatrix} \begin{pmatrix} x_1 \\ x_2 \\ x_3 \end{pmatrix} = \begin{pmatrix} 8 \\ 49 \\ 0 \end{pmatrix}$$

We can calculate the determinants:

$$\det A = \begin{vmatrix} 1 & 1 & 1 \\ 3 & 2 & 1 \\ 5 & -3 & 1 \end{vmatrix} = -12 \qquad \det A^{(1)} = \begin{vmatrix} 8 & 1 & 1 \\ 49 & 2 & 1 \\ 0 & -3 & 1 \end{vmatrix} = -156$$

$$\det A^{(2)} = \begin{vmatrix} 1 & 8 & 1 \\ 3 & 49 & 1 \\ 5 & 0 & 1 \end{vmatrix} = -180 \qquad \det A^{(3)} = \begin{vmatrix} 1 & 1 & 8 \\ 3 & 2 & 49 \\ 5 & -3 & 0 \end{vmatrix} = 240$$

From Cramer's rule, the solution is

$$x_1 = 13, \qquad x_2 = 15, \qquad x_3 = -20$$

EXAMPLE   Consider now the following set of non-homogeneous equations:

$$x_1 + 2x_2 + 3x_3 = 4$$
$$3x_1 - 7x_2 + x_3 = 13$$
$$4x_1 + 8x_2 + 12x_3 = 2$$

It can be written in matrix notation thus:

$$\begin{pmatrix} 1 & 2 & 3 \\ 3 & -7 & 1 \\ 4 & 8 & 12 \end{pmatrix} \begin{pmatrix} x_1 \\ x_2 \\ x_3 \end{pmatrix} = \begin{pmatrix} 4 \\ 13 \\ 2 \end{pmatrix}$$

We can calculate the determinant:

$$\det A = \begin{vmatrix} 1 & 2 & 3 \\ 3 & -7 & 1 \\ 4 & 8 & 12 \end{vmatrix} = 0$$

According to the above statement, this set of equations has either no unique solution or no solution at all. To decide which is the case, we use the Gauss–Jordan elimination and obtain, after the first step,

$$\begin{pmatrix} 1 & 2 & 3 \\ 0 & -13 & -8 \\ 0 & 0 & 0 \end{pmatrix} x = \begin{pmatrix} 4 \\ 1 \\ -14 \end{pmatrix}$$

The last equation $0 = -14$ is impossible. Thus the system has no solution at all. The same result follows if we look at the rank of the determinant of A. The rank is 2. Since the rank of $\det A^{(1)}$ is 3, there is no solution at all.

EXAMPLE   Consider again the set of homogeneous linear equations given in the Example on p. 383.

$$\begin{pmatrix} 1 & 4 & -1 \\ 4 & 16 & -4 \\ 2 & -3 & 1 \end{pmatrix} x = 0$$

The first and second equation differ only by the factor 4. Hence the equations are linearly dependent and

$$\begin{vmatrix} 1 & 4 & -1 \\ 4 & 16 & -4 \\ 2 & -3 & 1 \end{vmatrix} = \begin{vmatrix} 1 & 4 & -1 \\ 0 & 0 & 0 \\ 2 & -3 & 1 \end{vmatrix} = 0$$

Thus a non-trivial solution exists. Rewriting the first and third equations gives

$$x_1 + 4x_2 = x_3$$

$$2x_1 - 3x_2 = -x_3$$

From Cramer's rule, we have

$$x_1 = \frac{\begin{vmatrix} x_3 & 4 \\ -x_3 & -3 \end{vmatrix}}{\begin{vmatrix} 1 & 4 \\ 2 & -3 \end{vmatrix}} = -\frac{x_3}{11} \qquad x_2 = \frac{\begin{vmatrix} 1 & x_3 \\ 2 & -x_3 \end{vmatrix}}{\begin{vmatrix} 1 & 4 \\ 2 & -3 \end{vmatrix}} = \frac{3x_3}{11}$$

Hence we see that, as before, the solution contains one arbitrary parameter.

Vector Product in Determinant Notation

In Chapter 4, section 4.2.7, we defined the vector product of two vectors $\mathbf{a} = (a_x, a_y, a_z)$ and $\mathbf{b} = (b_x, b_y, b_z)$ as

$$\mathbf{a} \times \mathbf{b} = \mathbf{i}(a_y b_z - a_z b_y) + \mathbf{j}(a_z b_x - a_x b_z) + \mathbf{k}(a_x b_y - a_y b_x)$$

If we regard the expressions in brackets as two-row determinants, the RHS of the equation can be looked at as the evaluation of a three-row determinant:

$$\mathbf{a} \times \mathbf{b} = \begin{vmatrix} \mathbf{i} & \mathbf{j} & \mathbf{k} \\ a_x & a_y & a_z \\ b_x & b_y & b_z \end{vmatrix}$$

From the properties of determinants it follows that

$$\mathbf{a} \times \mathbf{b} = -\mathbf{b} \times \mathbf{a}, \qquad \text{since}$$

$$\begin{vmatrix} \mathbf{i} & \mathbf{j} & \mathbf{k} \\ a_x & a_y & a_z \\ b_x & b_y & b_z \end{vmatrix} = - \begin{vmatrix} \mathbf{i} & \mathbf{j} & \mathbf{k} \\ b_x & b_y & b_z \\ a_x & a_y & a_z \end{vmatrix}$$

Volume of a parallelepiped

Consider the parallelepiped defined by the three vectors $\mathbf{a}$, $\mathbf{b}$, and $\mathbf{c}$ (figure 15.1). From Chapter 4 we know that the value of the vector product $\mathbf{z} = \mathbf{a} \times \mathbf{b}$ represents the area of the base. Furthermore, $\mathbf{z}$ is a vector rectangular to the base.

The projection of $\mathbf{c}$ on to $\mathbf{z}$ represents the height of the parallelepiped. Thus the volume is

$$V = |\mathbf{c} \cdot \mathbf{z}| = |\mathbf{c}(\mathbf{a} \times \mathbf{b})|$$

Written as components:

$$V = |c_x(a_y b_z - a_z b_y) + c_y(a_z b_x - a_x b_z) + c_z(a_x b_y - a_y b_x)|$$

This can be expressed as a determinant (up to sign):

$$V = \begin{vmatrix} c_x & c_y & c_z \\ a_x & a_y & a_z \\ b_x & b_y & b_z \end{vmatrix} = \begin{vmatrix} a_x & a_y & a_z \\ b_x & b_y & b_z \\ c_x & c_y & c_z \end{vmatrix}$$

Note that the sign of the determinant is positive if **a**, **b** and **c** are oriented according to the right-hand screw rule (see Chapter 4).

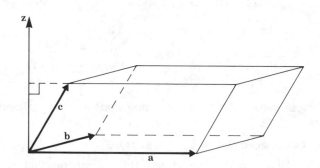

Figure 15.1

EXERCISES

15.2   Sets of Linear Equations

1. Solve the following equations using either Gaussian or Gauss–Jordan elimination. Use matrix notation.

    (a) $2x_1 + x_2 + 5x_3 = -21$
    $x_1 + 5x_2 + 2x_3 = 19$
    $5x_1 + 2x_2 + x_3 = 2$

    (b) $x - y + 3z = 4$
    $23x + 2y + 4z = 13$
    $11.5x + y + 2z = 6.5$

    (c) $x_1 + x_2 + x_3 = 8$
    $3x_1 + 2x_2 + x_3 = 49$
    $5x_1 - 3x_2 + x_3 = 0$

    (d) $1.2x - 0.9y + 1.5z = 2.4$
    $0.8x - 0.5y + 2.5z = 1.8$
    $1.6x - 1.2y + 2z = 3.2$

2. Obtain the inverse of the following matrices:

    (a) $\begin{pmatrix} 2 & 1 & 0 \\ 1 & 1 & -2 \\ 0 & 3 & -4 \end{pmatrix}$

    (b) $\begin{pmatrix} -4 & 8 \\ -6 & 7 \end{pmatrix}$

3. Investigate the following sets of homogeneous equations and obtain their solutions.

    (a) $x_1 + x_2 - x_3 = 0$
    $-x_1 + 3x_2 + x_3 = 0$
    $x_2 + x_3 = 0$

    (b) $2x - 3y + z = 0$
    $4x + 4y - z = 0$
    $x - \frac{3}{2}y + \frac{1}{2}z = 0$

15.3   Determinants

4. Evaluate the following determinants:

(a) $\begin{vmatrix} 4 & 3 & 2 \\ 1 & 0 & -1 \\ 5 & 2 & 2 \end{vmatrix}$

(b) $\begin{vmatrix} 1 & 7 & 4 & 12 \\ 5 & 5 & 4 & 3 \\ -2 & 6 & 25 & 3 \\ 5 & 35 & 20 & 60 \end{vmatrix}$

(c) $\begin{vmatrix} 3 & 4 & 0 & 2 \\ 6 & 1 & -3 & 1 \\ 0 & 0 & 4 & 0 \\ 5 & -1 & 2 & 4 \end{vmatrix}$

(d) $\begin{vmatrix} 4 & 6 & 0 & 7 \\ -3 & 0 & 2 & 8 \\ 10 & 1 & 0 & 2 \\ 5 & 2 & 0 & 1 \end{vmatrix}$

(e) $\begin{vmatrix} -1 & 0 & 2 & 3 \\ 2 & 1 & 8 & 5 \\ 0 & 0 & -4 & -2 \\ 1 & 0 & 1 & 4 \end{vmatrix}$

5. Determine the rank r of

(a) $A = \begin{pmatrix} -1 & 4 & 1 & 3 \\ 2 & -2 & -2 & 0 \\ 0 & 2 & 0 & 2 \end{pmatrix}$

(b) $B = \begin{pmatrix} 3 & 2 & 2 & 2 \\ 4 & 2 & 4 & 2 \\ 3 & 1 & 3 & 1 \\ 2 & 1 & 2 & 1 \end{pmatrix}$

6. Find out whether the sets of linear equations given in question 1 are uniquely solvable by examination of the determinant of the matrix of coefficients.

CHAPTER 16

Eigenvalues and eigenvectors of real matrices

16.1 TWO CASE STUDIES: EIGENVALUES OF 2 × 2 MATRICES

In Chapter 14 it was shown how a matrix $\mathbf{A}$ and a vector $\mathbf{r}$ can be multiplied to give a new vector $\mathbf{r}'$ (provided the dimensions of the vector and the matrix fit):

$$\mathbf{r}' = \mathbf{Ar}$$

Let us remember that each row of $\mathbf{A}$ is to be multiplied with $\mathbf{r}$, which is thought to be a column vector. As an example, we will consider a 2×2 matrix $\mathbf{A}$ and a 2-row vector $\mathbf{r}$; multiplication results in a new 2-row vector $\mathbf{r}'$.

If $\mathbf{A} = \begin{pmatrix} 0.5 & 0 \\ 0 & 2 \end{pmatrix}$ and $\mathbf{r} = \begin{pmatrix} 1 \\ 1 \end{pmatrix}$ then

$$\mathbf{r}' = \begin{pmatrix} 0.5 & 0 \\ 0 & 2 \end{pmatrix}\begin{pmatrix} 1 \\ 1 \end{pmatrix} = \begin{pmatrix} 0.5 \\ 2 \end{pmatrix}$$

Figure 16.1 shows both the old vector $\mathbf{r}$ and the new vector $\mathbf{r}'$. The result of applying $\mathbf{A}$ to $\mathbf{r}$ can be described as reducing the x-component by half and doubling the y-component.

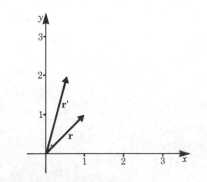

Figure 16.1

Generally, the new vector $\mathbf{r}'$ and the old vector $\mathbf{r}$ will point in different directions. However, there are some special vectors whose direction does not change when $\mathbf{A}$ is applied.

396

If **r** points along either axis, then for the matrix **A** under consideration the corresponding vector **r'** will point in the same direction (figure 16.2).

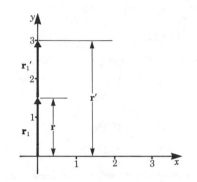

Figure 16.2

To give an example,

if $\qquad \mathbf{r}_1 = \begin{pmatrix} 0 \\ 1.5 \end{pmatrix} \qquad$ then $\qquad \mathbf{r}_1' = \begin{pmatrix} 0.5 & 0 \\ 0 & 2 \end{pmatrix}\begin{pmatrix} 0 \\ 1.5 \end{pmatrix} = \begin{pmatrix} 0 \\ 3 \end{pmatrix} = 2\mathbf{r}_1$

Instead of applying **A** to this special vector we could simply multiply $\mathbf{r}_1$ by the scalar 2. This is, of course, by no means true for any vector. Therefore a special nomenclature has been introduced.

> **DEFINITION**   Given an $n \times n$ matrix **A** and an n-vector **r**, if **r'** = **Ar** points in the same direction as **r**, i.e. $\mathbf{r}' = \lambda\mathbf{r}$ where λ is a real scalar, then **r** is called an *eigenvector* of **A** with real *eigenvalue* λ. The cases **r** = **0** or $\lambda = 0$ are excluded from this definition.

The last example could thus be rephrased as follows. The vector $\mathbf{r}_1$ is an eigenvector of **A** and the corresponding eigenvalue $\lambda_1 = 2$. In this case, there is also a second eigenvector, e.g. $\mathbf{r}_2 = \begin{pmatrix} 1 \\ 0 \end{pmatrix}$, with eigenvalue $\lambda_2 = 0.5$. Thus the matrix **A** possesses two real eigenvalues and we have found two corresponding eigenvectors.

Three questions now arise:

(1) What is the maximum number of real eigenvalues and eigenvectors for a given matrix?

(2) Does every matrix possess real eigenvalues and eigenvectors?

(3) How can these real eigenvalues and eigenvectors be computed?

We will restrict our examples to the case of 2×2 and 3×3 matrices, and, before discussing generalities, we should look at a second, slightly less trivial, case.

EXAMPLE   For $\mathbf{A} = \begin{pmatrix} 1.25 & 0.75 \\ 0.75 & 1.25 \end{pmatrix}$ find the eigenvalues and eigenvectors.

Clearly, vectors pointing in the direction of an axis do not solve this problem. We could embark on a trial and error search. But that could be tedious because real eigenvalues might not exist!

Therefore, let us start by reformulating the problem. We wish to find a number λ and a vector $\mathbf{r}$ such that

$$\mathbf{A}\mathbf{r} = \lambda \mathbf{r} \qquad (16.1)$$

Let us write this down as a set of two equations for the x and y components of $\mathbf{r}$:

$$\mathbf{A} = \begin{pmatrix} 1.25 & 0.75 \\ 0.75 & 1.25 \end{pmatrix} \qquad \mathbf{r} = \begin{pmatrix} x \\ y \end{pmatrix}$$

The equations are

$$1.25x + 0.75y = \lambda x$$
$$0.75x + 1.25y = \lambda y$$

By subtracting the RHS, a homogeneous set of two linear equations is obtained:

$$(1.25 - \lambda)x + 0.75y = 0$$
$$0.75x + (1.25 - \lambda)y = 0 \qquad (16.2)$$

By definition, the trivial solution $x = y = 0$ does not interest us. Are there any non-trivial solutions? We know from Chapter 15 that these indeed exist, if the determinant of the coefficients vanishes:

$$(1.25 - \lambda)^2 - 0.75^2 = 0 \qquad (16.3)$$

This is a quadratic equation in λ, and there are two distinct real roots:

$$\lambda_1 = 2, \qquad \lambda_2 = 0.5$$

The computed values are the only candidates for the eigenvalues of $\mathbf{A}$. Inserting them one after the other into the set of equations 16.2 in fact gives the following solutions:

For λ_1
$$\mathbf{r}_1 = \begin{pmatrix} 1 \\ -1 \end{pmatrix}$$

For λ_2
$$\mathbf{r}_2 = \begin{pmatrix} 1 \\ 1 \end{pmatrix}$$

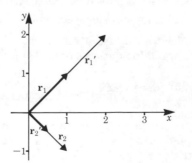

Figure 16.3

(Any scalar multiple would naturally do as well.) For the sake of clarity, we will check explicitly that these vectors do satisfy equation 16.1 with $\lambda = \lambda_1$ and $\lambda = \lambda_2$, respectively (figure 16.3):

$$r_1' = \mathbf{A}r_1 = \begin{pmatrix} 1.25 & 0.75 \\ 0.75 & 1.25 \end{pmatrix}\begin{pmatrix} 1 \\ 1 \end{pmatrix} = \begin{pmatrix} 2 \\ 2 \end{pmatrix} = 2\begin{pmatrix} 1 \\ 1 \end{pmatrix},$$

$$r_2' = \mathbf{A}r_2 = \begin{pmatrix} 1.25 & 0.75 \\ 0.75 & 1.25 \end{pmatrix}\begin{pmatrix} 1 \\ -1 \end{pmatrix} = \begin{pmatrix} 0.5 \\ -0.5 \end{pmatrix} = 0.5\begin{pmatrix} 1 \\ -1 \end{pmatrix}$$

Let us recapitulate. There are two eigenvalues for **A** and for each of them an eigenvector has been found. The eigenvalues were obtained as the roots of equation 16.3. That equation deserves some attention. It is called the *characteristic equation* of **A**.

We know full well that polynomial equations need not have any real roots, and, in general, some roots are complex and some are real. There are, at most, as many real roots as is the degree of the equation; in particular, a 2×2 matrix has, at most, two real eigenvalues. (Consider for example, the matrix given in question 3 of the exercise at the end of this chapter.) Also, any 2×2 matrix describing a rotation about an angle $\alpha \neq 0$ or π, has, evidently, no real eigenvalues.

Please note that we are dealing throughout this book with real matrices and real vectors, i.e. all entries must be real numbers. Accordingly, it would not be suitable to use complex scalars for multiplying vectors, and we do not consider complex eigenvalues. But you should be aware that in other situations it may be quite useful, or even unavoidable, to use the complex values.

16.2 GENERAL METHOD FOR FINDING EIGENVALUES

In order to find a general procedure for obtaining all eigenvalues and eigenvectors of a given matrix **A**, we will retrace the steps taken in the preceding section; however we will employ a somewhat more abstract notation.

Given a square $n \times n$ matrix **A**, we want to find all real eigenvalues of **A** (up to n distinct values) and an eigenvector for each of them.

Equation 16.1 still describes the general situation correctly:

$$\mathbf{A}r = \lambda r$$

Let us insert a unit matrix **I** on the RHS:

$$\mathbf{A}r = \lambda \mathbf{I}r$$

As before, the RHS is subtracted:

$$(\mathbf{A} - \lambda \mathbf{I})r = 0$$

This is again a set of linear equations and the condition for finding non-trivial solutions is that the determinant should vanish.

> **Theorem**   For the real scalar λ to be an eigenvalue of the matrix $\mathbf{A}$ it must be a real root of the characteristic equation:
>
> $$\det(\mathbf{A} - \lambda \mathbf{I}) = 0 \qquad (16.4)$$
>
> This is a polynomial equation of degree n if $\mathbf{A}$ is an $n \times n$ matrix.

For convenience, we give the explicit forms of the characteristic equation for dimensions 2 and 3:

If
$$\mathbf{A} = \begin{pmatrix} a_{11} & a_{12} \\ a_{21} & a_{22} \end{pmatrix}$$

then the characteristic equation is

$$\lambda^2 - (a_{11} + a_{22})\lambda + a_{11}a_{22} - a_{12}a_{21} = 0 \qquad (16.5)$$

If
$$\mathbf{A} = \begin{pmatrix} a_{11} & a_{12} & a_{13} \\ a_{21} & a_{22} & a_{23} \\ a_{31} & a_{32} & a_{33} \end{pmatrix}$$

then the characteristic equation is

$$-\lambda^3 + (a_{11} + a_{22} + a_{33})\lambda^2 - (a_{11}a_{22} + a_{11}a_{33} + a_{22}a_{33} - a_{12}a_{21} - a_{13}a_{31} - a_{23}a_{32})\lambda$$
$$+ \det \mathbf{A} = 0 \qquad (16.6)$$

For a square matrix of any dimension n the characteristic polynomial starts with $(-1)^n \lambda^n + (-1)^{n-1}\lambda^{n-1}(a_{11} + a_{22} + \ldots + a_{nn})$ and it always ends with $+\det \mathbf{A}$.

The first non-obvious coefficient is always the sum of the entries along the main diagonal of $\mathbf{A}$. It is called the *trace* of $\mathbf{A}$:

$$\text{tr}(\mathbf{A}) = a_{11} + a_{22} + a_{33} + \ldots + a_{nn}$$

After determining all real roots of the characteristic polynomial, we proceed to solve the homogeneous systems of linear equations in order to find eigenvectors.

16.3   WORKED EXAMPLE: EIGENVALUES OF A 3 × 3 MATRIX

This section goes step by step through the details of finding the eigenvalues and eigenvectors of a given 3×3 matrix.

$$\mathbf{A} = \begin{pmatrix} 2 & 1 & 3 \\ 1 & 2 & 3 \\ 3 & 3 & 20 \end{pmatrix}$$

(1) Find the characteristic equation. Its RHS = 0 and its LHS is given by the determinant

$$\det \begin{pmatrix} 2-\lambda & 1 & 3 \\ 1 & 2-\lambda & 3 \\ 3 & 3 & 20-\lambda \end{pmatrix} = -\lambda^3 + 24\lambda^2 - 65\lambda + 42 = 0$$

(2) Find the roots of the characteristic equation. This means solving a cubic equation — and we could try any of several approaches to this problem.

We can (a) use numerical methods; (b) refer to Cardan's formulae for third-order equations to find the solutions explicitly; (c) try to guess a first solution λ_1 and then divide the cubic polynomial by $(\lambda - \lambda_1)$ in order to obtain a quadratic polynomial.

For the given matrix $\mathbf{A}$, we use the third approach. It is not hard to see that $\lambda_1 = 1$ is a root. Therefore, we can split off the linear factor $(\lambda - 1)$ and the characteristic polynomial can be written thus:

$$-\lambda^3 + 24\lambda^2 - 65\lambda + 42 = (\lambda - 1)(-\lambda^2 + 23\lambda - 42) = 0$$

In order to find the two other eigenvalues, if the roots are real, we solve the quadratic equation

$$\lambda^2 - 23\lambda + 42 = 0$$

Its solutions are

$$\lambda_{2,3} = \frac{23}{2} \pm \sqrt{\left(\frac{23}{2}\right)^2 - 42} = \frac{23}{2} \pm \frac{19}{2}$$

Now we know that there are, in fact, three distinct real eigenvalues of the given matrix $\mathbf{A}$: These are

$$\lambda_1 = 1, \quad \lambda_2 = 2 \quad \text{and} \quad \lambda_3 = 21$$

(3) For each eigenvalue λ_i we must now find a non-trivial solution $\mathbf{r}_i$ of the respective homogeneous sets of linear equations

$$(\mathbf{A} - \lambda_i \mathbf{I})\mathbf{r}_i = \mathbf{0}$$

The vectors obtained will be eigenvectors of the matrix $\mathbf{A}$ to the respective eigenvalue λ_i.

When $\lambda = 1$. Set to solve:

$$\begin{pmatrix} 1 & 1 & 3 \\ 1 & 1 & 3 \\ 3 & 3 & 19 \end{pmatrix} \begin{pmatrix} x_1 \\ y_1 \\ z_1 \end{pmatrix} = \mathbf{0}$$

$$1x_1 + 1y_1 + 3z_1 = 0$$

$$1x_1 + 1y_1 + 3z_1 = 0$$

$$3x_1 + 3y_1 + 19z_1 = 0$$

A particular and non-trivial solution is obtained if $z_1 = 0$ and hence $x_1 = -y_1$. We can put, for example, $x_1 = 1$, $y_1 = -1$. Then the vector found is

$$\mathbf{r}_1 = \begin{pmatrix} 1 \\ -1 \\ 0 \end{pmatrix}$$

It is an eigenvector of $\mathbf{A}$ with eigenvalue 1.

When $\lambda = 2$. Set to solve:

$$0x_2 + 1y_2 + 3z_2 = 0$$

$$1x_2 + 0y_2 + 3z_2 = 0$$

$$3x_2 + 3y_2 + 18z_2 = 0$$

The third equation can be seen to be linearly dependent on the two other equations, so we can multiply each one of the two first equations by 3 and add. We need only consider the first two equations:

$$y_2 + 3z_2 = 0$$

$$x_2 + 3z_2 = 0$$

They give $x_2 = y_2 = -3z_2$. A particular solution is obtained by, e.g. letting $z_2 = -1$:

$$\mathbf{r}_2 = \begin{pmatrix} 3 \\ 3 \\ -1 \end{pmatrix}$$

It is an eigenvector of **A** with eigenvalue 2.

When $\lambda = 21$. Set to solve:

$$-19x_3 + 1y_3 + 3z_3 = 0$$

$$1x_3 - 19y_3 + 3z_3 = 0$$

$$3x_3 + 3y_3 - 1z_3 = 0$$

Again, the third equation can be seen to be linearly dependent on the two other equations, so we can add the first two equations and divide by -6. We need only consider the first two equations.

They give $6x_3 = 6y_3 = z_3$.

A particular solution is obtained by, e.g. letting $z_3 = 6$:

$$\mathbf{r}_3 = \begin{pmatrix} 1 \\ 1 \\ 6 \end{pmatrix}$$

It is an eigenvector of **A** with eigenvalue 21.

The problem of finding the eigenvalues and eigenvectors of the given matrix **A** is thereby solved exhaustively.

16.4 IMPORTANT FACTS ON EIGENVALUES AND EIGENVECTORS

The matrix **A** in the preceding section was chosen deliberately. It is symmetric, i.e. it equals its own transpose. It seems we were lucky in being confronted with a matrix which duly possesses three real eigenvalues and corresponding eigenvectors. But that was not a coincidence; it illustrates the following theorem, which we shall not prove.

> **Theorem**   A real non-singular symmetric $n \times n$ matrix possesses n real eigenvalues. Corresponding eigenvectors can be found such that each of them is orthogonal to each one of the others.

You should not find it too difficult to verify the second half of the assertion for the case of the matrix **A**. We can now answer the three questions posed in section 16.1 more explicitly. (We assume that the matrix is non-singular.)

(1) The maximum number of real eigenvalues and eigenvectors of a given $n \times n$ matrix is n. If the matrix happens to be symmetric, this maximum is attained.

(2) The following statement is relevant only for the case of non-symmetric matrices. If n is even there may be no real eigenvalues at all of a given $n \times n$ matrix.

If n is odd there must be at least one real eigenvalue of a given matrix, since the characteristic polynomial is of odd degree.

(3) Eigenvalues are found by solving the characteristic equation (equation 16.4). Eigenvectors are determined by finding a non-trivial particular solution of the resulting set of homogeneous linear equations. Please remember that the values $\lambda = 0$ and $\mathbf{r} = \mathbf{0}$, respectively, are not admitted.

EXERCISES

1. (a) For $\mathbf{A} = \begin{pmatrix} 4 & 2 \\ 1 & 3 \end{pmatrix}$ find the eigenvalues.

    (b) In a diagram draw two corresponding eigenvectors.

2. Is it possible for a real 2×2 matrix to have one real and one complex eigenvalue?

3. Prove that there are no real eigenvalues of the matrix

$$\mathbf{A} = \begin{pmatrix} 3 & 2 \\ -2 & 1 \end{pmatrix}$$

4. (a) For $\mathbf{A} = \begin{pmatrix} -1 & -1 & 1 \\ -4 & 2 & 4 \\ -1 & 1 & 5 \end{pmatrix}$ find all the eigenvalues.

    (*Hint*: they are all integers.)

    (b) Find corresponding eigenvectors.

5. In certain, rare, cases finding suitable eigenvectors may prove difficult. In order to illustrate what might happen, find the roots of the characteristic equation for

$$\mathbf{A} = \begin{pmatrix} 1 & 1 \\ 0 & 1 \end{pmatrix}$$

Then try to find corresponding eigenvectors.

CHAPTER 17

Numerical methods

The purpose of this chapter is to present, rather concisely, some basic numerical procedures. The applied scientist and engineer will certainly wish to solve standard problems on his or her own computer without resorting to the help of a specialist; and a program library may not be readily available. We have assumed that the user has an understanding of how an algorithm, given as a flow diagram or verbally, can be implemented on a computing device.

The methods described are quite straightforward. Their understanding requires nothing that has not been treated in the preceding chapters. If a very high level of precision must be attained, or if the problem warrants more advanced methods, the reader is referred to the specialist literature. Rounding errors are a new feature which arises with numerical calculations. Computers can handle only a limited number of digits and usually only approximate values can be found as solutions of a particular problem.

17.1 HORNER'S ALGORITHM

We will consider the problem of computing functional values of a polynomial with the least amount of effort.

Given that $f(x) = a_n x^n + a_{n-1} x^{n-1} + \ldots + a_1 x + a_0 = \sum_{i=n}^{0} a_i x^i,$

we wish to find $f(x_0)$ for some given value of x_0.

Horner's algorithm provides the optimal algorithm for this. It is based on the idea of rearranging the polynomial in order to facilitate the calculation. As an example, let us look at a polynomial of degree $n = 4$:

$$f(x) = a_4 x^4 + a_3 x^3 + a_2 x^2 + a_1 x + a_0$$

After rearranging, this becomes

$$f(x) = \{[(a_4 x + a_3)x + a_2]x + a_1\}x + a_0$$

Evaluating the polynomial in the factorised version is Horner's algorithm. There are two steps:

Step 1: Identify the coefficients $a_n, a_{n-1}, a_{n-2}, \ldots, a_1, a_0$. Vanishing coefficients must not be omitted.

Step 2: Evaluate the polynomial beginning with the innermost bracket and proceed to the outermost. Intermediate values are denoted by $b_n, \ldots, b_0$:

404

Horner's algorithm:

$$b_n = a_n$$

$$b_i = b_{i+1}x_0 + a_i \qquad \text{(with } i = n-1, \ldots, 1, 0)$$

$$b_0 = b_1 x_0 + a_0$$

(17.1)

The last value obtained, b_0, is the desired number: $f(x_0) = b_0$

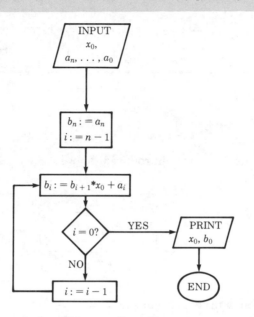

Figure 17.1

Figure 17.1 shows Horner's algorithm as a flow diagram.

Of course, in order to make the algorithm work on a computer, the cautious programmer would add some checks to ensure proper use and correct results. At first sight there is no reason why $n+1$ variables $b_n \ldots b_0$ should be defined. In fact, one single variable, say y, would suffice. We shall see very soon, however, that these b's can serve a useful purpose.

Let us now look at a concrete case.

EXAMPLE $f(x) = 4x^4 - x^2 + 2x + 5$. Evaluate $f(1.5)$. We follow the steps described above.

(1) $a_4 = 4$, $a_3 = 0$, $a_2 = -1$, $a_1 = 2$, $a_0 = 5$ (Note the value of a_3! It must not be omitted.)

(2) $b_4 = a_4 = 4$

$b_3 = 4 \times 1.5 + 0 = 6$

$b_2 = 6 \times 1.5 + (-1) = 8$

$b_1 = 8 \times 1.5 + 2 = 14$

$b_0 = 14 \times 1.5 + 5 = 26$

The value of the polynomial for $x_0 = 1.5$ is given by $b_0 = 26 = f(1.5)$

If the arithmetical operations are carried out by hand, then the following scheme will prove to be helpful. The arrows indicate the development of the computation. Starting with $b_n = a_n$, we write the products $b_k x_0 (k = n, \ldots, 1)$ in the adjoining column and add the respective coefficient *above* them. The last number, b_0, gives the required functional value $f(x_0)$.

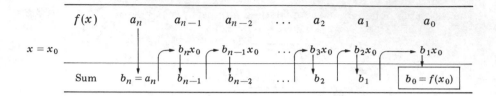

EXAMPLE   For the function introduced above, $f(x) = 4x^4 - x^2 + 2x + 5$, the pattern is as follows:

	4	0	−1	2	5
$x_0 = 1.5$		6	9	12	21
	4	6	8	14	$26 = f(1.5)$

We close this section with some comments:

If the polynomial $f(x)$ is divided by the linear factor $(x - x_0)$ then the values $b_n, \ldots, b_1$ are the coefficients of the quotient polynomial. Mathematically speaking

$$f(x) = \sum_{i=n}^{0} a_i x^i = (x - x_0) \sum_{i=n}^{1} b_i x^{i-1} + f(x_0)$$

For example: $f(x) = x^4 - x^2 + 2x + 5 = (x - 1.5)(4x^3 + 6x^2 + 8x + 14) + 26$

This piece of information is particularly useful if x_0 is a zero of $f(x)$. Horner's algorithm gives the quotient polynomial without further effort.

Furthermore, the value of the derivative $f'(x_0)$ is given by $f'(x_0) = \sum_{i=n}^{1} b_i x_0^{i-1}$

Regarding the question of reliability, unfortunately, in rare cases, rounding errors can accumulate drastically, especially if $|x_0|$ is large.

'Horner's method' was well known to the Chinese and Arab mathematicians half a millenium before G. W. Horner published his method in 1819.

17.2 ZEROS OF CONTINUOUS REAL FUNCTIONS

The solution z of a transcendental or a polynomial equation $f(x) = 0$ of more than degree 4 — called the zero of the equation — is in most cases not expressible explicitly. Therefore we calculate in a stepwise manner certain approximate values x_i. The following methods are valid for all types of differentiable or continuous functions; they need not be polynomial functions.

17.2.1 THE NEWTON–RAPHSON METHOD (TANGENT METHOD)
FOR DIFFERENTIABLE FUNCTIONS

Suppose we know an approximate solution x_1 of $f(z) = 0$. The basic idea is to replace, locally, the graph of $f(x)$ by a tangent. The zero x_2 of this tangent is easily evaluated, and it usually provides a better approximation than x_1 (see figure 17.2a).

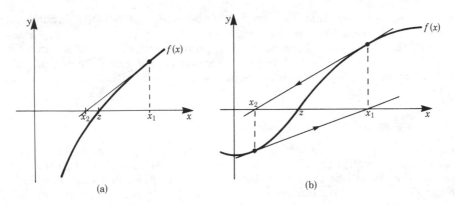

(a)   (b)

Figure 17.2

The formula for x_2 is

> Newton–Raphson method:
> $$x_2 = x_1 - \frac{f(x_1)}{f'(x_1)} \qquad (\text{with } f'(x_1) \neq 0) \tag{17.2}$$

By repeated application of this process, each time replacing the old x_1 by x_2, the true value of z can be approximated. Figure 17.3 shows the flow diagram. The process is terminated if $|f(x_2)|$ is less than some given δ. There is no exact information on how far the true value z is away. Please note also that this is *not* a foolproof algorithm: if the starting point is too far off its target then the sequence of alleged approximations might diverge or go round in circles in an infinite loop (see figure 17.2b).

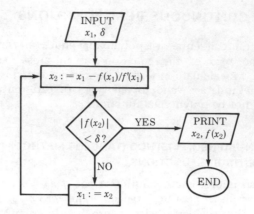

Figure 17.3

This method for finding a zero has already been introduced in Chapter 8. It can be interpreted as follows: the function $f(x)$ is replaced by the series expansion up to the first term, i.e. it is replaced by a linear function.

The snag is, however, that the given function may not be differentiable in the interval under consideration. An example is $f(x) = \sin|x^3 - 2x - 2|$. For this function, the Newton–Raphson method will not be appropriate, if we are close to the root of the polynomial ($1.769\ldots$), because $f(x)$ is not differentiable at this point. In many practical cases the function will be differentiable, but $f'(x)$ may be difficult to determine.

17.2.2   THE *REGULA FALSI* (SECANT METHOD)

This method for finding a zero is slightly less effective than the Newton–Raphson method, i.e. convergence to the true value z is slower. However, if $f'(x)$ is hard to evaluate, then it will be the right method to choose. Also, you may want to check the validity of the results obtained with one method by using a second method. The function $f(x)$ is assumed to be continuous, but it need not be differentiable.

Suppose we know of two points x_1 and x_2 which are close to a root z, and $f(x_1)$, $f(x_2)$ having opposite signs. A situation like this is depicted in figure 17.4.

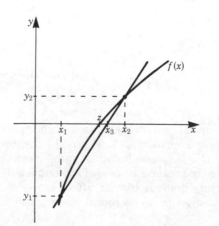

Figure 17.4

We know that z must lie in the interval $[x_1, x_2]$. In order to find an approximation x_3 for z the graph of $f(x)$ is replaced by the straight line segment (secant) between (x_1, y_1) and (x_2, y_2). Then x_3, the zero of the secant, is computed as follows:

Regula Falsi:

$$x_3 = x_1 - \frac{x_2 - x_1}{f(x_2) - f(x_1)} f(x_1) = \frac{x_1 f(x_2) - x_2 f(x_1)}{f(x_2) - f(x_1)} \qquad (17.3)$$

If $f(x_1)$ and $f(x_3)$ have opposite signs, then the root z must be in the interval $[x_1, x_3]$. Conversely, if $f(x_2)$ and $f(x_3)$ have opposite signs, then the root z must be in the interval $[x_3, x_2]$. By replacing either x_1 or x_2 accordingly by the new x_3, and by repeating this procedure, the uncertainty about z will be reduced step by step. Figure 17.5 shows a flow diagram. The process terminates if $|f(x_3)|$ is closer to 0 than some given δ.

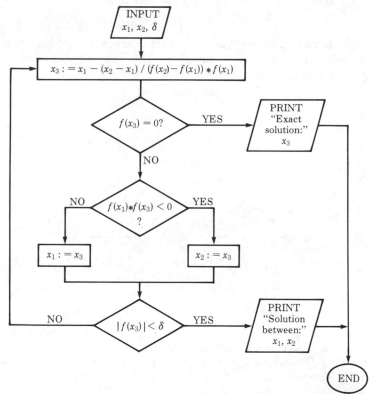

Figure 17.5

Again, the algorithm requires some standard precautions if it is implemented on a computer, e.g. an initial check that $f(x_1)$ and $f(x_2)$ have different signs. As a matter of fact, the algorithm *may* also work if both x_1 and x_2 lie on the same side of the root. It is sometimes suggested that there is no need to check the changing of the sign at all, replacing x_1 by x_2 and x_2 by x_3 in each step. This version of the algorithm is faster, but it is not guaranteed to produce sensible results.

17.2.3   HALVING THE INTERVAL (BISECTION METHOD)

Halving the interval is the crudest method for finding a zero, but it never fails. Given, as before, $f(x_1)$ and $f(x_2)$ with opposite signs, and hence with z in the interval $[x_1, x_2]$, define

$$x_3 = \tfrac{1}{2}(x_1 + x_2)$$

x_3 is the midpoint of the interval. Check if the changing of the sign takes place between x_1 and x_3 (case 1) or between x_3 and x_2 (case 2). Then replace either x_2 by x_3 (case 1) or x_1 by x_3 (case 2) and continue to bisect the interval until the true value z is enclosed in a small enough region.

17.2.4   ITERATION (METHOD OF SUCCESSIVE SUBSTITUTION)

The methods for finding zeros treated so far are prescriptions which must be carried out repeatedly, 'iterated', until the value obtained is sufficiently accurate.

The technical term *iteration*, however, is usually employed for the following situation: consider the problem of solving, for any given function $f^*(x)$,

$$f^*(x_0) = x_0$$

Geometrically speaking, we are looking for the intersection of the curve $y = f^*(x)$ with the straight line $y = x$ (see figure 17.6). This point of intersection x_0 is called *fixed point*. Applying f^* to x_0 does not produce any change.

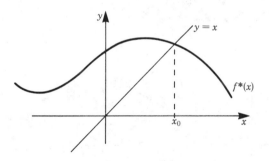

Figure 17.6

Let us choose a number x_1 which we know (or hope) is close enough to the solution x_0. Then $f^*(x_1) = x_2$ may be a better approximation for x_0. It therefore seems feasible that by applying f^* a number of times an approximation for x_0 to as good a degree as desired can be achieved:

Iteration:
$$x_1 = \text{estimated value}$$
$$x_2 = f^*(x_1)$$
$$x_3 = f^*(x_2) = f^*(f^*(x_1))$$
$$\ldots$$

(17.4)

In figure 17.7, the procedure is indeed shown to work, whereas in figure 17.8 it is shown to fail. The first figure depicts a *contracting* fixed point, the second a *repellent* fixed point.

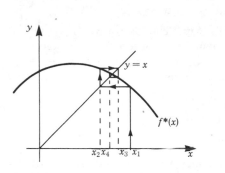 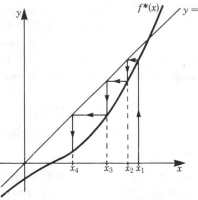

Figure 17.7   *Figure 17.8*

The moral to be learned from figure 17.8 is that iteration need not produce sensible approximations. But there is a handy criterion which guarantees that iteration does work under certain conditions:

Criterion for iteration:
If in some interval $[a, b]$ there exists a fixed point $x_0 = f^*(x_0)$ and a number $m < 1$ such that for all x in that interval

$$|f^{*\prime}(x)| \leqslant m$$

then any starting value in that interval gives rise to a sequence of values converging to x_0.

The flow diagram for the exceedingly simple algorithm of iteration is shown in figure 17.9.

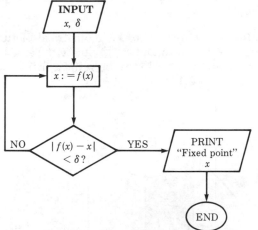

Figure 17.9

What still needs explaining is the connection between finding fixed points and finding zeros of functions. The answer is that if z solves $f(x) = 0$, then it also solves $Cf(x) + x = x$. For a given $f(x)$, let us introduce a new function: $f^*(x) = Cf(x) + x$.

Finding a zero of $f(x)$ is equivalent to finding a fixed point of $f*$. The constant C is to be chosen in such a way that the fixed point becomes contracting.

EXAMPLE   We wish to compute $\sqrt[3]{5}$ by solving $x^3 - 5 = 0$, i.e. by finding a zero of $f(x) = x^3 - 5$. Let us try to find a fixed point of $f*(x) = C(x^3 - 5) + x$.

We may concentrate on the interval $[1, 2]$, say. The derivative of $f*(x)$ is $f*'(x) = C3x^2 + 1$. If we choose $C = -0.1$, then the criterion for convergence is fulfilled and the iteration will contract to $x_0 = \sqrt[3]{5}$. Thus $f*(x) = -0.1x^3 + x + 0.5$ serves us well, and even so careless a starting value as $x = 1.0$ gives the desired result of $x_0 = 1.709\,975\,9\ldots$ within very few passes.

EXAMPLE   The following is known as Wallis' equation:

$$x^3 - 2x - 5 = 0$$

We wish to find its real root (there is only one such root). A crude sketch of the function $f(x) = x^3 - 2x - 5$ reveals that there must be a root in the interval $[2, 2.5]$. We will investigate the function $f*(x) = C(x^3 - 2x - 5) + x$. If we choose $C = -0.05$, then $|f*'(x)| \leqslant 0.5$ in this interval, and the criterion for convergence is fulfilled: the function $f*(x) = -0.05x^3 + 1.1x + 0.25$ has a contracting fixed point at $x_0 = 2.094\,551\,48\ldots$ This is the root of Wallis' equation.

17.3 NUMERICAL INTEGRATION

17.3.1 TRAPEZOIDAL RULE AND SIMPSON'S RULE FOR DEFINITE INTEGRALS

You will remember that the fundamental problem in the integral calculus is to compute the limit of infinite sums, e.g. in order to determine the area under a curve. That has been dealt with analytically in Chapter 6. In Chapter 8 a convenient device for numerical purposes was introduced, that of expanding a given complicated function into a power series and integrating term by term.

Trapezoidal Rule

For numerical purposes, let us return to the geometrical picture. We could follow the reasoning of Chapter 6 verbatim and approximate the area by a number of right quadrangles. But this method can obviously be refined, hence speeded up, by using trapeziums instead (cf. figure 17.10).

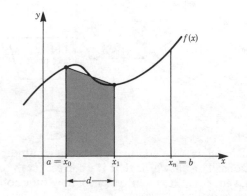

Figure 17.10

Thus the area is approximately

$$\int_a^b f(x)\,dx \;\approx\; \sum \text{trapezoidal areas}$$

It is convenient to use equidistant x values x_0, x_1, ..., x_n, the step width being $d = \dfrac{b-a}{n}$. The area of the kth trapezium is given by

$$\frac{d}{2}\left[f(x_k)+f(x_{k-1})\right]$$

The sum of all such values is an approximation for the definite integral. This is the *trapezoidal formula*:

> Trapezoidal formula:
> $$\int_a^b f(x)\,dx \;\approx\; \sum_{k=1}^{n}\frac{d}{2}\left[f(x_k)+f(x_{k-1})\right] \;=\; d\left(\frac{1}{2}\left[f(a)+f(b)\right]+\sum_{k=1}^{n-1}f(x_k)\right)\;(17.5)$$

In a given situation we will try to choose the subdivision fine enough to ensure a good approximation but coarse enough to avoid the accumulation of rounding errors.

If the function $f(x)$ is given empirically, i.e. through measurements, we may not have a choice at all: usually $f(x)$ will then be known only at finite points which should all be taken into account.

Simpson's Rule

It seems rather blunt to use straight line segments for approximating the curve to be integrated. Evidently, by using curved segments the accuracy of the approximation can be increased. This leads to a family of methods known as *Newton–Cotes methods*, where higher-order polynomials are employed instead of linear functions.

The formula obtained for polynomials of degree 2, i.e. parabolas, is *Simpson's rule* (see figure 17.11):

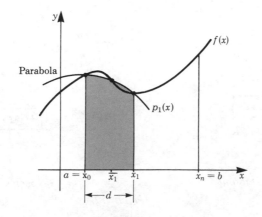

Figure 17.11

Simpson's rule:

$$\int_a^b f(x)\,dx \ \approx \ \frac{d}{3}\left(\frac{1}{2}\,[f(a)+f(b)] + \sum_{k=1}^{n-1} f(x_k) + 2\sum_{k=1}^{n} f(\bar{x}_k)\right), \qquad x_k = \frac{1}{2}(\bar{x}_k + x_{k-1})$$

(17.6)

With this formula good results are achieved with only a moderate amount of effort.

As before, in the trapezoidal formula, the interval $[a, b]$ has been divided into n equal strips of length d. But for Simpson's formula. in addition to the endpoints x_{k-1} and x_k, the midpoint $\bar{x}_k = \frac{1}{2}(x_k + x_{k-1})$ of the kth subinterval is needed. Observe that the y-values at midpoints have more weight assigned to them than the endpoints.

We will refrain from deriving Simpson's rule in detail. However, the general idea is firstly to find the parabola $p_k(x)$ that equals $f(x)$ at the three points $x_{k-1}, \bar{x}_k, x_k$, secondly to evaluate its integral

$$\int_{x_{k-1}}^{x_k} p_k(x)\,dx \ = \ \frac{d}{6}\left(f(x_{k-1}) + 4f(\bar{x}_k) + f(x_k)\right)$$

and thirdly to add up all the values.

In figures 17.12 and 17.13 the core of each algorithm for integration is shown.

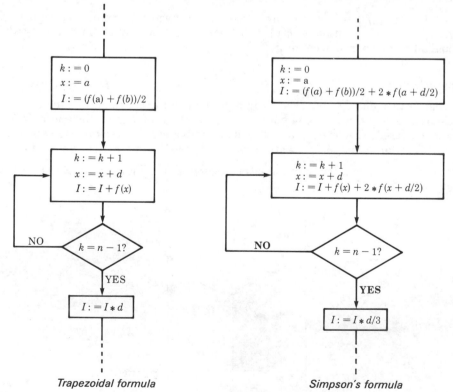

Trapezoidal formula   Simpson's formula

Figure 17.12   Figure 17.13

The trapezoidal rule uses as an approximation polynomials of degree 1 (straight lines), whereas Simpson's rule uses polynomials of degree 2 (parabolas) in order to increase the accuracy of the approximation. If higher-order polynomials are employed instead the accuracy of the approximation is only marginally improved. Therefore Simpson's rule is most frequently used.

EXAMPLE   Evaluate $\int_1^{11} \sqrt{x}\, dx$ by using the trapezoidal rule and Simpson's rule.

Apply formulas 17.5 and 17.6 with $x_0 = 1$, $x_5 = 11$, $n = 5$ and $d = 2$.

The solution is given in the following table:

x_i	y_i	Trapezoidal rule	Simpson's rule
$a = x_0 = 1$	1.000	0.500	0.500
$\bar{x}_1 = 2$	1.414		2.828
$x_1 = 3$	1.732	1.732	1.732
$\bar{x}_2 = 4$	2.000		4.000
$x_2 = 5$	2.236	2.236	2.236
$\bar{x}_3 = 6$	2.449		4.899
$x_3 = 7$	2.646	2.646	2.646
$\bar{x}_4 = 8$	2.828		5.657
$x_4 = 9$	3.000	3.000	3.000
$\bar{x}_5 = 10$	3.162		6.325
$b = x_5 = 11$	3.317	1.658	1.658
	Sum	11.772	35.481

$$\int_1^{11} x^{1/2}\, dx = 23.655 \qquad = 2 \times 11.772 \qquad = \tfrac{2}{3} \times 35.481$$

$$= 23.544 \qquad\qquad = 23.654$$

As an aside, we should like to point out that numerical differentiation, as opposed to integration, does not warrant special methods. To differentiate a given function we simply have to evaluate differential quotients. There is one general feature which should be taken into account when the curves under consideration are found empirically: differentiating usually makes a curve even more uneven, whereas integrating tends to smooth it out. Therefore, it may be advisable to find a smooth approximating curve to replace the empirically given curve before numerical differentiation is carried out.

17.3.2   FIRST-ORDER DEs: EULER'S METHODS

Euler's Method

We will indicate briefly how the formulae for definite integrals can be used to solve ordinary first-order DEs.

EXAMPLE   Given that $y' = y'(x, y)$, we wish to find the curve $y(x)$ through a given point (x_0, y_0).

The answer is obtained by constructing the curve step by step, starting from the given value $y_0 = y(x_0)$.

In order to compute the value of y at $x_0 + d$, we should evaluate an integral:

$$y(x_0 + d) = y(x_0) + \int_{x_0}^{x_0+d} y' \, dx$$

There are several ways the integral may be evaluated approximately. For qualitative considerations we may wish for a quick way to an approximate solution. This can be achieved by replacing the integral by the area of a rectangle:

$$\int_{x_0}^{x_0+d} y' \, dx \approx d \times y'(x_0, y_0)$$

The resulting formula for integrating a DE is called *Euler's method*:

Euler's method:
$$y(x_0 + d) \approx y(x_0) + d \times y'(x_0, y_0) \tag{17.7}$$

Improved Euler's Method

For better results, the integral in question should be approximated more faithfully. By using the trapezoidal rule we obtain, for one step,

$$\int_{x_0}^{x_0+d} y' \, dx \approx \frac{d}{2}\left(y'(x_0, y_0) + y'(x_0 + d, y(x_0 + d))\right)$$

We need y' at two points: $y'(x_0, y_0)$ and $y'(x_0 + d, y(x_0 + d))$. The latter value is approximately given by the following formula (note that Taylor's expansion is used for $y(x_0 + d)$):

$$y'(x_0 + d, y(x_0 + d)) \approx y'(x_0 + d, y(x_0) + d \times y'(x_0, y_0))$$

Putting it all together gives a good approximation for the next point. This method is called the *improved Euler's method*:

Improved Euler's method:
$$y(x_0 + d) \approx y(x_0) + \frac{d}{2}\left(y'(x_0, y_0) + y'(x_0 + d, y(x_0) + d \times y'(x_0, y_0))\right) \tag{17.8}$$

EXAMPLE   We will investigate the quality of both approximations by solving the equation

$$y' = 4x^2 - \frac{y}{x}$$

From Chapter 10, section 10.6.1, the analytic solution is known. It is

$$y(x) = x^3 + \frac{C}{x}$$

For $y_0 = y(0.1) = 10.001$, the special solution is

$$y(x) = x^3 + \frac{1}{x}$$

Using three different step widths, $d = 0.1$, $d = 0.01$ and $d = 0.001$, we compare results for the interval $[0.1, 1]$.

Euler's method for solving the DE $y' + \frac{y}{x} = 4x^2$, $f(0.1) = 10.001$

x	Step width $d = 0.1$	Step width $d = 0.01$	Step width $d = 0.001$	True value $x^3 + \frac{1}{x}$
0.1	10.001	10.001	10.001	10.001
0.2	0.0004	4.745	4.983	5.008
0.3	0.018	3.131	3.338	3.360
0.4	0.048	2.370	2.545	2.564
0.5	0.100	1.959	2.110	2.125
0.6	0.180	1.738	1.868	1.883
0.7	0.294	1.642	1.754	1.772
0.8	0.448	1.644	1.749	1.762
0.9	0.648	1.732	1.835	1.840
1.0	0.900	1.899	1.999	2.000

Note: The first column of y-values shows that the step width chosen for Euler's method in that case is too coarse.

Improved Euler's method for solving the DE $y' + \frac{y}{x} = 4x^2$, $f(0.1) = 10.001$

x	Step width $d = 0.1$	Step width $d = 0.01$	Step width $d = 0.001$	True value $x^3 + \frac{1}{x}$
0.1	10.001	10.001	10.001	10.001
0.2	5.009	5.007	5.007	5.008
0.3	3.363	3.360	3.359	3.360
0.4	2.568	2.563	2.564	2.564
0.5	2.130	2.124	2.126	2.125
0.6	1.889	1.881	1.883	1.883
0.7	1.779	1.771	1.768	1.772
0.8	1.770	1.762	1.762	1.762
0.9	1.849	1.841	1.846	1.840
1.0	2.010	2.001	2.009	2.000

Note: The third column of y-values shows that using too fine a step width introduces additional errors.

17.4 SETS OF LINEAR EQUATIONS

In Chapter 15 it was shown how a set of linear equations may be described by a matrix equation:

$$Ax = b$$

We will use matrix notation and describe a variation of the method presented in Chapter 15. The new feature of our algorithm is *pivoting*. Its rationale is as follows.

In practical applications the elements of the coefficient matrix and the elements of the vector **b** are often empirically obtained data. Their accuracy is limited as errors due to measurement or rounding are to be expected. The arithmetical operations may exacerbate this deficiency, especially when divisions by small numbers are performed. Pivoting is a prescription for avoiding some of the possible divisions by small numbers.

For eliminating x_1 we select the equation whose coefficient of x_1 has the largest absolute value. We assume that at least one element of the first column is $\neq 0$. This coefficient is called the *first pivot element* p_1. By rearranging we make this equation the first equation. (If it happens that there are two or more equally fitting candidates for the pivot element any one of them may be selected. For the sake of definiteness, we stipulate that the first one found should be chosen.) Now ordinary Gauss–Jordan elimination is carried out, i.e. the equation is divided by a_{11} and all other coefficients of x_1 are eliminated.

This procedure is now repeated for the other equations with respect to the second variable x_2. After $n-1$ such steps we are left with one single equation and no choice remains for the last pivot element p_n. This method is referred to as *partial pivoting*. The steps in the algorithm are as follows:

Gauss–Jordan elimination with partial pivoting:
(1) Augment the matrix **A** by the column vector **b**. (The new matrix **A*** = **A**|**b** is of size $n \times (n+1)$.)
(2) Starting with column $k = 1$ and proceeding to column $k = n$ (i.e. from the left to the right) do the following:
   (2.1) In column **k** in the lowest $n-k+1$ rows (i.e. below the diagonal) identify the entry of maximal absolute value. This is called pivot element p_k.
   (2.2) Interchange the row containing the pivot element with row k.
   (2.3) Divide all the coefficients of row k by p_k.
   (2.4) Subtract suitable multiples of that row from *all* the other rows to eliminate all entries but one in column k.
(3) Read the values in the appended column $k+1$. These are the values $x_1, \ldots, x_n$.

Partial pivoting is sometimes extended to general pivoting. This means we first select the coefficient with the largest absolute value, be it in any column whatsoever of the coefficient matrix. By rearranging the equations (rows) and variables (columns) we make this the first pivot element p_1 and the respective variable x_1. Then we carry out the first step of the Gauss–Jordan elimination. Second, we select the coefficient with the largest absolute value from the remaining equations (save the first) and make it the second pivot element p_2 by rearranging equations and variables again. This procedure is continued up to the last equation.

The added bonus of better numerical stability is, however, paid for by an increase in bookkeeping, because we must keep track of the swapping of variables. We do not propose to use this method in this book.

For the sake of concreteness, Gauss–Jordan elimination with partial pivoting will now be illustrated for a 4×4 set of linear equations.

EXAMPLE   Solve the following sets of equations:

$$\begin{pmatrix} 2 & -1 & 0 & 0 \\ 0 & -1 & 2 & -1 \\ -1 & 2 & -1 & 0 \\ 1 & 0 & 2 & -3 \end{pmatrix} \begin{pmatrix} x_1 \\ x_3 \\ x_2 \\ x_4 \end{pmatrix} = \begin{pmatrix} 1 \\ 3 \\ 2 \\ 2 \end{pmatrix}$$

Step 1: We construct the augmented matrix.

Step 2: This step consists of four passes:

Pass 1: Pivot element $p_1 = 2$ (in row 1). Steps (2.1) to (2.4) result in

$$\left(\begin{array}{cccc|c} 1 & -\frac{1}{2} & 0 & 0 & \frac{1}{2} \\ 0 & -1 & 2 & -1 & 3 \\ 0 & \frac{3}{2} & -1 & 0 & \frac{5}{2} \\ 0 & \frac{1}{2} & 2 & -3 & \frac{3}{2} \end{array}\right)$$

Pass 2: Pivot element $p_2 = \frac{3}{2}$ (in row 3). Steps (2.1) to (2.4) result in

$$\left(\begin{array}{cccc|c} 1 & 0 & -\frac{1}{3} & 0 & \frac{4}{3} \\ 0 & 1 & -\frac{2}{3} & 0 & \frac{5}{3} \\ 0 & 0 & \frac{4}{3} & -1 & \frac{14}{3} \\ 0 & 0 & \frac{7}{3} & -3 & \frac{2}{3} \end{array}\right)$$

Pass 3: Pivot element $p_3 = \frac{7}{3}$ (in row 4). Steps (2.1) to (2.4) result in

$$\left(\begin{array}{cccc|c} 1 & 0 & 0 & -\frac{3}{7} & \frac{10}{7} \\ 0 & 1 & 0 & -\frac{6}{7} & \frac{13}{7} \\ 0 & 0 & 1 & -\frac{9}{7} & \frac{2}{7} \\ 0 & 0 & 0 & \frac{5}{7} & \frac{30}{7} \end{array}\right)$$

Pass 4: There is no choice left for the pivot element. It must be the one in row 4, i.e. $p_4 = \frac{5}{7}$. Steps (2.1) to (2.4) result in

$$\left(\begin{array}{cccc|c} 1 & 0 & 0 & 0 & 4 \\ 0 & 1 & 0 & 0 & 7 \\ 0 & 0 & 1 & 0 & 8 \\ 0 & 0 & 0 & 1 & 6 \end{array}\right)$$

Step 3: The last column on the right gives the result $x_1 = 4$, $x_2 = 7$, $x_3 = 8$, $x_4 = 6$, and it easy to verify that these numbers are indeed correct.

The method described works well for almost any set of linear equations which possesses a unique solution. We know from Chapter 15 that this implies a non-vanishing determinant. If, however, $\det A = 0$, then one step, step (2.3), must fail, because there is no non-zero pivot element to be found. In fact, knowning the pivot elements $p_1, \ldots, p_n$ is (almost) the same as knowing $\det A$:

$$\det A = p_1 \times p_2 \times \ldots \times p_n \qquad \text{(up to sign)} \qquad (17.9)$$

The sign of the determinant is given by the number of times p_k is not found in row k, i.e. the number of times rows have to be interchanged: $+$ for an even number, $-$ for an odd number.

EXAMPLE   In the last example interchanging of rows took place in passes 2 and 3. Hence the determinant is positive:

$$\det A = 2 \times \tfrac{3}{2} \times \tfrac{7}{3} \times \tfrac{5}{7} = 5$$

Of course, the elimination method can also be used to compute the inverse matrix A^{-1}. This requires augmenting A by the unit matrix I: $A^* = A|I$ and proceeding as before.

EXAMPLE   Find the inverse of the matrix A of the previous example:

$$A|I = \begin{pmatrix} 2 & -1 & 0 & 0 & | & 1 & 0 & 0 & 0 \\ 0 & -1 & 2 & -1 & | & 0 & 1 & 0 & 0 \\ -1 & 2 & -1 & 0 & | & 0 & 0 & 1 & 0 \\ 1 & 0 & 2 & -3 & | & 0 & 0 & 0 & 1 \end{pmatrix}$$

Gauss–Jordan elimination turns this matrix into

$$\begin{pmatrix} 1 & 0 & 0 & 0 & | & 1 & 0.6 & 0.8 & -0.2 \\ 0 & 1 & 0 & 0 & | & 1 & 1.2 & 1.6 & -0.4 \\ 0 & 0 & 1 & 0 & | & 1 & 1.8 & 1.4 & -0.6 \\ 0 & 0 & 0 & 1 & | & 1 & 1.4 & 1.2 & -0.8 \end{pmatrix}$$

The right-hand part equals A^{-1}.

We know that the uniqueness of the solution of a set of linear equations is, in theory, a very clear-cut matter if $\det A \neq 0$. But in practice this is more complicated.

The first problem is that $\det A$ may be almost zero. If, in the elimination procedure, a pivot element turns out to be extremely small, then the matrix is 'almost singular', i.e. the solution will be unreliable. The algorithm should, therefore, include the following condition:

> If $|p_k| < \epsilon$ for some given positive bound ϵ then terminate
> and print out "Matrix is (almost) singular".

The second problem is that even if $\det A$ is not very small in absolute terms, it may be so in terms of its coefficients.

EXAMPLE

If
$$A = \begin{pmatrix} 10 & 99 \\ 1 & 10 \end{pmatrix} \quad \text{then} \quad \det A = 1$$

An error of measurement of 1% of a_{11} could change this to

$$A' = \begin{pmatrix} 10.1 & 99 \\ 1 & 10 \end{pmatrix}, \quad \text{and} \quad \det A' = 2$$

For $b = \begin{pmatrix} 1 \\ 0 \end{pmatrix}$ the first set of equations reads

$$\begin{pmatrix} 10 & 99 \\ 1 & 10 \end{pmatrix}\begin{pmatrix} x \\ y \end{pmatrix} = \begin{pmatrix} 1 \\ 0 \end{pmatrix} \quad \text{with solution} \quad \begin{pmatrix} x \\ y \end{pmatrix} = \begin{pmatrix} 10 \\ -1 \end{pmatrix}$$

The second set of equations reads

$$\begin{pmatrix} 10.1 & 99 \\ 1 & 10 \end{pmatrix}\begin{pmatrix} x' \\ y' \end{pmatrix} = \begin{pmatrix} 10 \\ -1 \end{pmatrix} \quad \text{with solution} \quad \begin{pmatrix} x' \\ y' \end{pmatrix} = \begin{pmatrix} 5 \\ -0.5 \end{pmatrix}$$

It is to be seen that an uncertainty of 1% in one single coefficient may, in this case, lead to a change of 50% in the solution. The reader should note that this would not be detected by the Gauss–Jordan method; the values of the pivot elements are 10 and 0.1 which are not very small values.

A matrix such as the one shown in the example is called *ill conditioned*. The result of the elimination procedure must then be interpreted with great care, because very small changes in the coefficients may result in drastic changes in the solution vector. Other techniques must be employed, e.g. iterative methods.

For the sake of completeness and as a guideline for deciding whether a given matrix is ill conditioned, we mention *Turing's numbers* for an $n \times n$ matrix, i.e.

$$\nu(A) = \frac{1}{n}\sqrt{\text{tr}\,(A^tA) \times \text{tr}\,[(A^{-1})^tA^{-1}]}$$

$$\mu(A) = n \times (\text{maximum element of } A) \times (\text{maximum element of } A^{-1})$$

The larger the values obtained the worse the condition of the given matrix. The following relationship always holds true:

$$1 \leqslant \nu(A) \leqslant \mu(A)$$

Remember that A^t denotes the transpose of the matrix A and that the trace of a square matrix is defined as the sum of all the elements on the leading diagonal, i.e.

$$\text{tr}\,(A) = \sum_{k=1}^{n} a_{kk}$$

17.5 HINTS FOR YOUR COMPUTATIONAL WORK

Brute Force Methods

Many algorithms in numerical mathematics were developed long ago, when calculations had essentially to be performed by hand. Then it did indeed make a difference whether the method chosen gave answers to the prescribed degree of accuracy within five or within fifty passes.

Today, with personal computers proliferating and computing facilities becoming ever cheaper, the situation is quite different. The time required for solving a given problem is determined mainly by the task of programming the machine. Therefore, there are cases when a very simple, a brute force, algorithm (e.g. bisecting the interval) is to be preferred to an elaborate one (e.g. the Newton–Raphson method). The time lost by using a less effective method may well be compensated for by the time gained when programming and debugging the software. It is quite reasonable to use the most simple-minded approach if it meets our needs. Computing is usually not a matter of aesthetics but a matter of obtaining results in as efficient a manner as possible.

Rounding Errors

For every machine there is an optimal degree of accuracy to be obtained by a given method. At first sight, Simpson's rule, for example, should give results of ever increasing validity if the number n of steps is increased. But this is not the case. Each summand is affected by a rounding error, and there is no guarantee that these errors will level out. In general, this will not be the case. Increasing n means a higher degree of theoretical accuracy, but this is counterbalanced by more computational inaccuracy (cf. figure 17.14 showing a rough sketch).

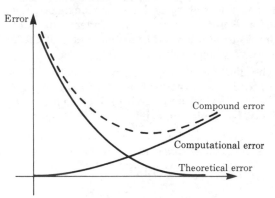

Figure 17.14

A good way of investigating the limits of a computer is to test it by evaluating formulae the results of which are known.

EXERCISES

17.2   Zeros of Real Functions

1. Use all four methods (Newton–Raphson, secant, halving the interval, iteration) to solve the following equations:

   (a) $\cosh x_0 = 2$ $(x_0 > 0)$ (b) $2x_0^3 + x_0^2 = 1$ $(0.5 < x_0 < 0.7)$
   (c) $2 \sin x_0 = x_0$ $(x_0 > 0)$

17.3   Numerical Integration

2. Use both the trapezoidal rule and Simpson's rule in the following:

   (a) It is claimed that $\tan^{-1}x$ is an integral of $\dfrac{1}{1+x^2}$. Compare $\tan^{-1}s$ and

   $\displaystyle\int_0^s \frac{dx}{1+x^2}$ for $s = 0.1, 0.2, \ldots, 2.0$. Use as step width $d = 0.05$.

(b) Compute $\int_1^s \dfrac{1}{x}\,dx$ for $s = 2$, $s = 3$, $s = 4$, $s = 5$, using $d = 0.025$. Compare your results with $\ln s$.

(c) Evaluate $\int_0^{10} e^{-x^2}\,dx$. (*Hint:* See Chapter 8, section 8.6.2). Since the integral from 10 to ∞ does not contribute much, your results should be good approximations of $\dfrac{\sqrt{\pi}}{2}$. Use $d = 1$, $d = 0.5$ and $d = 0.1$.

17.3.2   First Order DEs

3. Determine the solution curves for the following differential equations. First solve them analytically, and second, use both Euler's methods.

(a) $y' = C(y - y_1)(y - y_2)$
(*Remark:* This DE is called the 'logistic equation'.)
Use $C = -10^{-3}$, $y_1 = 10^2$, $y_2 = 0$. The starting point is $(0, 200)$, the step width is $d = 0.2$, and the interval is $[0, 20]$.

(b) $y' + \dfrac{y}{x} = xy^4$

The starting point is $(\tfrac{1}{3}, \tfrac{3}{2})$, the first step width is $2/300$, the subsequent step widths are $d = 0.01$, and the interval is $[\tfrac{1}{3}, 2.9]$.

17.4   Sets of Linear Equations

4. Solve the following sets of linear equations and compute the inverses of the given matrices:

(a)
$$\begin{pmatrix} 1 & 1 & 4 & -1 \\ 4 & 0 & 13 & -2 \\ 0 & 1 & -4 & 0 \\ -2 & 0 & -6 & 1 \end{pmatrix} \begin{pmatrix} x_1 \\ x_2 \\ x_3 \\ x_4 \end{pmatrix} = \begin{pmatrix} 33 \\ 111 \\ -27 \\ -52 \end{pmatrix}$$

(b)
$$\begin{pmatrix} 3 & 5 & 4 & 5 & 1 \\ 4 & 7 & 5 & 6 & 2 \\ 2 & 4 & 3 & 2 & 0 \\ 4 & 7 & 6 & 7 & 1 \\ 3 & 6 & 4 & 5 & 2 \end{pmatrix} \begin{pmatrix} x_1 \\ x_2 \\ x_3 \\ x_4 \\ x_5 \end{pmatrix} = \begin{pmatrix} 57 \\ 78 \\ 34 \\ 79 \\ 66 \end{pmatrix}$$

5. Compute the inverse matrices of
$$A = \begin{pmatrix} 10 & 99 \\ 1 & 10 \end{pmatrix} \quad \text{and} \quad A' = \begin{pmatrix} 10.1 & 99 \\ 1 & 10 \end{pmatrix}$$

6. Compute the inverse matrices of
$$A = \begin{pmatrix} 27 & 10 & -17 & -7 \\ -16 & -6 & 10 & 5 \\ -7 & -3 & 5 & 0 \\ 4 & 2 & -3 & 0 \end{pmatrix} \quad \text{and} \quad A' = \begin{pmatrix} 27.2 & 10 & -17 & -7 \\ -16 & -6 & 10 & 5 \\ -7 & -3 & 5 & 0 \\ 4 & 2 & -3 & 0 \end{pmatrix}$$

CHAPTER 18

* Fourier series; harmonic analysis

18.1 EXPANSION OF A PERIODIC FUNCTION INTO A FOURIER SERIES

In Chapter 8 we showed that a function $f(x)$ which may be differentiated any number of times can usually be expanded in an infinite series in powers of x, i.e.

$$f(x) = \sum_{n=0}^{\infty} a_n x^n$$

The advantage of the expansion is that each term can be differentiated and integrated easily and, in particular, it is useful in obtaining an approximate value of the function by taking the first few terms.

We now ask whether a function can be expanded in terms of functions other than power functions, and especially whether a periodic function may be expanded in terms of periodic functions, say trigonometric functions. Many problems in engineering and science involve periodic functions, particularly in electrical engineering, vibrations, sound and heat conduction. A *periodic function* $f(x)$ is a function such that $f(x) = f(x+L)$, where L is the smallest value for which the relationship is satisfied.

Fourier's theorem relates to periodic functions and states that any periodic function can be expressed as the sum of sine functions of different amplitudes, phases and periods. The periods are of the form L divided by a positive integer.

Thus, however irregular the curve representing the function may be, as long as its ordinates repeat themselves after equal intervals, it is possible to resolve it into a number of sine curves, the ordinates of which when added together give the ordinates of the original function. This resolution of a periodic curve is known as *harmonic analysis*.

To simplify the mathematics we will start by considering functions whose period is 2π; this implies (see figure 18.1) that

$$f(x) = f(x + 2\pi)$$

Expressed mathematically, Fourier's theorem states that

$$y = f(x) = \sum_{n=0}^{\infty} A_n \sin(nx + \phi_n) \tag{18.1}$$

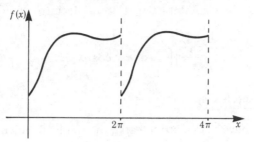

Figure 18.1

Since $\sin(nx + \phi_n) = \sin nx \cos \phi_n + \cos nx \sin \phi_n$, we can express the function in terms of sine and cosine functions. We have

$$y = f(x) = \frac{a_0}{2} + \sum_{n=1}^{\infty} (a_n \cos nx + b_n \sin nx) \qquad (18.2)$$

This series is called a *Fourier series*. The terms in Fourier series differ in period (or frequency). The nth term has the period $\dfrac{2\pi}{n}$ $\left(\text{or the frequency } \dfrac{n}{2\pi}\right)$. The $1/2$ in $a_0/2$ is to make a_0 fit the general equation.

18.1.1   EVALUATION OF THE COEFFICIENTS

Before actually proceeding with the evaluation of the coefficients, we will state the results of some definite integrals in the range $-\pi$ to π, where n and m are positive integers. It can be shown that the same results are obtained in the range from 0 to 2π.

$$\int_{-\pi}^{\pi} \cos nx \, dx = \int_{-\pi}^{\pi} \sin nx \, dx = 0 \qquad [1]$$

$$\int_{-\pi}^{\pi} \cos mx \, \cos nx \, dx = \int_{-\pi}^{\pi} \sin mx \, \sin nx \, dx = \begin{cases} 0, & m \neq n \\ \pi, & m = n \end{cases} \qquad [2]$$

$$\int_{-\pi}^{\pi} \sin mx \, \cos nx \, dx = 0 \qquad [3]$$

The integrals in equation 1 are standard.
Let us evaluate the first integral in equation 2. We integrate by parts:

$$\int_{-\pi}^{\pi} \cos mx \, \cos nx \, dx = \left[\frac{1}{m} \sin mx \, \cos nx \right]_{-\pi}^{\pi} + \frac{n}{m} \int_{-\pi}^{\pi} \sin mx \, \sin nx \, dx$$

The first term is zero. The second term can be integrated by parts once more:

$$\frac{n}{m} \int_{-\pi}^{\pi} \sin mx \, \sin nx \, dx = \left[-\frac{n}{m^2} \cos mx \, \sin nx \right]_{-\pi}^{\pi} + \frac{n^2}{m^2} \int_{-\pi}^{\pi} \cos mx \, \cos nx \, dx$$

Again the first term is zero. Inserting this result into the original integral and rearranging gives

$$\left(1 - \frac{n^2}{m^2}\right)\int_{-\pi}^{\pi} \cos mx \cos nx \, dx = 0$$

Thus the integral is zero except for $\dfrac{n^2}{m^2} = 1$, i.e. $n = m$.

In the latter case we have a standard integral, the result of which is known:

$$\int_{-\pi}^{\pi} \cos^2 mx \, dx = \pi$$

The integral with sine functions may be solved by the reader. The solution can be obtained in exactly the same way.

The integral in equation 3 may be solved in the same way too. In the case of $m = n$ we have

$$\int_{-\pi}^{\pi} \sin mx \cos mx \, dx = \frac{1}{2}\int_{-\pi}^{\pi} \sin 2mx \, dx = 0$$

Evaluation of a_o

To find a_0 we integrate the Fourier series from $-\pi$ to π:

$$\int_{-\pi}^{+\pi} f(x) \, dx = \frac{1}{2}\int_{-\pi}^{+\pi} a_0 \, dx + \sum_{n=1}^{\infty}\left(a_n\int_{-\pi}^{+\pi} \cos nx \, dx + b_n\int_{-\pi}^{+\pi} \sin nx \, dx\right)$$

According to equation 1 above, all integrals in the infinite sum vanish. Hence

$$\int_{-\pi}^{+\pi} f(x) \, dx = \frac{1}{2}\int_{-\pi}^{\pi} a_0 \, dx = \pi a_0$$

Therefore, we have obtained a_0:

$$a_0 = \frac{1}{\pi}\int_{-\pi}^{+\pi} f(x) \, dx$$

Note that $\dfrac{a_0}{2}$ is the average value of the function in the range $-\pi$ to $+\pi$.

Evaluation of a_n

The coefficients a_n have to be evaluated one by one. For a given $n = k$ multiply the Fourier series by $\cos kx$ and integrate from $-\pi$ to $+\pi$:

$$\int_{-\pi}^{+\pi} f(x) \cos kx \, dx = \int_{-\pi}^{+\pi} \frac{a_0}{2} \cos kx \, dx + \sum_{n=1}^{\infty}\int_{-\pi}^{\pi} a_n \cos nx \cos kx \, dx$$

$$+ \sum_{n=1}^{\infty}\int_{-\pi}^{+\pi} b_n \sin nx \cos kx \, dx$$

By virtue of equations 1 and 3 above, all integrals on the right vanish except the one for $n = k$. We thus obtain

$$\int_{-\pi}^{+\pi} f(x) \cos nx \, dx = \int_{-\pi}^{+\pi} a_n \cos nx \cos nx \, dx = a_n \int_{-\pi}^{+\pi} \cos^2 nx \, dx = a_n \pi$$

Hence

$$a_n = \frac{1}{\pi} \int_{-\pi}^{\pi} f(x) \cos nx \, dx$$

Evaluation of b_n

We proceed in the same way: we multiply the Fourier series by $\sin kx$ and integrate in the range $-\pi$ to π. All integrals on the right vanish except for $n = k$.

$$\int_{-\pi}^{\pi} b_n \sin^2 nx = b_n \pi$$

Hence

$$b_n = \frac{1}{\pi} \int_{-\pi}^{\pi} f(x) \sin nx \, dx$$

The result is:

If a function $f(x)$ of period 2π can be represented in a Fourier series then

$$f(x) = \frac{a_0}{2} + \sum_{n=1}^{\infty} a_n \cos nx + \sum_{n=1}^{\infty} b_n \sin nx$$

where

$$a_0 = \frac{1}{\pi} \int_{-\pi}^{\pi} f(x) \, dx \qquad\qquad (18.3)$$

$$a_n = \frac{1}{\pi} \int_{-\pi}^{\pi} f(x) \cos nx \, dx \qquad n = 1, 2, \ldots \qquad (18.4)$$

$$b_n = \frac{1}{\pi} \int_{-\pi}^{\pi} f(x) \sin nx \, dx \qquad n = 1, 2, \ldots \qquad (18.5)$$

Since $f(x)$ is a periodic function of period 2π we could, if we wished, use the range 0 to 2π instead, or any other interval of length 2π.

The terms $\cos x$, $\sin x$ are known as the *fundamental* or *first harmonic*, $\cos 2x$, $\sin 2x$ as the *second harmonic*, $\cos 3x$, $\sin 3x$ as the *third harmonic* and so on.

We have not yet discussed the conditions that must be satisfied by $f(x)$ for the expansion to be possible. There are, in fact, several sufficient conditions which guarantee that the Fourier expansion is valid, and most functions the applied scientist is likely to meet in practice will be Fourier expandable.

We should mention one criterion which is connected with the name of the eminent mathematician Peter G. L. Dirichlet (1805–59).

Dirichlet's lemma states that a periodic function $f(x)$ which is bounded (i.e. there is a constant B such that $|f(x)| < B$ for all x) and which has a finite number of maxima and minima and a finite number of points of discontinuity in the interval $[-L; L]$ has a convergent Fourier series. This series converges towards the value of the function $f(x)$ at all points where it is continuous. At points of discontinuity the value of the Fourier series is equal to the arithmetical mean of the left-hand and right-hand limit of the function $f(x)$, i.e. it is equal to

$$\tfrac{1}{2}[\lim_{\substack{\Delta x \to 0 \\ \Delta x > 0}} f(x + \Delta x) + \lim_{\substack{\Delta x \to 0 \\ \Delta x > 0}} f(x - \Delta x)]$$

The proof of this lemma is beyond the scope of this book, and the reader should refer to advanced books on mathematics.

18.1.2   FOURIER SPECTRUM

A periodic function of time, such as a vibratory motion, expressed as a Fourier series is often represented by a Fourier spectrum. It consists of the values of the amplitudes and phases of the different terms as a function of the frequency. This is shown in figure 18.2. a_1 is the amplitude of the fundamental frequency ω; a_2, a_3 etc. are the amplitudes of the harmonics 2ω, 3ω, etc and ϕ_1, ϕ_2, etc are the phase angles when the function is expressed in the form of equation 18.1. For example, if s is a vibration or a signal in an electrical system then equation 18.1 takes on the following form:

$$s = \sum_{n=1}^{\infty} a_n \sin(n\omega t + \phi_n) \qquad (\omega = \text{frequency}, \ t = \text{time})$$

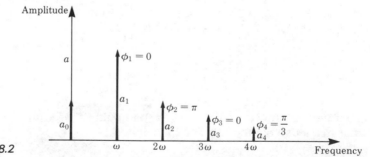

Figure 18.2

A periodic electrical signal introduced into a transducer, filter or amplifier will be modified; it will be damped and distorted unless special precautions are taken in the design of the equipment. By expressing the signal as a Fourier series it is easy to find out how each Fourier component is affected.

For sine and cosine functions these modifications are often easy to find either empirically by measurement or theoretically by calculation. As a rule, modifications depend on the frequency of the function.

Thus the modification of an arbitrary periodic signal can be obtained by expressing the signal as a Fourier series, finding the modifications of the Fourier components and reconstructing the modified signal.

In a transmission line, for example, it is most important to ensure that a signal is not distorted. This implies that the relative amplitudes of the terms making up the signal, as well as the phase angles of the harmonics, are faithfully reproduced. In other words, it is necessary to ensure that the time shift due to the transmission remains the same for all harmonics.

18.1.3   ODD AND EVEN FUNCTIONS

Even Functions

A function is even when $f(x) = f(-x)$. In this case all the coefficients b_n vanish. Since $f(x) \sin nx$ is an odd function, its integral from $-\pi$ to π is zero.

For an even function the Fourier series is

$$f(x) = \frac{a_0}{2} + \sum_{n=1}^{\infty} a_n \cos nx$$

Odd Functions

A function is odd when $f(x) = -f(-x)$. In this case all the coefficients a_n vanish. Since $f(x) \cos nx$ is an odd function, its integral from $-\pi$ to π is zero.

For an odd function the Fourier series is

$$f(x) = \sum_{n=1}^{\infty} b_n \sin nx$$

Thus the Fourier series for an even function consists of cosine terms only, whereas that for an odd function consists of sine terms only.

18.2   EXAMPLES OF FOURIER SERIES

Sawtooth Waveform

The sawtooth function is shown in figure 18.3 with a period of 2π. It is defined by

$$f(x) = \begin{cases} \dfrac{1}{\pi}x + 1, & -\pi \leqslant x \leqslant 0 \\[3mm] \dfrac{1}{\pi}x - 1, & 0 \leqslant x \leqslant \pi \end{cases}$$

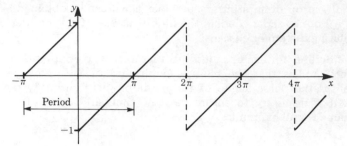

Figure 18.3

Since the function is odd, only the coefficients b_n are required. Because there are two branches of the function we have to split the interval of integration. We then have

$$b_n = \frac{1}{\pi} \int_{-\pi}^{0} \left(\frac{x}{\pi} + 1\right) \sin nx \; dx + \frac{1}{\pi} \int_{0}^{\pi} \left(\frac{x}{\pi} - 1\right) \sin nx \; dx$$

$$= \frac{1}{\pi^2} \int_{-\pi}^{\pi} x \sin nx \; dx + \frac{1}{\pi} \int_{-\pi}^{0} \sin nx \; dx - \frac{1}{\pi} \int_{0}^{\pi} \sin nx \; dx$$

The first integral can be solved by parts; the other two are standard.

Integrating gives

$$b_n = -\left[\frac{1}{\pi^2 n} x \cos nx\right]_{-\pi}^{\pi} + \underbrace{\left[\frac{1}{\pi^2 n^2} \sin nx\right]_{-\pi}^{\pi}}_{= 0} - \left[\frac{1}{\pi n} \cos nx\right]_{-\pi}^{0} + \left[\frac{1}{\pi n} \cos nx\right]_{0}^{\pi}$$

$$b_n = -\frac{2}{\pi n}$$

Hence, for the sawtooth waveform the Fourier series is

$$f(x) = -\frac{2}{\pi} \sum_{n=1}^{\infty} \frac{\sin nx}{n}$$

Figure 18.4 shows the first six terms of the expansion. As the number of terms is increased the series gets closer and closer to the function.

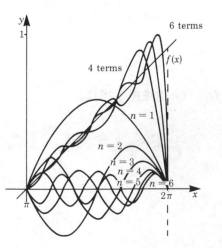

Figure 18.4

Triangular Waveform

The triangular function is shown in figure 18.5. Its period is 2π. It is defined by

$$f(x) = \begin{cases} -x, & -\pi < x \leqslant 0 \\ x, & 0 \leqslant x \leqslant \pi \end{cases}$$

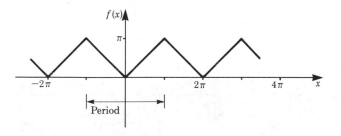

Figure 18.5

Since it is an even function, we only need to calculate the coefficients a_n. Because there are two branches of the function we have to split the interval of integration.

$$a_0 = \frac{1}{\pi}\int_{-\pi}^{0} (-x)\,dx + \frac{1}{\pi}\int_{0}^{\pi} x\,dx = \pi$$

$$a_n = \frac{1}{\pi}\int_{-\pi}^{0} (-x)\cos nx\,dx + \frac{1}{\pi}\int_{0}^{\pi} x\cos nx\,dx$$

Integrating by parts gives

$$a_n = -\left[\frac{1}{\pi n^2}\cos nx\right]_{-\pi}^{0} - \underbrace{\left[\frac{x}{\pi n}\sin nx\right]_{-\pi}^{0}}_{= 0} + \left[\frac{1}{\pi n^2}\cos nx\right]_{0}^{\pi} + \underbrace{\left[\frac{x}{\pi n}\sin nx\right]_{0}^{\pi}}_{= 0}$$

$$a_n = \frac{2}{\pi n^2}(\cos n\pi - 1)$$

If n is even $\qquad\qquad \cos n\pi = +1 \qquad$ hence $\qquad a_n = 0$

If n is odd $\qquad\qquad \cos n\pi = -1 \qquad$ hence $\qquad a_n = -\frac{4}{\pi n^2}$

The Fourier series of the triangular waveform is

$$f(x) = \frac{\pi}{2} - \frac{4}{\pi}\sum_{n=0}^{\infty} \frac{\cos(2n+1)x}{(2n+1)^2}$$

Rectangular Waveform

The function is shown in figure 18.6. It is defined in the interval $-\pi$ to π by

$$f(x) = \begin{cases} -1, & -\pi < x \leqslant -\dfrac{\pi}{2} \\[2mm] 1, & -\dfrac{\pi}{2} < x \leqslant \dfrac{\pi}{2} \\[2mm] -1, & \dfrac{\pi}{2} < x \leqslant \pi \end{cases}$$

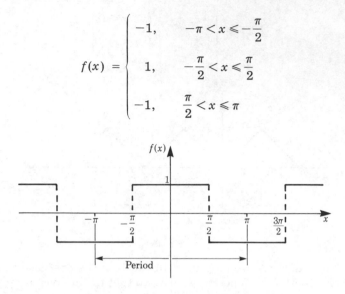

Figure 18.6

Since the function is even, we need only calculate the coefficients a_n.

$$a_0 = \frac{1}{\pi}\int_{-\pi}^{\pi} f(x)\,\mathrm{d}x = 0 \qquad \text{(we can see this by inspection)}$$

$$a_n = \frac{1}{\pi}\left(-\int_{-\pi}^{-\pi/2}\cos nx\,\mathrm{d}x + \int_{-\pi/2}^{\pi/2}\cos nx\,\mathrm{d}x - \int_{\pi/2}^{\pi}\cos nx\,\mathrm{d}x\right)$$

$$a_n = \frac{2}{\pi n}\sin\frac{n\pi}{2}$$

The Fourier series for the rectangular waveform is

$$f(x) = \frac{2}{\pi}\sum_{n=1}^{\infty}\frac{1}{n}\sin\frac{n\pi}{2}\cos nx$$

Figure 18.7 shows approximations to the function. We see that as we take more and more terms we approach the original function $f(x) = f_1 + f_2 + f_3 + \ldots$.

There is, however, a snag to approximating a function at a point of discontinuity such as $x = \pi/2$ in our example. If a lot of terms are added, $f_1 + f_2 + f_3 + \ldots + f_k$, then the graph overshoots before and after the discontinuity.

This is known as the *Gibbs phenomenon*. For most practical purposes this effect can usually be neglected; but if special attention is to be paid to the function at a discontinuity, the Gibbs phenomenon must be kept in mind.

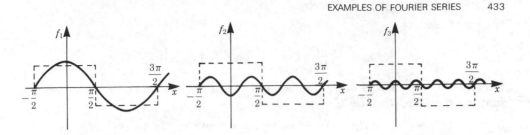

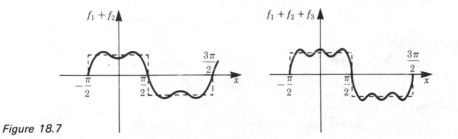

Figure 18.7

Half sine wave, such as a rectified alternating current.

The function is shown in figure 18.8. It is defined by

$$f(x) = \begin{cases} 0, & -\pi < x \leqslant 0 \\ I_0 \sin x, & 0 < x \leqslant \pi \end{cases}$$

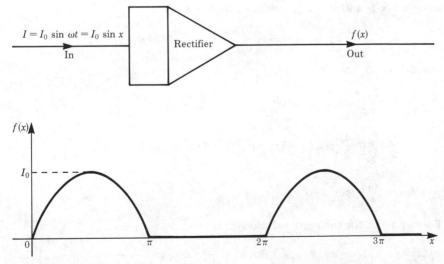

Figure 18.8

The coefficient a_0 is

$$a_0 = \frac{I_0}{\pi} \left(\int_{-\pi}^{0} 0 \, dx + \int_{0}^{\pi} \sin x \, dx \right) = 2\frac{I_0}{\pi}$$

To find the coefficients a_n and b_n we use two identities which are based on the addition formulae (Chapter 1, section 1.5.6):

$$\sin(n+1)x = \sin nx \cos x + \sin x \cos nx$$
$$\sin(n-1)x = \sin nx \cos x - \sin x \cos nx$$

Subtracting gives

$$\sin(n+1)x - \sin(n-1)x = 2\sin x \cos nx$$

Similarly, we obtain

$$\cos(n+1)x = \cos nx \cos x - \sin nx \sin x$$
$$\cos(n-1)x = \cos nx \cos x + \sin nx \sin x$$

Subtracting gives

$$\cos(n-1)x - \cos(n+1)x = 2\sin nx \sin x$$

Now we evaluate a_n:

$$a_n = \frac{I_0}{\pi} \int_{0}^{\pi} \sin x \cos nx \, dx = \frac{I_0}{\pi} \int_{0}^{\pi} \frac{1}{2} \left(\sin(n+1)x - \sin(n-1)x \right) dx$$

$$= \frac{I_0}{2\pi} \left[-\frac{\cos(n+1)x}{n+1} + \frac{\cos(n-1)x}{n-1} \right]_{0}^{\pi} \qquad (n \neq 1)$$

If $n = 1$, $\qquad a_1 = \frac{I_0}{\pi} \int_{0}^{\pi} \sin x \cos x \, dx = 0$; $\qquad$ hence $a_1 = 0$

If n is odd $n+1$ and $n-1$ are even, so that

$$a_n = 0 \qquad\qquad (n \text{ odd})$$

If n is even $n+1$ and $n-1$ are odd, so that

$$a_n = -\frac{2I_0}{\pi(n+1)(n-1)} \qquad\qquad (n \text{ even})$$

The coefficients b_n are

$$b_n = \frac{I_0}{\pi} \int_{0}^{\pi} \sin x \sin nx \, dx$$

$$= \frac{I_0}{\pi} \int_{0}^{\pi} \frac{1}{2} \left(\cos(n-1)x - \cos(n+1)x \right) dx = 0 \qquad \text{if } n \neq 1$$

If $n = 1, b_1 = \frac{I_0}{\pi} \int_{0}^{\pi} \sin^2 x \, dx = \frac{I_0}{2}$

The Fourier series for the rectified waveform is

$$f(x) = \frac{I_0}{\pi} \left(1 + \frac{\pi}{2} \sin x - \frac{2}{1 \times 3} \cos 2x - \frac{2}{3 \times 5} \cos 4x - \frac{2}{5 \times 7} \cos 6x - \ldots \right)$$

18.3 EXPANSION OF FUNCTIONS OF PERIOD 2*L*

A periodic function $f(x)$ of period $2L$ is repeated when x increases by $2L$, i.e.

$$f(x + 2L) = f(x)$$

If we put $z = \dfrac{\pi}{L}x$, then the new function $f(z)$ is a periodic function of period 2π. As x increases from $-L$ to L, z increases from $-\pi$ to π, and equation 18.2 holds true for the new variable z. We have

$$f(z) = \frac{a_0}{2} + \sum_{n=1}^{\infty} (a_n \cos nz + b_n \sin nz) \tag{18.6}$$

To get back to the original function, we simply replace z by $\dfrac{\pi}{L}x$ and obtain

$$f(x) = \frac{a_0}{2} + \sum_{n=1}^{\infty} \left(a_n \cos\frac{n\pi}{L}x + b_n \sin\frac{n\pi}{L}x \right) \tag{18.7}$$

The coefficients of a Fourier series of a function with period $2L$ are

$$a_0 = \frac{1}{L}\int_{-L}^{L} f(x)\,dx \tag{18.8}$$

$$a_n = \frac{1}{L}\int_{-L}^{L} f(x)\cos\frac{n\pi}{L}x\,dx \quad n = 1, 2 \ldots \tag{18.9}$$

$$b_n = \frac{1}{L}\int_{-L}^{L} f(x)\sin\frac{n\pi}{L}x\,dx \quad n = 1, 2 \ldots \tag{18.10}$$

Rectangular Waveform of Period 4

The function is defined by

$$f(x) = \begin{cases} 0, & -2 < x \leqslant 0 \\ 1, & 0 < x \leqslant 2 \end{cases}$$

In this case, we have $L = 2$. All integrals from -2 to 0 vanish.

From equation 18.8: $a_0 = \dfrac{1}{2}\displaystyle\int_{0}^{2} dx = 1$

From equation 18.9: $a_n = \dfrac{1}{2}\displaystyle\int_{0}^{2} \cos\frac{n\pi x}{2}\,dx = 0$

From equation 18.10: $b_n = \dfrac{1}{2}\displaystyle\int_{0}^{2} \sin\frac{n\pi x}{2}\,dx = \dfrac{1}{n\pi}(1 - \cos n\pi)$

We therefore have $f(x) = \dfrac{1}{2} + \dfrac{2}{\pi}\left(\sin\frac{\pi x}{2} + \frac{1}{3}\sin\frac{3\pi x}{2} + \frac{1}{5}\sin\frac{5\pi x}{2} + \ldots \right)$

EXERCISES

1. Obtain the Fourier series for the function (figure 18.9) defined by

$$f(x) = \begin{cases} 0, & -\pi \leqslant x < -\dfrac{\pi}{2} \\[2mm] 1, & -\dfrac{\pi}{2} \leqslant x < \dfrac{\pi}{2} \\[2mm] 0, & \dfrac{\pi}{2} \leqslant x \leqslant \pi \end{cases}$$

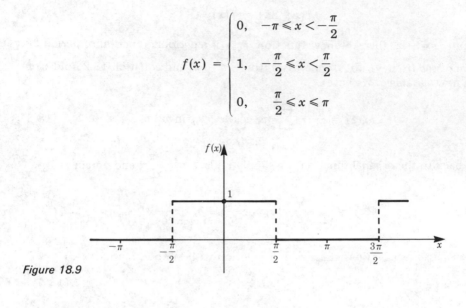

Figure 18.9

2. Obtain the Fourier series for the function (figure 18.10) defined below. The function is periodic with period 2π.

$$f(x) = \begin{cases} 1, & -\pi \leqslant x < 0 \\ -1, & 0 \leqslant x \leqslant \pi \end{cases}$$

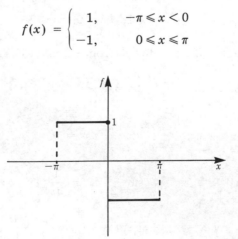

Figure 18.10

3. Obtain the Fourier series for a rectified waveform (figure 18.11) given by

$$f(x) = |\sin x| = \begin{cases} -\sin x, & -\pi < x < 0 \\ \sin x, & 0 < x < \pi \end{cases}$$

Use the results obtained in the example on page 433.

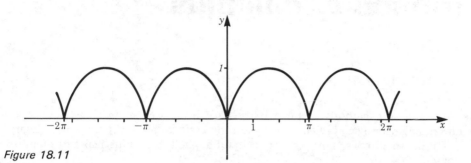

Figure 18.11

4. Obtain the Fourier series for the function (figure 18.12) defined below. Its period is 4π.

$$f(x) = \begin{cases} 0, & -2\pi \leqslant x < -\pi \\ 1, & -\pi \leqslant x < \pi \\ 0, & \pi \leqslant x < 2\pi \end{cases}$$

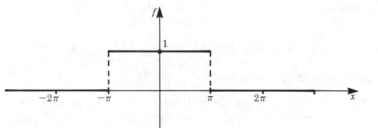

Figure 18.12

CHAPTER 19

Probability calculus

19.1 INTRODUCTION

The concepts and methods of probability have been used increasingly in the last decades in engineering and science. They are the basis for an understanding of a large part of modern physics, both theoretical and applied, e.g. statistical and quantum mechanics.

Statistical mechanics describes physical systems such as gases, solid bodies, and fluids, all of which consist of many atoms and molecules. The properties which refer to a single element of a system are called microscopic, e.g. state of motion, position, kinetic and potential energy.

Those which refer to the whole system are called macroscopic, e.g. pressure, volume, temperature, magnetism, electrical conductivity and so on.

Statistical mechanics attributes the macroscopic properties of the whole system to the microscopic properties of the constituent elements. Statistical mechanics uses, above all, the fact that the whole system consists of a very large number of elements, e.g. 1 litre of air contains about 10^{23} gas molecules.

Quantum mechanics describes physical objects like atoms, atomic nuclei etc. In quantum mechanics statistics plays a major role, since only probability statements can be made about the properties of these objects.

New mathematical methods are needed for the quantitative treatment of physical systems with a large number of elements; and it has been found that the theory of probability provides the necessary methods for the mathematical treatment of such systems. One important fact is that the larger the number of elements of a system under consideration is, the more closely will the results of the theory of probability agree with experiments.

A further field of application is the theory of errors. All physical measurements are in principle liable to errors. Before we can draw useful conclusions about an experiment, we must first estimate the errors. In Chapter 21 we will show how to infer the accuracy of measurements from the scatter of points, i.e. the variance of experimental data.

An important application is quality control in the small or large-scale industrial production of almost any product. Manufacturers need the sound knowledge of quality control which statistics can provide.

19.2 CONCEPT OF PROBABILITY

19.2.1 RANDOM EXPERIMENT, OUTCOME SPACE AND EVENTS

The concept of probability in mathematics has been derived from the colloquial word 'probable'. If the sky is cloudy we say: 'It will probably rain today'. Of course, we are not sure that it will rain, i.e. whether the event will occur or not, but there is a likely chance that it will. It is the aim of probability theory to calculate and express quantitatively the degree of certainty or uncertainty in the possible occurrence of an event.

The concept of probability can be deduced from an analysis of well-known games of chance. Suppose a die is thrown. This can be repeated arbitrarily often. A process which can be repeated arbitrarily many times in accordance with certain rules is called an *experiment*. In this case the experiment can have six different outcomes. The possible outcomes are the number of spots on the faces of the die, i.e. the numbers 1, 2, 3, 4, 5, 6. The outcome cannot be forecast with certainty. Such an experiment is called a *random experiment*.

The outcome of a random experiment depends on the statement of the problem. When throwing a die, as in our example, we could ask: 'What is the number of spots?' or 'is the number even or odd?'.

In the first case, the set of outcomes is

$$\{1, 2, 3, 4, 5, 6\}$$

In the second case the set of outcomes is

$$\{\text{ even, odd }\} \qquad \text{or} \qquad \{\{2, 4, 6\}\{1, 3, 5\}\}$$

The set of outcomes is called the *outcome space* or the *sample space* of the experiment.

Possible outcomes with equal probability are called *elementary events*. Equal probability means that no outcome is more favoured than any other for physical or logical reasons.

In the case of throwing a die, we have six elementary events, the number of spots:

$$R = \{1, 2, 3, 4, 5, 6\}$$

Any set of elementary events is called an *event*.

In the case of throwing a die and regarding the outcomes as 'number even' or 'number odd' we have two events, each of which is a subset of the set of elementary events:

$$\{2, 4, 6\}\{1, 3, 5\}$$

Thus an event can be any outcome of a random experiment, depending on the problem:

(1) an elementary event;
(2) any combination of elementary events.

For the sake of completeness, an *impossible event* is defined as an event which is not an element of the outcome space. For example, when throwing a die, the event 'number of spots = 7' is impossible.

19.2.2   THE CLASSICAL DEFINITION OF PROBABILITY

Let a random experiment consist of N equally possible outcomes, i.e. elementary events. Let us consider an event A consisting of N_A elementary events. The probability P of outcome A is given by the classic definition of probability:

DEFINITION   Classical definition of probability

$$P(A) = \frac{N_A}{N} = \frac{\text{number of elementary events contained in event } A}{\text{total number of possible elementary events}}$$

Some authors refer to the number N_A as 'favoured'.

EXAMPLE   What is the probability that an even number will appear when throwing a die?

There are 6 possible ways a die can fall; hence $N = 6$. Of these 6 there are 3 which will show an even number $\{2, 4, 6\}$ and 3 an odd number $\{1, 3, 5\}$; therefore $N_A = 3$. Consequently the probability that an even number will appear is

$$P(\text{even}) = \frac{N_A}{N} = \frac{3}{6} = \frac{1}{2}$$

EXAMPLE   A pack of 52 cards contains 4 jacks. What is the probability of drawing one jack?

It is possible to draw 52 different cards; thus $N = 52$ (possible elementary events). Since there are 4 jacks in the pack, we have 4 possibilities to realise the event 'one jack'; hence $N_A = 4$. Therefore the probability of drawing one jack is

$$P(\text{jack}) = \frac{N_A}{N} = \frac{4}{52} = \frac{1}{13}$$

The basis of the classic definition of probability is the assumption that all elementary events are equally probable. In reality, all experiments carried out with real dice, playing cards and other games approach this assumption only to a certain degree. A perfect or fair die, for example, is one whose material is absolutely homogeneous and whose shape is perfect.

19.2.3   THE STATISTICAL DEFINITION OF PROBABILITY

In many cases it does not make sense in practice to assume that all elementary events are equally probably. For example, a die becomes deformed after long use, or the centre of mass may be deliberately altered so that it is 'loaded' with the intent to defraud.

In cases of this kind we can also state the probability that a particular event will occur.

Let N experiments be carried out under identical conditions and let N_A of them have the outcome A. The expression

$$h_A = \frac{N_A}{N} \qquad \text{is called the } \textit{relative frequency}.$$

The relative frequency is a quantity which has to be determined experimentally. It should not be confused with the classic definition of probability. N_A is the actual number of experiments with outcome A.

If the values of the relative frequency do not change for a large number N of experiments actually carried out, then we can interpret the relative frequency as the probability. We call the probability obtained from the relative frequency the *statistical probability* $P(A)$.

> **DEFINITION**   The statistical definition of probability is the limit of the relative frequency of an event A.
>
> $$P(A) = h_A = \frac{N_A}{N} \quad \text{as} \quad N \to \infty$$

In practice, of course, it is not possible to carry out an arbitrarily large number of experiments.

EXAMPLE   The probability of getting a 1 when throwing a die can be determined empirically, as shown in figure 19.1 which is the result of an actual test. The relative frequency approaches the value of 1/6 as the number of experiments increases. If we repeat the test, the shape of the curve will be different at first but will again approach the value 1/6 for a large number of experiments.

If the curve approaches a different value we can deduce that the die is faulty or loaded. In this way we can check the die.

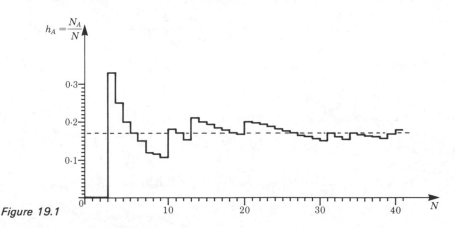

Figure 19.1

The concept of the statistical definition of probability refers to actual experiments. It can be applied to the result of scientific measurements (from observations of the radioactive decay process we can derive the probability of the decay of a radium atom, for instance). In the case of throwing ideal dice or similar games, both definitions of probability coincide for large values of N. Unless we state otherwise, we will use the classic definition; this means that we will consider ideal experiments. In chapter 21, which deals with the theory of errors, we shall concern ourselves with the analysis of the results of real experiments.

19.2.4 GENERAL PROPERTIES OF PROBABILITIES

We will now derive a few general properties of probabilities which will be useful in practical calculations by means of an example.

EXAMPLE   A box contains N balls. N_1 balls have the number 1 written on them, N_2 the number 2, ... and N_k the number k. k is the largest number on a ball.

In our case let $N = 9$, $k = 4$

$$N_1 = 2 \qquad N_2 = 1 \qquad\qquad N_3 = 4 \qquad\qquad N_4 = 2$$

① ①	②	③ ③ ③ ③	④ ④
↑	↑	↑	↑
2 balls with the number 1	1 ball with the number 2	4 balls with the number 3	2 balls with the number 4

We take one ball out of the box. Let this ball carry the number j. This event is denoted by j whose probability is

$$P(j) = \frac{N_j}{N}$$

Now we want to find what the probability is of our taking out a ball with the number i on it *or* a ball with the number j. This event is represented by the balls carrying the number i or j. There are N_i and N_j balls with the number i and j respectively.

The probability of our taking out one ball in this subset is

$$P(i \text{ or } j) = \frac{N_i + N_j}{N} = \frac{N_i}{N} + \frac{N_j}{N} = P(i) + P(j)$$

This is the *addition law* for probabilities and is only valid if the events are mutually exclusive or *disjoint*, i.e. independent.

> **Rule**   Addition law for independent probabilities
> $$P(i \text{ or } j) = P(i) + P(j)$$

This theorem can be generalised for any number of mutually exclusive events.

> **Rule**
> $$P(1 \text{ or } 2 \text{ or } \ldots \text{ or } k) = \sum_{i=1}^{k} P(i)$$

If the set of events is equal to the set of all possible events, then the probability $P(1 \text{ or } 2 \text{ or } \ldots \text{ or } k)$ is a certainty. In such a case, the *normalisation condition* must be satisfied.

> **Rule**   Normalisation condition for probabilities
> $$\sum_{i=1}^{k} P(i) = 1$$

PROOF   Let the total number of events be N, then $\displaystyle\sum_{i=1}^{k} N_i = N$

and hence $$\sum_{i=1}^{k} P(i) = \sum_{i=1}^{k} \frac{N_i}{N} = \frac{1}{N} \sum_{i=1}^{k} N_i = \frac{1}{N} N = 1$$

An event with the probability $P = 1$ occurs with certainty.

What is the probability of taking out of the box a ball with the number m, where $m > k$? Since k is the largest number on a ball, we know that there are no balls with the number m. Therefore

$$P(m) = \frac{0}{N} = 0$$

This means that an event which cannot take place has the probability zero: it is an impossible event.

EXAMPLE   Each group of spots from 1 to 6 on a die has the probability $1/6$ of appearing when the die is thrown.

The probability of throwing either a 1 or a 2 is

$$P(1 \text{ or } 2) = \tfrac{1}{6} + \tfrac{1}{6} = \tfrac{1}{3}$$

The probability of throwing an even number is

$$P(\text{even}) = \tfrac{1}{6} + \tfrac{1}{6} + \tfrac{1}{6} = \tfrac{1}{2}$$

The probability of obtaining any number between 1 and 6 is a certainty:

$$P(1 \text{ or } 2 \text{ or } 3 \text{ or } 4 \text{ or } 5 \text{ or } 6) = \tfrac{1}{6} + \tfrac{1}{6} + \tfrac{1}{6} + \tfrac{1}{6} + \tfrac{1}{6} + \tfrac{1}{6} = 1$$

The probability of throwing a number greater than 6 is zero, an impossible event.

We will consider one more relationship of this kind. Let the probability P for the occurrence of an event i be known and let N be the total number of possible events.

To obtain the number N_i, we need only solve the equation $P(i) = \dfrac{N_i}{N}$ for N_i. Hence

$$N_i = P(i) N$$

EXAMPLE   When throwing a die, let $N = 60$. The probability of throwing a 1 is $P(1) = 1/6$. The number N_1 for the event of 'getting a 1' is

$$N_1 = P(1) N = \tfrac{1}{6} \times 60 = 10$$

This means that, on average, after 60 throws we can expect the number 1 to appear ten times.

19.2.5   **PROBABILITY OF STATISTICALLY INDEPENDENT EVENTS.
COMPOUND PROBABILITY**

In this section we will determine the probability of the simultaneous occurrence of two different events.

For example, a person throws a die and at the same time tosses a coin. We are interested in the probability of the die showing a 6 and the coin showing a tail. The experiments and their outcomes are independent of each other.

Such composite events are called *compound events*.

Suppose we have two groups of independent events, the number of spots on the die (group A) and the faces of the coin (group B). We require the probability of the simultaneous occurrence of these two events. The table lists all events and their possible combinations.

Coin: 2 elementary events	Die: 6 elementary events					
	1	2	3	4	5	6
Head (H)	H1	H2	H3	H4	H5	H6
Tail (T)	T1	T2	T3	T4	T5	T6
	Compound events					

We see that there are 12 possible elementary compound events.

The total number of elementary compound events is equal to the product of the number of elementary events of group A (the die) and the number of elementary events of group B (the coin). Hence

$$N = N_1 \times N_2$$

In the example above $N_1 = 6$, $N_2 = 2$, $N = 12$

From the table above we can determine the probability of a particular compound event when the die is thrown and the coin is tossed.

EXAMPLE   What is the probability that the die will show a 6 and the coin a tail?

In this case $N_{6T} = 1$ and $N = 12$. Hence $P(6T) = \frac{1}{12}$.

EXAMPLE   What is the probability that the die will show an even number and the coin a head?

The number of events is $N = 6 \times 2 = 12$.

For group A (die), $N_1 = 3$.

For group B (coin), $N_2 = 1$.

Probability $P(\text{even, head}) = \dfrac{3 \times 1}{12} = \dfrac{1}{4}$.

The concept of statistically independent events is important. In our example occurrence of the number 6 on the die is in no way influenced by the coin showing head or tail. The two events are independent.

The compound event 'number even' on the die and 'head' on the coin is made up of the events 'number even' and 'head'. Hence the total number of compound events is given by the product of elementary events Λ and B. Thus

$$N_{\text{number even, head}} = N_{\text{number even}} \times N_{\text{head}}$$

The required probability of both events occurring is

$$P_{\text{(even, head)}} = \frac{N_{\text{even}} \times N_{\text{head}}}{N_1 N_2} = \frac{N_{\text{even}}}{N_1} \times \frac{N_{\text{head}}}{N_2}$$

Rule   The probability of the occurrence of a compound event $-A$ and $B-$ is for statistically independent events, A and B, the product of the probability of the occurrence of A and of B

$$P(A \text{ and } B) = P(A) \times P(B)$$

The probability of compound events is generally less than the probability of singular events:

$$P(A \text{ and } B) \leqslant P(A)$$

$$P(A \text{ and } B) \leqslant P(B)$$

EXAMPLE   The probability that a thirty-year-old man will reach the age of 65 is, according to statistics, $P(\text{man}) \approx 0.75$. The probability that his wife (who is the same age) will live through these 30 years is $P(\text{woman}) \approx 0.80$. What is the probability that both will be living 30 years hence?

$$P(\text{man and woman}) = 0.75 \times 0.80 = 0.6$$

Note that

$$P(\text{man and woman}) \leqslant P(\text{man})$$

$$P(\text{man and woman}) \leqslant P(\text{woman})$$

19.3   PERMUTATIONS AND COMBINATIONS

To calculate probabilities the standard problem is to calculate the number of elementary events contained in an event A. To simplify the solution of these problems we introduce the concept of permutations and combinations.

19.3.1   PERMUTATIONS

Given n objects, we may place them in a row in any order; each such arrangement is called a *permutation*. Thus two elements, a and b can be arranged in two different ways: ab and ba. Three elements, a, b and c, can be arranged in six different ways:

$$abc, \ bac, \ cab, \ acb, \ bca, \ cba$$

How many permutations are there for n different elements $a_1, a_2, \ldots, a_n$? The following reasoning leads us to the solution. First of all, we assume that the n possible places are empty. The element a_1 can be put in any place, i.e. there are n possibilities. The element a_2 can be put in any one of the remaining $(n-1)$ places which are still empty, i.e. there are $(n-1)$ possibilities. Thus the number of possibilities for placing the first two elements is given by

$$n(n-1)$$

For element a_3 there are $(n-2)$ possibilities, etc. until we reach element a_n for which there is only one possibility left.

Thus the total number of arrangements of the n elements $a_1, a_2, \ldots, a_n$ is

$$n(n-1)(n-2) \ldots 1 = n!$$

Rule   The number n_P of permutations of n different elements is

$$n_\text{p} = 1 \times 2 \times 3 \times \ldots \times (n-1)n = n!$$

Remember from Chapter 8, section 8.2, that $n!$ is referred to as 'factorial n'. We also defined factorial zero as $0! = 1$.

For example, $3! = 1 \times 2 \times 3 = 6$ is read as '3 factorial equals 6'.

It is a little more difficult to obtain the number of permutations of n elements if some of them are equal as, for example, in the case of the three elements a, b, b. We can write the number of permutations of these 3 elements thus:

$$abb, \ bab, \ bba, \ abb, \ bab, \ bba$$

But note that there are three pairs of identical permutations; hence there are only three different permutations. The identical permutations are formed by rearranging equal elements. If we are interested in the number of different permutations, then we must divide the total number of permutations by the number of permutations of equal elements.

$$N_{abb} = \frac{3!}{2!} = \frac{6}{2} = 3$$

Rule   If out of n elements $n_1, n_2, \ldots, n_m$ elements are equal to each other, then the number of different permutations is

$$n_p = \frac{n!}{n_1! n_2! \ldots n_m!}$$

PROOF   Let n_p be the required number of permutations. From any of these, if the n_1-like elements were different, we could make $n_1!$ new permutations. Thus if the n_1-like elements were all different, we would get $n_p n_1!$ permutations.

Similarly, if the n_2-like elements were different, we would get $n_2!$ new permutations from each of the second set of permutations.

Thus, if the n_1-like elements and the n_2-like elements were all different, we would get $n_p n_1! n_2!$ permutations in all.

The process is continued until all the sets of like elements are dealt with, and we then get the number of permutations of n elements which are all different. This is $n!$

Therefore $n_p n_1! n_2! \ldots n_m! = n!$

Hence
$$n_p = \frac{n!}{n_1! n_2! \ldots n_m!}$$

19.3.2   **COMBINATIONS**

We now alter the problem considered in the previous section to ask the number of different ways in which k elements can be selected out of n elements.

EXAMPLE   Consider the four letters a, b, c, d $(n = 4)$. They can be arranged in groups of two $(k = 2)$ as follows:

$$ab, ac, ad, bc, bd, cd$$

That is, there are six groups in all.

EXAMPLE   Now consider the five letters, a, b, c, d, e $(n = 5)$. They can be arranged in groups of three $(k = 3)$ as follows:

$$abc, abd, abe, acd, ace, ade, bcd, bce, bde, cde$$

That is, there are 10 groups in all.

DEFINITION   Each of the groups which can be made by taking k different elements of a number of elements n is called a *combination* of these elements.

Generally, if there are given n different elements and each group contains k of them, the symbol $_nC_k$ or $\binom{n}{k}$ is used to denote the number of all possible groups.

In other words, $_nC_k$ is the number of combinations of n different elements taken k at a time *without repetition*. This is also referred to as the kth class. The number obtained is given here without proof:

Binomial coefficient:

$$\binom{n}{k} = {_nC_k} = \frac{n!}{k!(n-k)!}$$

Note: This is called the *binomial coefficient*. These coefficients occur in the binomial

expansion $(a+b)^n = \sum_{k=0}^{n} \binom{n}{k} a^{n-k} b^k$.

Note the special case $\binom{n}{0} = \binom{n}{n} = \frac{n!}{0!n!} = 1$ since $0! = 1$.

EXAMPLE   A club has 20 members. The managing committee is formed by 4 members having equal rights. How many combinations are possible in choosing a committee?

We must choose 4 members out of the 20. Thus there are $\binom{20}{4}$ combinations:

$$\binom{20}{4} = \frac{20!}{4!16!} = \frac{20 \times 19 \times 18 \times 17}{4 \times 3 \times 2 \times 1} \approx 5000$$

So far we have considered combinations containing different elements only. Now we will consider combinations where several elements are allowed to recur. We will see that, in this case, more combinations are possible.

EXAMPLE   Given the four letters a, b, c, d, how many combinations are there when taking 2 letters at a time?

(i) Without repetition: ab, ac, ad, bc, bd, cd

These are 6 combinations $6 = \binom{4}{2}$.

(ii) With repetition: there are 6 combinations as in (i) plus aa, bb, cc, dd, giving 10 combinations, i.e. 4 more than when repetition is not allowed.

We give the rule without proof.

Rule   There are $\binom{n+k-1}{k}$ combinations of n different elements taken k at a time with repetition.

Up to now we have not made any allowance for the order in determining the number of combinations. If we distinguish between combinations with the same elements but with different orders, we are considering *variations*. We give the rule for the number of variations without proof.

> **Rule**   There are $\dfrac{n!}{(n-k)!}$ variations of n different elements taken k at a time with repetitions and taking account of different order.

EXERCISES

1. A menu contains five dishes from which two can be chosen freely. Give the sample space.

2. A set of cards consists of 16 red and 16 black cards. What is the probability of drawing a black card out of the pile?

3. What is the probability that when tossing a die the number that appears will be divisible by 3?

4. An experiment is carried out 210 times. Outcome A has been measured 7 times. What is the relative frequency of outcome A?

5. A box contains 20 balls. Of these, 16 are blue and 4 are green. Evaluate the probability of taking out a blue ball and, after putting it back, the probability of drawing a green ball.

6. What is the probability that when tossing two dice we will obtain a sum of 2 spots after the first throw and a sum of 5 after the second throw?

7. A player casts with two dice. What is the probability that a 2, a 3 or a 4 will be thrown?

8. A sales representative has to visit 6 towns. In how many ways can the route be fixed so that the journey always starts from town A?

9. A teacher has a class of 15 pupils. Of these, 3 must be selected for a special task. How many possibilities are there?

CHAPTER 20

Probability distributions

20.1 DISCRETE AND CONTINUOUS PROBABILITY DISTRIBUTIONS

20.1.1 DISCRETE PROBABILITY DISTRIBUTIONS

In the practical treatment of statistical problems, it is wise to characterise individual outcomes of a random experiment with numerical values. A simple method is to number the individual outcomes consecutively. For example, the outcomes of throwing a die can be described by the number of the spots on the faces.

The set of numerical values can be regarded as the range of definition of a variable, which is called a *random variable* or a *variate*. The outcomes of a random experiment are therefore assigned to the values of the random variable within its range of definition.

If the range of definition consists of discrete values, we call the variable a *discrete random variable*. We are then in a position to assign to each value of the random variable the probability of its outcome in a random experiment.

This can be expressed symbolically as

Outcome of random experiment j → random variable x_j → probability $P(j)$

The complete set of probabilities for each value of the discrete random variable in a random experiment is called a *discrete probability distribution*.

> **DEFINITION** A *discrete probability distribution* is the complete set of probabilities of the discrete values of the random variable in a random experiment.

EXAMPLE Consider an ideal or 'fair' die. There are six possible outcomes when throwing a die. The discrete random variable, in this case, 'number of spots', assumes the values from 1 to 6. Each value of the random variable has the probability

$$P(x) = \tfrac{1}{6} \qquad (x = 1, 2, 3, 4, 5, 6)$$

The probability distribution for this somewhat trivial example is shown in figure 20.1.

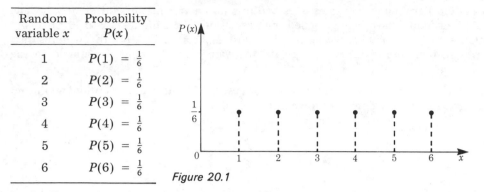

Random variable x	Probability $P(x)$
1	$P(1) = \frac{1}{6}$
2	$P(2) = \frac{1}{6}$
3	$P(3) = \frac{1}{6}$
4	$P(4) = \frac{1}{6}$
5	$P(5) = \frac{1}{6}$
6	$P(6) = \frac{1}{6}$

Figure 20.1

In the figure, the values of the discrete random variable are plotted on the abscissa and their respective probabilities along the ordinate.

EXAMPLE   Two dice are thrown. We choose as the random variable the sum of the number of spots. Thus the random variable assumes the values 2, 3, 4, ..., 12. We now seek the probability distribution for this random variable.

The outcome $x = 2$, for example, can only be realised if each die shows a '1'.

The probability for this event is

$$P(2) = \tfrac{1}{6} \times \tfrac{1}{6} = \tfrac{1}{36}$$

The outcome $x = 5$, as another example, can be realised by 4 elementary events:

First die	Second die	Sum
1	4	5
4	1	5
2	3	5
3	2	5

The probability distribution in this example is shown as figure 20.2.

Random variable x	Probability $P(x)$
2	$\frac{1}{36}$
3	$\frac{2}{36}$
4	$\frac{3}{36}$
5	$\frac{4}{36}$
6	$\frac{5}{36}$
7	$\frac{6}{36}$
8	$\frac{5}{36}$
9	$\frac{4}{36}$
10	$\frac{3}{36}$
11	$\frac{2}{36}$
12	$\frac{1}{36}$

Figure 20.2

We could have numbered the values of the random variable in a different way. The outcomes could have been assigned, just as easily, to the numbers 1 to 11 or to any other set of numbers consisting of 11 values, e.g. 100 to 110, but this would surely have made it more difficult to relate the values of the random variable to the numerical results of the experiment.

EXAMPLE   Consider a cylinder, having a base of 1 square unit, containing air and subdivided into 5 regions, as shown in figure 20.3.

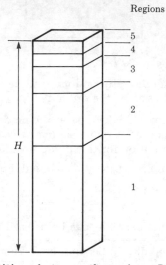

Figure 20.3

There are no material partitions between the regions. Let us consider an arbitrary air molecule. The purpose of the experiment is to measure the position (height) of this air molecule at a particular instant when the air has been thoroughly mixed. We wish to evaluate the probability that we will find the air molecule in a particular region. We number the regions 1 to 5; these are our random variables. The volume of each region is given in figure 20.4. The probability $P(x)$ of the air molecule being at a particular position is given by the ratio of the partial volume V_x to the total volume V of the cylinder. Hence

$$P(x) = \frac{V_x}{V}$$

Random variable x	Volume of region V_x	Probability $P(x)$
1	8	$\frac{1}{2}$
2	4	$\frac{1}{4}$
3	2	$\frac{1}{8}$
4	1	$\frac{1}{16}$
5	1	$\frac{1}{16}$

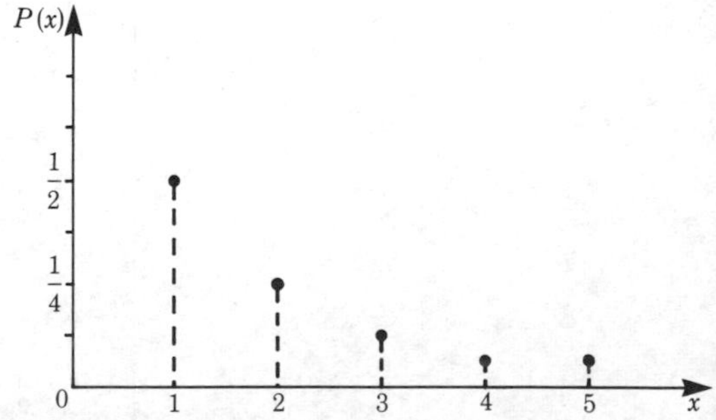

Figure 20.4

The table and the diagram show this discrete probability distribution. In this example, we assumed that the positional probability in a region was proportional to the volume of the region.

20.1.2 CONTINUOUS PROBABILITY DISTRIBUTIONS

The random experiments considered so far had discrete values of the random variable. There are, however, many random experiments whose results are best expressed by means of a continuous variable.

Let us consider again the example of the cylinder filled with air and an arbitrarily chosen air molecule.

The purpose of the experiment is to measure the position of this air molecule at a certain instant when the air is thoroughly mixed, and we now wish to evaluate the probability that we will find this air molecule at a position defined by the height h (figure 20.5). The external conditions of this random experiment remain unchanged; all that has changed is the formulation of the problem. The result is the position h of the air molecule which can take on any value between 0 and H.

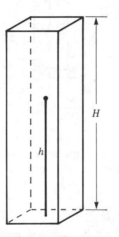

Figure 20.5

The position (reading the height) h is now a *continuous* quantity; it is a logical way of describing the outcome of the random experiment with h as the random variable. We have thus obtained a new type of random experiment. The outcomes and the random variable have a continuous set of values. This is always the case when we consider a measurement as a random experiment in which the measured quantity varies continuously.

An important and perhaps a surprising conclusion for continuous random variables can be drawn from our example. The probability of finding the air molecule somewhere between 0 and H is 1 (by the normalising condition). The number of possible readings of its position is infinite for an arbitrarily small subdivision of the height H. It follows, therefore, that the probability at a definite position h_0 has to approach the value zero.

For continuous random variables it is not possible to specify a probability different from zero for an *exactly* defined value of a random variable.

A value for the probability of the outcome of a random experiment for a continuous variable can, however, be given if we formulate the problem differently, i.e. by asking for the probability that the air molecule will be inside a given *interval*.

What then is the probability of finding the air molecule in the interval h_0 to $h_0 + \Delta h$? (see figure 20.6).

The total height of the cylinder is H. The probability $P(h_0 \leqslant h \leqslant h_0 + \Delta h)$ of finding the air molecule in the interval h_0 and $h_0 + \Delta h$ is proportional to the interval Δh, i.e.

$$P(h_0 \leqslant h \leqslant h_0 + \Delta h) = \frac{1}{H} \Delta h$$

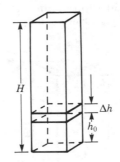

Figure 20.6

In general, the probability will also depend on the value of h_0 considered. This will be the case when, for example, we take the influence of gravity into account.

Hence, more generally, we have

$$P(h_0 \leqslant h \leqslant h_0 + \Delta h) = F(h_0) \Delta h$$

$F(h_0)$ is called the *probability density function* and is as yet unknown. If the probability density is known as a function of h, then we can evaluate the probability of finding the air molecule in the interval Δh for each value h_0.

This concept of a probability density function has to be very clearly distinguished from the probability of a discrete distribution. The probability density is not quite a probability, but a probability per unit of the random variable. The probability itself can only be given for an interval of the random variable. The probability for an interval is then the product of the probability density function by the magnitude of the interval (for small intervals).

Let us now determine the probability distribution in the case of the air molecule.

Case 1: We will assume for the time being that the influence of gravity can be neglected, in which case the probability density function will not depend on h_0. The probability of finding the air molecule in an interval of length Δh was

$$\frac{\Delta h}{H} = F(h) \Delta h$$

Hence the probability density function is $F(h) = \frac{1}{H}$; it is independent of h.

Case 2: We now take the influence of gravity into account. The density of the air decreases with height in accordance with the barometric equation (see Chapter 13, section 13.2).

It is shown in statistical mechanics that the probability density function for a gas molecule, taking gravity into account is given by

$$F(h) = \frac{1}{c} e^{-\alpha h} \quad \text{for} \quad 0 \leqslant h \leqslant H$$

The constant c depends on the conditions of the gas inside the column:

$$\alpha = \frac{\rho_0}{P_0} g$$

ρ_0 = density at $h = 0$, P_0 = pressure at $h = 0$, and g = acceleration due to gravity.

c is a constant which is obtained by the *normalisation condition*, i.e.

$$\int_0^H F(h)\, dh = 1$$

and hence

$$\int_0^H \frac{1}{c} e^{-\alpha h}\, dh = 1$$

from which we get

$$c = \frac{1}{\alpha}(1 - e^{-\alpha H})$$

In this case, the probability density function is a function of the altitude h.

Let us denote the probability density function for an arbitrary random variable x by $f(x)$. In order to determine the probability that x will assume any value between x_1 and x_2, we have to add up the probabilities of all the intervals between x_1 and x_2. When the lengths of the intervals Δx tend to zero the sum becomes an integral and we have

$$P(x_1 \leqslant x \leqslant x_2) = \int_{x_1}^{x_2} f(x)\, dx$$

All probability distributions have to satisfy the normalisation condition; for discrete probability distributions this is

$$\sum_{i=1}^{n} P(i) = 1$$

For a continuous distribution, the *normalisation condition* is

$$\int_{-\infty}^{\infty} f(x)\, dx = 1$$

20.2 MEAN VALUES OF DISCRETE AND CONTINUOUS VARIABLES

Mean Value of Discrete Random Variables

A class of students is given a test with the following results:

Grade	Number of students
1	4
2	4
3	6
4	2
5	1
6	0

It is often very useful to characterise the performance of a class by means of a single number, although by doing so a lot of information is lost, since the individual grades of the pupils are not included. Nevertheless, such a number, if carefully chosen, is sufficient in many cases. Such a number is the *arithmetical mean value*.

The arithmetical mean value is defined as the sum of all individual grades divided by the number of students. In our example, we have

$$\frac{(1 \times 4 + 2 \times 4 + 3 \times 6 + 4 \times 2 + 5 \times 1 + 6 \times 0)}{17} = \frac{43}{17} = 2.53$$

We encounter the same type of problem when we measure a physical quantity several times in succession and the measured values do not agree completely with each other, which is frequently the case in practice. In this case we also take the arithmetical mean value as the most probable value of the physical quantity.

Now let us generalise. If we carry out a random experiment n times, the random variable may take on the values $x_1, x_2, x_3, \ldots, x_n$.

The *arithmetical mean value* $\bar{x}$ of discrete random variables is defined as

$$\bar{x} = \frac{1}{n} \sum_{i=1}^{n} x_i$$

Let the range of definition of a discrete random variable x be $x_1, x_2, \ldots, x_k$. We then carry out a random experiment n times. If the random value x_i appears with a frequency n_i, then the mean value $\bar{x}$ of the random variable is defined as

$$\bar{x} = \frac{1}{n} \sum_{i=1}^{k} n_i x_i$$

If we know the probability distribution $P(i)$ of a discrete random variable, then the mean value is defined as

$$\bar{x} = \sum_{i=1}^{k} P(i) \, x_i$$

Mean Value of Continuous Random Variables

If a continuous random variable is defined between the values x_1 and x_2 and the probability density function is $f(x)$, then the *mean value* of x is given by

$$\bar{x} = \int_{x_1}^{x_2} x f(x)\, dx$$

EXAMPLE   The probability density function of a gas in a cylinder is given by

$$f(h) = \frac{e^{-\alpha h}}{\frac{1}{\alpha}(1 - e^{-\alpha H})} \qquad (0 \leqslant h \leqslant H)$$

The mean value of the random variable h (the height of the gas molecule in the cylinder shown in figure 20.7) is

$$\bar{h} = \int_0^H \frac{h\, e^{-\alpha h}}{\frac{1}{\alpha}(1 - e^{-\alpha H})}\, dh = \frac{1}{\alpha} - \frac{H\, e^{-\alpha H}}{1 - e^{-\alpha H}}$$

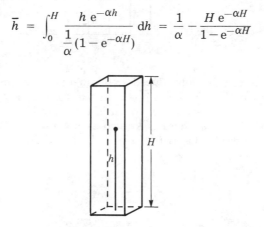

Figure 20.7

In the case of an air molecule in the Earth's atmosphere, α has the value $0.000\,18\,\mathrm{m}^{-1}$. Figure 20.8 shows the mean height h of a gas molecule plotted against the altitude H.

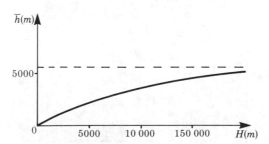

Figure 20.8

We notice that for small values of H, $\bar{h} = \dfrac{H}{2}$. This means that the mean value of the random variable h is in the middle of the height of the column. For $H \to \infty$, $\bar{h}$ approaches the limiting value of $5400\,\mathrm{m}$, which means that for arbitrarily large heights of the column $\bar{h}$ remains finite.

20.3   THE NORMAL DISTRIBUTION AS THE LIMITING VALUE OF THE BINOMIAL DISTRIBUTION

A coin is tossed 10 times in succession. The probability of the outcome being a head k times and a tail $(10-k)$ times is given by the *binomial distribution* for a random experiment carried out n times with two possible outcomes.

For $n = 10$ and equally likely outcomes, the probabilities have been plotted as a function of k in figure 20.9.

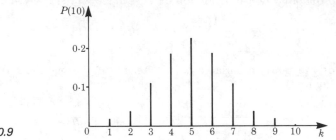

Figure 20.9

Such a distribution can be reproduced experimentally in the following manner. n rows of nails are positioned on a board in the form of a pyramid, as shown in figure 20.10. Each nail is exactly positioned in the middle of two nails belonging to the next lower row. Such a board is known as a Galtonian board, and the figure illustrates two such boards, one with 4 rows of nails and the other with 8 rows of nails.

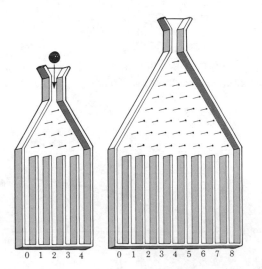

Figure 20.10

A marble (or a small steel sphere) is introduced in the tunnel at the top of the board and allowed to drop freely so that it impinges on the top nail centrally. It is then deflected to the right or left with equal probability, after which it impinges on a nail in the next row and so on through every row. Each time it hits a nail it is deflected to the left or right with equal probability. The sphere is then collected in one of the partitions shown in the figure.

If there are n rows of nails, then this corresponds exactly to n experiments with two equally probable outcomes. This leads to the binomial distribution. A marble reaches the kth partition if it has been deflected on k nails to the left and $(n-k)$ nails to the right. The probability $P(n;k)$ for this (cf. section 20.3.2) is

$$P(n;k) = \binom{n}{n_k} (\tfrac{1}{2})^k (\tfrac{1}{2})^{n-k} = \binom{n}{n_k} (\tfrac{1}{2})^n$$

If we allow many marbles to drop through the Galtonian board, we then find that they are distributed in accordance with the probability $P(n;k)$. The number of marbles in the different partitions approaches the binomial distribution. (From a practical point of view, we have to select the radius of the marbles and the distance between the nails in such a way that ideal conditions are obtained.)

Figure 20.11 shows results obtained empirically with $n = 4$ and $n = 8$.

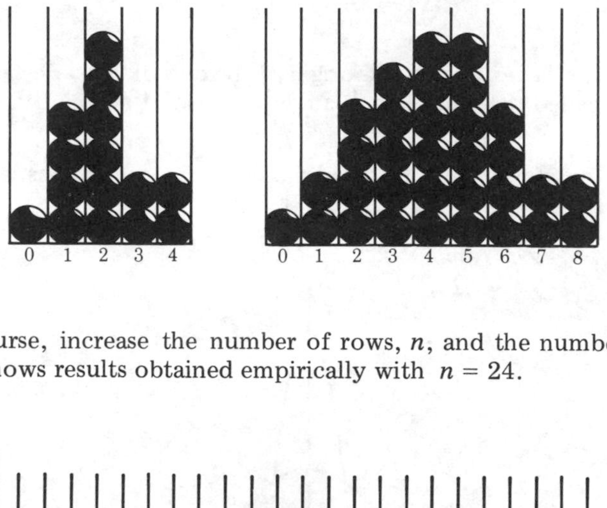

Figure 20.11

We can, of course, increase the number of rows, n, and the number of partitions. Figure 20.12 shows results obtained empirically with $n = 24$.

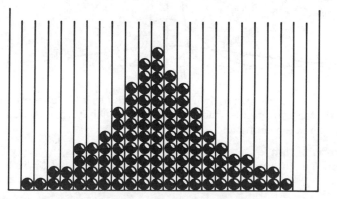

Figure 20.12

In the limit, as $n \to \infty$, we will obtain a continuous function, as shown in figure 20.13. It is known as the *normal distribution*.

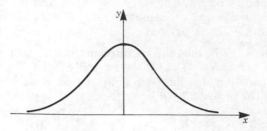

Figure 20.13

The proof that the binomial distribution is transformed into the normal distribution for $n \to \infty$ is quite complex and beyond the scope of this book.

20.3.1 PROPERTIES OF THE NORMAL DISTRIBUTION

The analytical expression for the normal distribution, given without proof is

$$f(x) = \frac{1}{\sigma\sqrt{2\pi}}\, e^{-\frac{1}{2}\left(\frac{x}{\sigma}\right)^2}$$

It is symmetrical with respect to the origin of the coordinates. Because of its shape, it is frequently called the bell-shaped curve. Because of symmetry, it has its maximum at $x = 0$.

The parameter σ defines the particular shape of the normal distribution. This is illustrated in figure 20.14 for $\sigma = 1$, 2 and 3.

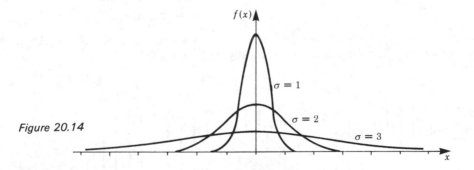

Figure 20.14

When σ is small the curve is narrow and elongated, i.e. it has a sharp maximum. The larger σ the flatter and broader the curve becomes. Whatever the value of σ the area under the curve remains constant.

The normal distribution is a probability distribution and has to satisfy the normalisation condition, i.e. the sum of all probabilities must be equal to 1. In this case, we have

$$\int_{-\infty}^{\infty} f(x)\,dx = \int_{-\infty}^{\infty} \frac{1}{\sigma\sqrt{2\pi}}\, e^{-\frac{1}{2}\left(\frac{x}{\sigma}\right)^2}\,dx = 1$$

The proof is given in the appendix to this chapter.

If we consider the normal distribution as the probability distribution of a random variable x, we can then compute its mean value. We have

$$\bar{x} = \int_{-\infty}^{\infty} x f(x)\, dx$$

$$= \int_{-\infty}^{\infty} \frac{x}{\sigma\sqrt{2\pi}} e^{-\frac{1}{2}\left(\frac{x}{\sigma}\right)^2}\, dx = \left[\frac{-2\sigma}{\sigma\sqrt{2\pi}} e^{-\frac{1}{2}\left(\frac{x}{\sigma}\right)^2} \right]_{-\infty}^{+\infty}$$

$$\bar{x} = 0$$

Here the normal distributed random variable has zero as its mean value.

A new normal distribution is obtained if we shift it along the x-axis by amount μ (figure 20.15). This is achieved by replacing x by $(x - \mu)$ so that

$$f(x) = \frac{1}{\sigma\sqrt{2\pi}} e^{-\frac{1}{2}\left(\frac{x-\mu}{\sigma}\right)^2}$$

This function, of course, has its maximum value at $x = \mu$. The mean value of the random variable in this case is $\bar{x} = \mu$, which should be fairly obvious.

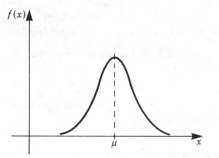

Figure 20.15

The parameter σ is called the *standard deviation*. It is given by

$$\sigma^2 = \int_{-\infty}^{\infty} (x - \mu)^2 f(x)\, dx$$

σ determines the width of the normal distribution curve. It is a measure of the variation of the random variable x about its mean value. Within the interval $\mu + \sigma$ and $\mu - \sigma$ lies 68% of the area beneath the normal distribution curve (Figure 20.16). Its meaning will be discussed further in Chapter 21 which deals with the theory of errors.

The normal distribution is symmetrical about its maximum value at $x = \mu$.

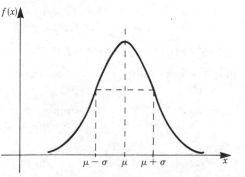

Figure 20.16

20.3.2   **DERIVATION OF THE BINOMIAL DISTRIBUTION**

There are some random experiments for which there are only two possible outcomes, e.g. tossing a coin.

Let us denote the two possible outcomes of an experiment by A and B, and let $P(A)$ and $P(B)$ be the probabilities for the occurrence of events A and B, respectively. Since A and B are the only two elementary events in this random experiment, we must have

$$P(A) + P(B) = 1$$

We now carry out this experiment N times in succession, e.g. we toss a coin N times. What is the probability of event A occurring N_A times in N experiments? We call this probability $P(N; N_A)$. We are not concerned here with the order in which the N_A events occur.

The solution can be divided into three steps.

(1) We compute the probability for the occurrence of a particular outcome with N_A events A and N_B events B, so that $N = N_A + N_B$.

(2) We obtain the number of particular outcomes which differ only in arrangement, not in N_A. This is the permutation discussed in Chapter 19, section 19.3.1.

(3) We evaluate, from the probability of one particular outcome and the number of equally probable outcomes, the probability that out of N experiments N_A will lead to the event A.

Let us start with step 1. We pick out one particular outcome, perhaps the one for which event A occurs N_A times, and after that event B occurs $N_B = (N - N_A)$ times, i.e.

$$\underbrace{A\,A\,A\dots A\,A}_{N_A \text{ times}} \qquad \underbrace{B\,B\,B\dots B\,B}_{N_B \text{ times}}$$

The elementary event A has the probability $P(A)$. The compound probability for the occurrence of N_A events A is

$$P(N_A) = P(A) \times P(A) \times \dots \times P(A) = P(A)^{N_A}$$

The elementary event B has the probability $P(B)$. The compound probability for the occurrence of $N_B = N - N_A$ events B is

$$P(N_B) = P(B) \times P(B) \times \dots \times P(B) = P(B)^{N_B} = P(B)^{N-N_A}$$

The compound probability for the simultaneous occurrence of N_A events A and $N_B = N - N_A$ events B is then the product

$$P(A)^{N_A} P(B)^{N_B} = P(A)^{N_A} P(B)^{N-N_A}$$

We have thus determined the probability for the occurrence of one particular outcome of the test.

Step 2: In Chapter 19, section 19.3.1, we determined that the number of permutations of N elements in which N_A and N_B are equal is

$$\frac{N!}{N_A! N_B!} = \frac{N!}{N_A!(N-N_A)!} = \binom{N}{N_A}$$

Step 3: Each of the permutations in Step 2 has the same probability of occurring. The probability that one of these permutations may occur is (in accordance with the addition theorem for probabilities) obtained by summing the probabilities of all the permutations.

Hence, with the results obtained in steps 1 and 2, the probability $P(N; N_A)$ that out of N random experiments N_A will have the outcome A is given by

$$P(N; N_A) = \frac{N!}{N_A!(N-N_A)!} P(A)^{N_A} P(B)^{N-N_B} = \binom{N}{N_A} P(A)^{N_A} P(B)^{N-N_A}$$

This also means that out of N experiments those that remain, namely N_B, will have the outcome B. For a preassigned N, $P(A)$ and $P(B)$, this probability is a function of N_A. The range of definition of this function is the set of integers from 0 to N. For each N_A, we can calculate the accompanying probability using the above equation. The totality of these probabilities is called the *binomial distribution*.

Note: Repeating an experiment N times can be replaced by carrying out N equal experiments simultaneously, e.g. tossing one coin N times is equivalent to tossing N coins simultaneously.

Summary

Suppose an experiment has two possible outcomes, A and B, which occur with probabilities $P(A)$ and $P(B)$. If we carry out N such experiments, then the probability for the occurrence of N_A events A is given by the binomial distribution

$$P(N; N_A) = \binom{N}{N_A} P(A)^{N_A} P(B)^{N-N_A}$$

APPENDIX

Evaluation of the integral

$$I = \frac{1}{\sigma\sqrt{2\pi}} \int_{-\infty}^{\infty} e^{-\frac{1}{2}\left(\frac{x-\mu}{\sigma}\right)^2} dx = 1$$

First substitute

$$z = \frac{1}{\sqrt{2}}\left(\frac{x-\mu}{\sigma}\right)$$

Then

$$dx = \sqrt{2}\sigma\, dz$$

Hence the integral becomes

$$I = \frac{1}{\sqrt{\pi}} \int_{-\infty}^{\infty} e^{-z^2} dz$$

To integrate this expression we use a 'trick': we multiply the expressions

$$I = \frac{1}{\sqrt{\pi}} \int_{-\infty}^{\infty} e^{-z^2} dz \quad \text{and} \quad I = \frac{1}{\sqrt{\pi}} \int_{-\infty}^{\infty} e^{-u^2} du$$

So obtaining the double integral

$$I^2 = \frac{1}{\pi} \int_{-\infty}^{\infty} \int_{-\infty}^{\infty} e^{-(z^2 + u^2)} \, dz \, du$$

Now we introduce polar coordinates:

$$r^2 = z^2 + u^2, \qquad z = r \cos \phi$$
$$u = r \sin \phi$$

The differential area $dz \, du$ becomes $r \, dr \, d\phi$. We integrate ϕ from 0 to 2π and r from 0 to ∞ and obtain

$$I^2 = \frac{1}{\pi} \int_0^{\infty} \int_0^{2\pi} e^{-r^2} r \, dr \, d\phi$$

The integration with respect to ϕ gives

$$\int_0^{2\pi} d\phi = 2\pi$$

so that

$$I^2 = 2 \int_0^{\infty} r \, e^{-r^2} \, dr$$

The solution can be verified easily

$$I^2 = 2 \left[-\frac{1}{2} e^{-r^2} \right]_0^{\infty} = 2 \left[0 + \frac{1}{2} \right] = 1$$

Thus $I^2 = 1$, and hence

$$I = \frac{1}{\sqrt{2\pi}} \int_{-\infty}^{\infty} e^{-\frac{1}{2}\left(\frac{x-\mu}{\sigma}\right)} \, dx = 1$$

(The value $I = -1$ is ruled out, since the integrand assumes positive values only.)

If we now insert in the integral $\mu = 0$ and $\sigma = \dfrac{1}{\sqrt{2}}$ we obtain

$$\frac{1}{\sqrt{\pi}} \int_{-\infty}^{\infty} e^{-x^2} \, dx = 1 \qquad \text{or} \qquad \int_{-\infty}^{\infty} e^{-x^2} \, dx = \sqrt{\pi}$$

Evaluation of the integral

$$\frac{1}{\sigma\sqrt{2\pi}} \int_{-\infty}^{\infty} (x-\mu)^2 \, e^{-\frac{1}{2}\left(\frac{x-\mu}{\sigma}\right)^2} \, dx = \sigma^2$$

First substitute

$$z^2 = \frac{1}{2} \left(\frac{x-\mu}{\sigma} \right)^2$$

Then $(x - \mu)^2 = 2\sigma^2 z^2$ and $dx = \sqrt{2}\,\sigma \, dz$.

The integral

$$\frac{1}{\sigma\sqrt{2\pi}}\int_{-\infty}^{\infty}(x-\mu)^2\, e^{-\frac{1}{2}\left(\frac{x-\mu}{\sigma}\right)}\, dx$$

is transformed into

$$\frac{2\sigma^2}{\sqrt{\pi}}\int_{-\infty}^{\infty}z^2 e^{-z^2}\, dz$$

Integrating by parts gives

$$\frac{2\sigma^2}{\sqrt{\pi}}\int_{-\infty}^{\infty}z^2 e^{-z^2}\, dz = \underbrace{\frac{2\sigma^2}{\sqrt{\pi}}\left[-z\frac{1}{2}e^{-z^2}\right]_{-\infty}^{+\infty}}_{=\ 0} - \frac{2\sigma^2}{\sqrt{\pi}}\int_{-\infty}^{\infty}\left(-\frac{1}{2}e^{-z^2}\right)dz$$

$$= \frac{\sigma^2}{\sqrt{\pi}}\int_{-\infty}^{\infty}e^{-z^2}\, dz$$

From above we know that

$$\int_{-\infty}^{\infty}e^{-z^2}\, dz = \sqrt{\pi}$$

Hence

$$\frac{1}{\sigma\sqrt{2\pi}}\int_{-\infty}^{\infty}(x-\mu)^2 e^{-\frac{1}{2}\left(\frac{x-\mu}{\sigma}\right)}\, dx = \sigma^2$$

EXERCISES

1. Two dice are thrown. Calculate the mean value of the random variable 'sum of the number of spots'.

2. A random variable has the probability distribution

$$f(x) = \begin{cases} \dfrac{x}{2}, & 0 \leqslant x \leqslant 2 \\ 0, & \text{otherwise} \end{cases}$$

Compute the mean value of the random variable.

3. 60% of the students who start to study for an engineering degree complete their studies and obtain a degree. What is the probability that in a group of 10 arbitrarily chosen students in the first term of study 8 will obtain a degree?

4. Evaluate the mean values of the random variable x which have the following normal distributions:

    (a) $f(x) = \dfrac{1}{3\sqrt{2\pi}} e^{-(x-2)^2/18}$ (b) $f(x) = \dfrac{1}{\sqrt{2\pi}} e^{-(x+4)^2/2}$

CHAPTER 21

Theory of errors

21.1 PURPOSE OF THE THEORY OF ERRORS

The theory of errors is a part of mathematical statistics and deals with the following facts.

Given the results of measurements carried out in a laboratory, we require statements about the 'true' value of the measured quantity and a prediction of the accuracy of the measurements.

There are two types of errors which arise when we carry out a measurement: *systematic* or *constant errors* and *random errors*.

Constant errors are errors generated in the measuring instruments or in the method of measurement. They always bias the result in a particular direction so that it is either too large or too small; they arise through wrong calibration of the measuring instrument or not paying attention to secondary effects. An example of a constant error is often found in the speedometer of a car, sometimes as a result of design. The speed indicated is frequently found to be 5% above the 'true' speed of the car, but this can vary from 0 to 7% in practice.

Constant errors can only be avoided by a critical analysis of the measuring technique and of the instruments, and such errors cannot be discovered with the help of the theory of errors.

Random errors are due to interference during measurements, so that a repetition of measurements does not give exactly the same results, i.e. the measured values vary. For example, if we weigh a body repeatedly we will always obtain a different result. Although we take great care on each occasion, we are not able to read each time exactly the same position of a pointer between two very fine marks. Furthermore, the pointer itself does not always settle at the same position. Random errors are the result of a multiplicity of interference factors like the fluctuation of the boundary conditions which had initially been assumed to be controllable (temperature, air pressure, voltage fluctuations, shocks and errors of observation).

To avoid random errors we must naturally improve the method of measurement. This leads to more reliable measurements but does not solve the fundamental problem. The influence of random errors can be limited, and the accuracy of measurement may even be improved in powers of 10; however, each instrument has limited accuracy and hence random errors will always appear.

The purpose of the theory of errors can now be formulated more precisely.

From the measured values we want to be able to infer the 'true' value of the measured quantity and estimate the reliability of the measurement. Each reading in an experiment is made up of a hypothetical 'true' value of the measured quantity and an error component:

$$x = T + E$$

where x = measured value, T = 'true' value free of errors, and E = error component.

Furthermore,

$$E = E_1 + E_2$$

where E_1 = random error which can be estimated by repeated measurements, and E_2 = constant error.

21.2 MEAN VALUE AND VARIANCE

21.2.1 MEAN VALUE

The acceleration due to gravity, g, is to be determined experimentally. The time of fall of a sphere is measured with a stopwatch and the distance with a tape measure. In order to increase the reliability, the measurements are repeated in a series of readings. A series of readings comprising 20 measurements is considered to be a *random sample* of all possible measurements for this experimental set-up. The arithmetic mean value of n measurements is taken as the best estimate for the 'true' value. We have

$$\bar{x} = \frac{1}{n} \sum_{j=1}^{n} x_j$$

If individual measured values occur repeatedly, the mean value can be expressed in terms of the frequency h_i with which they occur, in which case we have

$$\bar{x} = \sum_{i=1}^{k} h_i x_i$$

where n_i is the frequency and $h_i = \frac{n_i}{n}$ is the relative frequency of the measured value x_i. If $n \to \infty$, the relative frequencies become the probabilities $P(i)$ (see Chapter 19, section 19.2).

The sum of all deviations from the arithmetic mean value vanishes, i.e.

$$\sum_{i=1}^{n} \Delta x_i = \sum_{i=1}^{n} (x_i - \bar{x}) = 0$$

21.2.2 VARIANCE AND STANDARD DEVIATION

Individual measurements deviate from the mean value partly because of random errors. These deviations become smaller as the measurements become more reliable and more exact. They thus enable us to predict the reliability of the measurements and the magnitude of the random errors. In order to draw this conclusion, we have to define a measure of dispersion. Let us start by considering the deviation of the individual measurements from the mean. Since positive and negative deviations from the mean cancel out, their sum is zero. However, to obtain positive values we square the deviations. A suitable measure of dispersion is obtained by taking the mean value of the square of the deviations. Such a measure is called *variance*.

DEFINITION   The *variance* is the mean value of the square of the deviation:

$$S^2 = \frac{1}{n} \sum_{i=1}^{n} (x_i - \bar{x})^2 \tag{21.1}$$

If a frequency distribution exists, then the variance becomes

$$S^2 = \sum_{j=1}^{k} h_j (x_j - \bar{x})^2$$

The unit of variance is the square of a physical quantity.

The measure of dispersion of the physical quantity is obtained by taking the square root of the variance and is called the *standard deviation*.

DEFINITION   Standard deviation:

$$S = \sqrt{\frac{1}{n} \sum_{i=1}^{n} (x_i - \bar{x})^2} = \sqrt{\sum_{j=1}^{k} h_j (x_j - \bar{x})^2} \tag{21.2}$$

Meaning of Standard Deviation

For a large number of measurements, about 68% of all measured values of x will lie within the interval

$$\bar{x} - S < x < \bar{x} + S$$

The mean value and the variance are related to each other. We demonstrate below that for the arithmetic mean the variance (and with it the standard deviation) assume a minimum value.

PROOF   We take the variance as the mean value of the squares of the deviations from some independent reference value $\tilde{x}$ so that

$$S^2 = \frac{1}{n} \sum (x_i - \tilde{x})^2$$

We wish to determine $\tilde{x}$ in such a way that the variance assumes a minimum. According to the rules of differential calculus, we find the minimum if we set

$$\frac{d}{d\tilde{x}}(S^2) = 0$$

i.e.

$$\frac{d}{d\tilde{x}}\left(\frac{1}{n}\sum(x_i - \tilde{x})^2\right) = \frac{2}{n}\sum(x_i - \tilde{x})(-1) = 0$$

or

$$\sum x_i - n\tilde{x} = 0$$

Hence

$$\tilde{x} = \frac{1}{n}\sum x_i = \bar{x}, \qquad \text{the arithmetic mean value.}$$

Now we have to check whether there is, in fact, a minimum for this value of x. We find

$$\frac{d^2}{dx^2}(S^2) = 2, \qquad \text{a positive value.}$$

Hence the variance is a minimum for $\tilde{x} = \bar{x}$.

21.2.3 MEAN VALUE AND VARIANCE IN A RANDOM SAMPLE AND PARENT POPULATION

A random sample contains n measurements of one quantity. We consider the measurements as a random selection out of the set of all the possible measurements in an experiment. This set is called the *parent population*. It is always larger than the random sample. The parent population, like the random sample, is characterised by a mean value and a variance. The mean value for the parent population is the hypothetical 'true' value. The values which relate to the parent population can be estimated on the basis of random sample data. The larger the random sample is, the more reliable the estimate becomes.

Quantities which relate to the random sample are denoted by latin letters, i.e.

$$\bar{x} = \text{mean value}$$
$$S^2 = \text{variance}$$

and those which relate to the parent population are denoted by greek symbols, i.e.

$$\mu = \text{mean value}$$
$$\sigma^2 = \text{variance}$$

The estimates for the 'true' values for the parent population are the ones we are interested in, and the results of measurements are estimates of the unknown 'true' values. We give below, without proof, the most important formulae for these estimates. You will find detailed explanations and proofs in textbooks on mathematical statistics.

Best *estimate of the arithmetic mean value*

$$\mu \approx \bar{x} = \frac{1}{n} \sum_{i=1}^{n} x_i \qquad (21.3a)$$

Best *estimate of the variance*

$$\sigma^2 \approx \frac{n}{n-1} S^2 = \frac{1}{n-1} \sum_{i=1}^{n} (x_i - \bar{x})^2 \qquad (21.3b)$$

This estimate for the parent population is larger than that for the sample by the factor $\frac{n}{(n-1)}$; for sufficiently large n this factor tends to the value 1, and hence S^2 can be used as an estimate for σ^2.

Best *estimate of the standard deviation*

$$\sigma = \sqrt{\frac{n}{n-1}} \qquad S = \sqrt{\frac{n}{n-1}} \sqrt{\sum_{i=1}^{n} (x_i - \bar{x})^2}$$

EXAMPLE   The diameter of a wire has been measured a number of times, as shown in the first column of the table below. Calculate the mean value and the standard deviation for the parent population. Columns 2 and 3 show the required workings.

	Diameter x_i (mm)	$(x_i - \bar{x})$ (mm)	$(x_i - \bar{x})^2$ (mm)^2
	14.1×10^{-2}	-0.1×10^{-2}	0.01×10^{-4}
	13.8×10^{-2}	-0.4×10^{-2}	0.16×10^{-4}
	14.3×10^{-2}	0.1×10^{-2}	0.01×10^{-4}
	14.2×10^{-2}	0	0
	14.5×10^{-2}	0.3×10^{-2}	0.09×10^{-4}
	14.1×10^{-2}	-0.1×10^{-2}	0.01×10^{-4}
	14.2×10^{-2}	0	0
	14.4×10^{-2}	0.2×10^{-2}	0.04×10^{-4}
	14.3×10^{-2}	0.1×10^{-2}	0.01×10^{-4}
	13.9×10^{-2}	-0.3×10^{-2}	0.09×10^{-4}
	14.4×10^{-2}	0.2×10^{-2}	0.04×10^{-4}
Sum	156.2×10^{-2}	0	0.46×10^{-4}

(1) We calculate the mean value of the sample:

$$\bar{x} = \bar{d} = \frac{156.2 \times 10^{-2}}{11} = 0.142 \, \text{mm}$$

(2) Now we calculate the variance and standard deviation of the sample:

$$S^2 = \frac{0.46 \times 10^{-4}}{11} = 0.042 \times 10^{-4}\,\text{mm}^2$$

$$S = 0.20 \times 10^{-2} = 0.002\,\text{mm}$$

(3) Mean value, variance and standard deviation for the parent population:

$$\bar{x} = 0.142\,\text{mm}, \qquad \text{as for the sample.}$$

$$\sigma^2 \approx S^2 \frac{n}{n-1} = \frac{11}{10} \times 0.46 \times 10^{-4} = 0.046 \times 10^{-4}$$

$$\sigma = 0.21 \times 10^{-2} = 0.0021\,\text{mm}$$

21.3 MEAN VALUE AND VARIANCE OF CONTINUOUS DISTRIBUTIONS

The concepts of variance and mean value can be applied to continuous distributions.

Let the probability density function $p = f(x)$ of a distribution be given.

We know that for discrete samples the mean value is

$$\bar{x} = \sum_{i=1}^{n} \frac{x_i}{n} = \sum_{j=1}^{k} h_j x_j$$

We replace the discrete frequencies h_j by the probability density function $p = f(x)$ and proceed to the limit as $k \to \infty$. The sum becomes an integral:

Mean value μ of a continuous distribution

$$\mu = \int_{-\infty}^{\infty} x\, f(x)\, \mathrm{d}x \tag{21.4}$$

The variance for a given frequency distribution of the sample gives the best estimate of the variance for the parent population. We have seen that it is given by

$$\sigma^2 = \frac{n}{n-1} \sum_{i=1}^{k} h_i (x_i - \bar{x})^2$$

In the limit, *the variance for a continuous distribution* becomes

$$\sigma^2 = \int_{-\infty}^{\infty} (x - \mu)^2 f(x)\, \mathrm{d}x \tag{21.5}$$

EXAMPLE   A factory manufactures bolts for use in the construction industry. Because manufacturing processes are not perfect not all manufactured parts are absolutely identical. In the case of the bolts, their diameter x is a random variable. We *assume* that the density function has the following distribution:

$$f(x) = \begin{cases} A(x-0.9)(1.1-x), & 0.9 < x < 1.1 \\ 0, & \text{otherwise} \end{cases}$$

The constant A is unknown.

Determine the value of the constant A, plot $f(x)$ graphically and calculate the mean value μ and the variance σ^2.

First let us calculate A

From section 20.1.2 we know that all probability distributions have to satisfy the normalisation condition, i.e. the sum of all probabilities has to equal 1. Hence

$$\int_{0.9}^{1.1} f(x)\, dx = 1$$

i.e.

$$\int_{0.9}^{1.1} (x-0.9)(1.1-x)\, dx = \frac{1}{A}$$

Integrating and solving for A gives $A = 750$

Figure 21.1 shows tabulated values of the function and its graph.

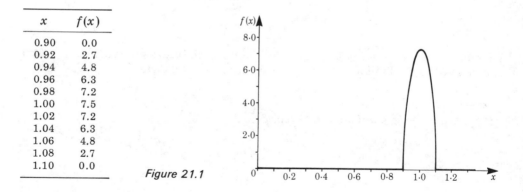

x	$f(x)$
0.90	0.0
0.92	2.7
0.94	4.8
0.96	6.3
0.98	7.2
1.00	7.5
1.02	7.2
1.04	6.3
1.06	4.8
1.08	2.7
1.10	0.0

Figure 21.1

The function is symmetrical; in fact it is a parabola. We see from the graph that μ must be equal to 1. Let us check this with our definition:

$$\mu = 750 \int_{0.9}^{1.1} x(x-0.9)(1.1-x)\, dx$$

$$= 750 \left[\frac{2}{3}x^3 - \frac{1}{4}x^4 - \frac{0.99}{2}x^2 \right]_{0.9}^{1.1} = 1.0$$

The variance is $\sigma^2 = 750 \int_{0.9}^{1.1} (x-1)^2(x-0.9)(1.1-x)\, dx = 0.002$

21.4   **ERROR IN MEAN VALUE**

Up to now we have considered the mean value of one random sample of n measurements to be the best estimate for the 'true' value. We have not yet answered the question concerning the error in such an estimate.

Consider now several random samples (series of readings) of the same number n in the same parent population. The mean values of these random samples are also scattered around the 'true' value. We need to know whether the dispersion of the mean values is smaller than that of the individual values. This dispersion of the mean is most important to us, since it determines the reliability of the result of a series of readings.

Let $\sigma_M{}^2$ be the variance of the mean values of random samples, the variance of the individual values being σ^2.

The variance of the mean values is then given by $\sigma_M{}^2 = \dfrac{\sigma^2}{n}$

The standard deviation of the mean values is $\sigma_M = \dfrac{\sigma}{\sqrt{n}}$

The standard deviation of the mean values is a measure of the accuracy of the mean values of a series of readings. It is referred to as the *sampling error* or the mean error of the mean values.

> **DEFINITION**   The *standard deviation of the mean value*, the sampling error (mean error of the mean value) is
>
> $$\sigma_M = \frac{\sigma}{\sqrt{n}} = \sqrt{\frac{\Sigma(x-\bar{x})^2}{n(n-1)}} \qquad (21.6)$$

The accuracy of the measurements can be increased by increasing the number of independent measurements. If the sampling error in the mean value is to be halved, for example, then the number of measurements has to be quadrupled.

EXAMPLE   The example in section 21.2.3 concerning the diameter of a wire gave the following results:

$$\text{Arithmetic mean value } \bar{x} = 0.142 \text{ mm.}$$

$$\text{Standard deviation} \quad \sigma = 0.0021 \text{ mm.}$$

Mean error of the mean values:

$$\sigma_M = \frac{\sigma}{\sqrt{n}} = \frac{0.0021}{11} = 0.0006 \text{ mm}$$

Hence $\bar{d} = 0.142 \pm 0.0006 \text{ mm.}$

Confidence intervals

Although the mean value $\bar{x}$ obtained is an estimate of the 'true' value μ, we also need to take into account the sampling error.

Since it is reasonable to suppose that the sample means are normally distributed, we can expect the 'true' value to lie with a probability of 68% in the interval $\bar{x} \pm 1\sigma_M$ and with a probability of 95% in the interval $\bar{x} \pm 2\sigma_M$ (see section 21.5).

These intervals are called *confidence intervals*. The results of measurements are usually presented by quoting the mean value and the sampling error.

21.5   NORMAL DISTRIBUTION: DISTRIBUTION OF RANDOM ERRORS

We have assumed that random errors are normally distributed. This assumption can be rendered plausible on the basis that random errors arise as a result of the super-position of many very small sources of errors referred to as *elementary errors*. The individual errors in measurements arise as a result of these elementary accidental errors like the deviations of the marbles in the Galtonian board. Starting with the hypothesis of a large number of uncontrollable statistical interference factors, Gauss showed that the probability function of the distribution of measured values can be described by the normal distribution:

$$f(x) \; = \; \frac{1}{\sigma\sqrt{2\pi}} \; e^{-\frac{1}{2}\left(\frac{x-\mu}{\sigma}\right)^2}$$

The usefulness of this model has also been proved empirically. In many cases, the errors in measurements are actually distributed in accordance with this normal distribution around their corresponding mean value. The larger the number of measurements the more closely the distribution approaches the Gaussian normal distribution. The measurements are dispersed around the 'true' value μ. The standard deviation is σ.

Using the value of the standard deviation, we can specify the percentage of the measured values which can be expected to fall within a given interval in the neigh-bourhood of the mean value, as shown in figure 21.2.

A deviation of more than $\pm 3\sigma$ from the mean value can only, on the average, occur once in 300 measurements by chance. This means that practically all measured results fall inside the interval $\mu \pm 3\sigma$.

The mean values of random samples are always normally distributed, the dispersion being of course smaller. As already explained in section 21.4, the variance of the mean values decreases as n, the number of measured values, is increased since

$$\sigma_M \; = \; \frac{\sigma}{\sqrt{n}} \qquad \text{(standard deviation of the mean value)}$$

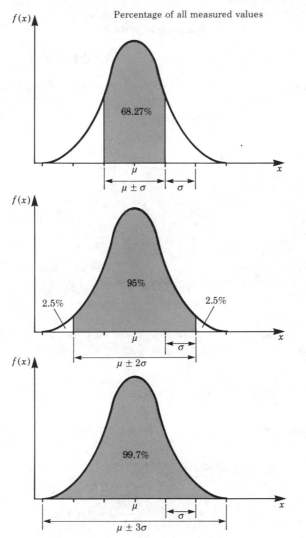

Figure 21.2

21.6 LAW OF ERROR PROPAGATION

In science and engineering, we have to distinguish between fundamental quantities (length, mass, time) and derived quantities (specific weight, velocity, acceleration, force, pressure etc.). Many physical quantities cannot be measured directly, since they are calculated from several other measured quantities. For example, specific weight is calculated from the measured quantities weight and volume. Thus the sampling error in the specific weight is made up of the sampling errors which arise in the measurement of the weight and the volume.

If we assume that some physical quantity g cannot be measured directly but is a function of two measurable quantities x and y we have

$$g = f(x, y), \qquad x \text{ and } y \text{ being independent.}$$

We can then measure x and y and obtain the mean values $\bar{x}$ and $\bar{y}$ and the standard deviations σ_{Mx}, σ_{My}. The standard deviation σ_{Mg} of the quantity g is then calculated by the application of the Gaussian *law of error propagation* given here without proof.

Law of error propagation

$$\sigma_{Mg} = \sqrt{\left(\frac{\partial f}{\partial x}\right)^2 {\sigma_{Mx}}^2 + \left(\frac{\partial f}{\partial y}\right)^2 {\sigma_{My}}^2} \qquad (21.7)$$

The partial derivatives are calculated at the values $\bar{x}$ and $\bar{y}$.

EXAMPLE   A coil having 500 windings is made of copper wire having a diameter $D = 0.142 \pm 0.0006$ mm and a total length $L = 94\,290 \pm 30$ mm. The specific resistance of copper is

$$\rho = 1.7 \times 10^{-5} \ \Omega \, \text{mm}$$

If A is the cross sectional area of the wire, then the total resistance R of the coil is

$$R = \rho \frac{L}{A} = \frac{4\rho L}{\pi D^2}$$

$$= \frac{4 \times 1.7 \times 10^{-5} \times 9.4290}{\pi (0.142)^2} = 101.2 \, \Omega$$

We need to know the standard deviation σ_{MR} of this total resistance. We have

$$\frac{\partial R}{\partial L} = \frac{4\rho}{\pi D^2} = 9.32 \times 10^{-3}$$

$$\frac{\partial R}{\partial D} = -\frac{8\rho L}{\pi D^3} = 1425.6$$

Thus

$$\sigma_{MR} = \sqrt{(9.32 \times 10^{-3} \times 30)^2 + (1425.6 \times 6 \times 10^{-4})^2} \approx 1\,\Omega$$

Hence the total resistance of the coil is $R = (101 \pm 1)\,\Omega$

This value would then be quoted by the manufacturer.

21.7 WEIGHTED AVERAGE

It frequently happens that a physical quantity can be determined by two different measuring methods. The results of the measurements usually deviate from each other. We then have two results whose true value lies with a probability of 0.68 in the confidence intervals

$$\bar{x}_1 \pm \sigma_{M_1} \quad \text{and} \quad \bar{x}_2 \pm \sigma_{M_2}$$

A better estimate for the 'true' value is obtained if we combine both measurements. We do not calculate the arithmetical mean value of the individual values but 'weigh' the measurements. A simple arithmetic average would not give the best value because it would mean that we were giving equal importance to both values.

The measurement with the smaller sampling error has a larger weight. The following expression is called the *weight*.

$$g_i = \frac{1}{\sigma_{M_i}^2}$$

The weighted average (weighted mean value) is given by

$$\bar{x} = \frac{g_1 \bar{x}_1 + g_2 \bar{x}_2}{g_1 + g_2}$$

It can be shown that in this expression the sampling error of x has a minimum. The expression can be generalised for more than two readings.

Weighted mean value:	$\bar{x} = \dfrac{g_1 \bar{x}_1 + g_2 \bar{x}_2}{g_1 + g_2}$
Weight:	$g_i = \dfrac{1}{\sigma_{M_i}^2}$

EXAMPLE The diameter of the wire in the example in section 21.2.3 is measured by a second method with the following results:

$$\bar{d}_1 = 0.1420 \pm 0.0006 \text{ mm}$$

$$\bar{d}_2 = 0.1410 \pm 0.001 \text{ mm}$$

The weighted mean value is obtained in the following way:

$$g_1 = \frac{1}{\sigma_1^2} = \frac{1}{(6 \times 10^{-4})^2} = 2.778 \times 10^6$$

$$g_2 = \frac{1}{\sigma_2^2} = \frac{1}{(10^{-3})^2} = 10^6$$

Hence
$$\bar{d} = \frac{2.778 \times 10^6 \times 0.142 + 10^6 \times 0.141}{3.778 \times 10^6} = 0.1417 \text{ mm}$$

21.8 CURVE FITTING: METHOD OF LEAST SQUARES, REGRESSION LINE

We have so far considered measurements which were related to one physical quantity. We will now consider experiments which investigate the relationships between two physical quantities, x and y, where the quantity y is measured and the quantity x is varied.

EXAMPLE   The graph of the cooling curve for a liquid has to be determined.

We measure the temperature at intervals of 30 seconds and plot a graph of temperature versus time. The experiment is repeated n times and we then calculate the mean value and the standard deviation at each time interval and plot these, as shown in figure 21.3.

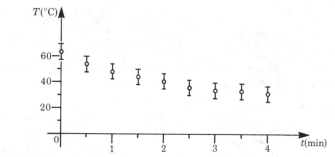

Figure 21.3

We now draw the best curve through the measured values, i.e. the curve which is as close as possible to the measured values. This method is called curve fitting.

To obtain the best fit we use the method of least squares. In this method the squares of the distances of the measured points from the curve are a minimum. Before we can begin, however, we must decide on the type of function that will best fit the experimental data. In general; this function will depend on the underlying theory. The types of functions most frequently used are straight lines, log functions, parabolas and exponential functions.

We propose to demonstrate the method using a straight line relationship. The task is to determine, from the given data, the coefficients a and b in the equation

$$y = ax + b$$

assuming it to be a first approximation to the experimental data. It is called a *regression line*.

Let $\qquad y = ax + b$

Then $\qquad a = \dfrac{\displaystyle\sum_{i=1}^{n} x_i y_i - n\bar{x}\bar{y}}{\displaystyle\sum_{i=1}^{n} x_i^2 - n\bar{x}^2}$ $\qquad$ and $\qquad b = \bar{y} - a\bar{x}$ $\qquad\qquad$ (21.8)

The straight line passes through the point $(\bar{x}, \bar{y})$ which is the 'centroid' of the test data.

PROOF

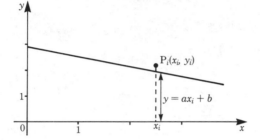

Figure 21.4

The vertical distance d_i from the test point $P_i(x_i y_i)$ to the straight line is

$$d_i = |y_i - (ax_i + b)| = |y_i - ax_i - b|$$

Summing up the square of the distances for all the points, $i = 1, 2, 3, \ldots, n$, gives

$$S = \sum d_i^2 = \sum (y_i - ax_i - b)^2$$

We now consider a and b as variable quantities and determine their values such that S will be a minimum.

For S to be a minimum we must set

$$\frac{\partial S}{\partial a} = \frac{\partial S}{\partial b} = 0$$

Hence $\qquad \dfrac{\partial S}{\partial a} = 0 = \sum 2(y_i - ax_i - b)(-x_i)$

$\qquad\qquad 0 = \sum (x_i y_i - ax_i^2 - bx_i) = \sum x_i y_i - a\sum x_i^2 - \sum bx_i$

$$\frac{\partial S}{\partial b} = 0 = \sum 2(y_i - ax_i - b)(-1)$$

$$0 = \sum (y_i - ax_i - b) = \sum y_i - a \sum x_i - bn$$

By definition of the arithmetic mean value: $\sum x_i = n\bar{x}$ and $\sum y_i = n\bar{y}$

We thus have the two following equations in the two unknowns a and b:

$$\sum x_i y_i - a \sum x_i^2 - bn\bar{x} = 0$$

$$n\bar{y} - an\bar{x} - bn = 0 \tag{21.9}$$

Substituting and solving gives

$$b = \bar{y} - a\bar{x}$$

$$a = \frac{\sum x_i y_i - n\bar{x}\bar{y}}{\sum x_i^2 - n\bar{x}^2} \tag{21.10}$$

Thus we have determined the coefficients a and b of the regression line.

The straight line $y = ax + b$ is the best fit for the experimental data points.

EXAMPLE   The values of the temperature during a cooling experiment of a liquid at intervals of 0.5 min are as follows:

Time, t (min)	Temperature, T (°C)
0	62
0.5	55
1	48
1.5	46
2	42
2.5	39
3	37
3.5	36
4	35

As a crude approximation, we propose to fit a straight line, i.e. $T = at + b$.

To obtain the values of a and b according to the method of least squares, we proceed in accordance with the following scheme.

(1) Tabulate the test data and calculate the necessary sums and products according to equation 21.10, as shown in the table.

t_i (min)	t_i^2 (min^2)	T_i (°C)	$t_i T_i$
0	0	62	0
0.5	0.25	55	27.5
1	1	48	48
1.5	2.25	46	69
2	4	42	84
2.5	6.25	39	97.5
3	9	37	111
3.5	12.25	36	126
4	16	35	140
$\sum$ 18.0	51.0	400	703

(2) Calculate the mean value $\bar{t}$ and $\bar{T}$:

$$\bar{t} = 2\,\text{min}, \qquad \bar{T} = 44.44°C$$

(3) We substitute the values in the equations for a and b. This gives

$$a = \frac{\sum t_i T_i - n\bar{t}\bar{T}}{\sum t_i^2 - n\bar{T}^2} = \frac{703 - 9 \times 2 \times 44.44}{51 - 9 \times 22}$$

$$= -6.47°C/\text{min}$$

and
$$b = \bar{T} - a\bar{t} = 44.44 + 6.47 \times 2$$
$$= 57.4°C$$

Hence the regression curve is

$$T = (-6.47t + 57.4)°C$$

It is shown in figure 21.5.

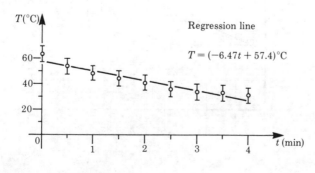

Figure 21.5

21.9 CORRELATION AND CORRELATION COEFFICIENT

Let us consider two sets of measurements of two variables x and y which are graphically represented in figure 21.6. The first impression is that the variables x and y are correlated. In fact, if we calculate the regression lines they both lead to the same line. But it is obvious that the first set, A, suggests a stronger dependence between the variables than the second set, B. In the first set, we might assume a clearcut functional dependence between x and y which is distorted by random errors of measurement.

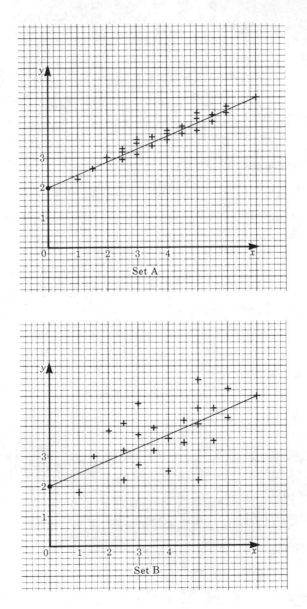

Figure 21.6

In the second set of measurements, we might say that there is a correlation between x and y, but there might be other variables too which influence y. Empirical data of this kind often occur in biology, psychology, meteorology, and economics, but they also occur in technology.

Our task is to find a measure which describes the magnitude of the correlation between two variables, while a variance is superimposed, the details of which we do not know.

This measure is given by the *correlation* r^2, or the *correlation coefficient r*.

They are defined by

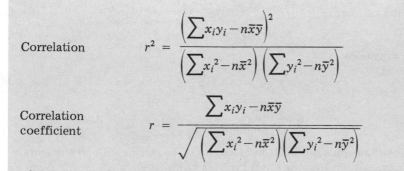

Correlation

$$r^2 = \frac{\left(\sum x_i y_i - n\bar{x}\bar{y}\right)^2}{\left(\sum x_i^2 - n\bar{x}^2\right)\left(\sum y_i^2 - n\bar{y}^2\right)}$$

Correlation coefficient

$$r = \frac{\sum x_i y_i - n\bar{x}\bar{y}}{\sqrt{\left(\sum x_i^2 - n\bar{x}^2\right)\left(\sum y_i^2 - n\bar{y}^2\right)}}$$

where x_i, y_i = measured values, and $\bar{y}, \bar{x}$ = mean values. All sums are formed from $i = 1$ to $i = n$. n = number of measurements.

Note that the correlation coefficient can take on values between -1 and $+1$. We get $r = 1$ if all points (x_i, y_i) lie on a straight line with a positive slope, i.e. the regression line.

We get $r = 0$ if x_i and y_i are statistically independent, e.g. if the dots lie in a circular-shaped cluster.

r is negative if the regression line has a negative slope, indicating that when x increases y decreases.

The definition of the correlation coefficient may seem quite arbitrary. It may be useful to know how we arrive at this concept.

We established the regression line by means of the condition that the sum of the differences between y_i-values and the adjunct points of the regression line $\hat{y}_i$ should be a minimum. There remains a variance $\frac{1}{n}\sum(y_i - \hat{y}_i)^2$ which is only then zero if all y_i-values coincide with the regression line. Consider the variance of the y_i-values with respect to the mean value $\bar{y}$. It is obviously greater.

A measure which describes the quality of the regression line as an approximation of the given values is the relative reduction of variance achieved by the regression line. The amount of the reduction of variance is called the *explained variance*. The fraction of variance thus explained relating to the regression line with the original variance relating to the mean value is called the *correlation* r^2. Its root is the *correlation coefficient r* (often called the *product moment correlation*).

The correlation can be obtained in the following way.

Let the variance with respect to the mean value be

$$S_M = \frac{\sum (y_i - \bar{y})^2}{n}$$

The remaining variance with respect to the regression line is

$$S_r = \frac{\sum (y_i - \hat{y}_i)^2}{n}$$

Thus the explained variance is

$$S_e = v_M - v_{rl}$$

The correlation is given by

$$r^2 = \frac{S_e}{S_M} = \frac{\sum (y_i - \bar{y})^2 - \sum (y_i - \hat{y}_i)^2}{\sum (y_i - \bar{y})^2}$$

If the formula for the regression line is inserted, it can be shown with an elementary but cumbersome calculation that the given formula holds true.

EXERCISES

1. In the following examples, state whether the error is systematic (constant) S or random R.

   (a1) A 100-metre race is held in a school during a sports day. The judges start their stopwatches when the sound of the starting pistol reaches them. What type of error arises in this case?

   (a2) The timing of the start and end of the 100-metre race is subject to individual fluctuations, e.g. reaction time.

   (b) The zero point of a voltmeter has been wrongly set. The measurements are therefore subject to an error. What kind of error is it?

   (c) The resistance of a copper coil is obtained by measuring the current flowing through when a voltage is applied to it. As the coil warms up the resistance increases. What kind of error will ensue?

2. (a) Nine different rock samples are taken from a crater on the moon whose densities are then determined with the following results:

   $$3.6, 3.3, 3.2, 3.0, 3.2, 3.1, 3.0, 3.1, 3.3 \, g/cm^3.$$

   Calculate the mean value and standard deviation of the parent population.

   (b) The velocity of a body travelling along a straight line is measured 10 times. The results are

   $$1.30, 1.27. 1.32, 1.25, 1.26, 1.29, 1.31, 1.23, 1.33, 1.24 \, m/s.$$

   Calculate the mean value and the standard deviation.

3. A continuous random variable has the following density function:

$$f(x) = \begin{cases} 1, & 0 \leqslant x \leqslant 1 \\ 0, & \text{otherwise} \end{cases}$$

Calculate the mean value and the variance.

4. Determine the confidence intervals $\bar{x} \pm \sigma_M$ and $\bar{x} \pm 2\sigma_M$ for exercises 2(a) and (b).

5. A measured variable is normally distributed with a mean value $\mu = 8$ and a standard deviation $\sigma = 1$. What percentage of all test data are smaller than 7?

6. (a) The sides of a rectangle are $x = 120 \pm 0.2$ cm and $y = 90 \pm 0.1$ cm. Calculate the area and the standard deviation.

    (b) Calculate the density and the standard deviation of a sphere of diameter 6.2 ± 0.1 mm and mass 1000 ± 0.1 g.

7. The sensitivity of a spring balance is to be determined. To achieve this we place different masses m on the balance and record the deflection S. The results are

mass (mg):	2000	3000	4000	5000	6000
deflection (mm)	16	27	32	35	40

If a straight line is to be fitted through these data points, i.e. $S = am + b$, calculate the values of a and b.

8. An angle has been measured several times with two theodolites and the following values were obtained: $73°2'7'' \pm 10'$ and $73°2'12'' \pm 20''$. Calculate the weighted average.

Answers

Chapter 3

1. Vectors are: acceleration, centripetal force, velocity, momentum, magnetic intensity.

2. The order in which the vectors are added is immaterial.

   (a)

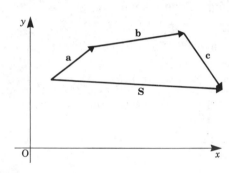

   Figure 3.36

   (b)

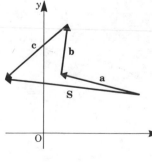

   Figure 3.37

3. (a)

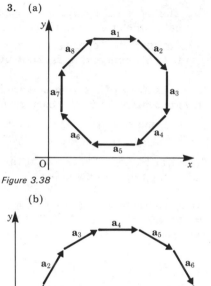

   Figure 3.38

   (b)

   Figure 3.39

4. (a)

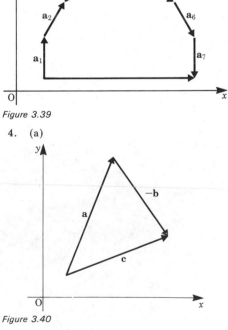

   Figure 3.40

(b)

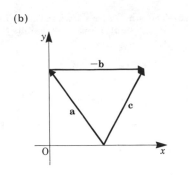

Figure 3.41

5. (a)

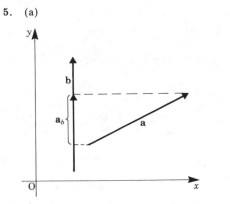

Figure 3.42

(b)

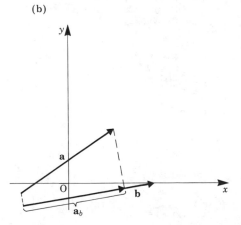

Figure 3.43

6. (a) $|a_b| = |a| \cos 60° = 5 \times \frac{1}{2} = 2.5$
   (b) $|a_b| = 0$
   (c) $|a_b| = 4$
   (d) $|a_b| = |\frac{3}{2} \cos(\pi - \frac{\pi}{3})| = |\frac{3}{2}(-\cos\frac{\pi}{3})| = \frac{3}{4}$

7. $P = (10, -2)$ Steps in the computation:
   $a = \overrightarrow{P_1P_2} = P_2 - P_1 = (5, 2)$
   $b = P_3 - P_1 = (3, -5)$
   $P_4 = P_1 + a + b = (10, -2)$

8. $a = \overrightarrow{P_1P_2} = (x_2 - x_1, \ y_2 - y_1)$,
   $b = (x_3 - x_2, \ y_3 - y_2)$;
   $c = (x_4 - x_3, \ y_4 - y_3)$,
   $d = (x_1 - x_4, \ y_1 - y_4)$,
   $S = (x_2 - x_1 + x_3 - x_2 + x_4 - x_3 + x_1 - x_4,$
   $\quad y_2 - y_1 + y_3 - y_2 + y_4 - y_3 + y_1 - y_4)$
   $= (0, 0) = 0$

9. $F = (90\,\text{N}, \ 10\,\text{N})$

10. (a) $(4, 3, 0)$ (b) $(5, 3, 7)$

11. (a) $a = (-14\frac{1}{2}, -1, 0)$
    (b) $a = (-7, -22, 12)$

12. (a) $e_a = \dfrac{3}{\sqrt{14}}, \dfrac{-1}{\sqrt{14}}, \dfrac{2}{\sqrt{14}}$
    (b) $e_a = \frac{2}{3}, -\frac{1}{3}, -\frac{2}{3}$

13. (a) $|a| = \sqrt{24} = 4.90$
    (b) $|a| = \sqrt{53} = 7.28$

14. (a) $V_1 + V_2 = (0, 250)\,\text{km/h}$
    (b) $V_1 + V_3 = (50, 300)\,\text{km/h}$
    (c) $V_1 + V_4 = (0, 350)\,\text{km/h}$
    (d) $|V_1 + V_2| = 250\,\text{km/h}$
    (e) $|V_1 + V_3| = 304\,\text{km/h}$
    (f) $|V_1 + V_4| = 350\,\text{km/h}$

Chapter 4

1. (a) $a \cdot b = ab \cos \alpha = 3 \times 2 \times \frac{1}{2} = 3$
   (b) $a \cdot b = 10$
   (c) $a \cdot b = 2 \times \sqrt{2} \approx 2.828$
   (d) $a \cdot b = -3.75$

2. (a) $\alpha = \dfrac{\pi}{2}$ Vectors are perpendicular
   (b) $\alpha = 0$ Vectors are parallel
   (c) $\alpha = \dfrac{\pi}{3}$
   (d) $\dfrac{\pi}{2} < \alpha < \pi$

3. (a) $a \cdot b = -3 - 2 + 20 = 15$
   (b) $a \cdot b = -1.25$
   (c) $a \cdot b = -\dfrac{11}{12}$
   (d) $a \cdot b = 4$

4. (a) $a \cdot b = 0$ Vectors are perpendicular to each other
   or one vector is zero

(b) $\mathbf{a} \cdot \mathbf{b} = -12$ a is not perpendicular to $\mathbf{b}$

(c) $\mathbf{a} \cdot \mathbf{b} = 0$ a is perpendicular to $\mathbf{b}$

(d) $\mathbf{a} \cdot \mathbf{b} = 0$ a is also perpendicular to $\mathbf{b}$

(e) $\mathbf{a} \cdot \mathbf{b} = -1$ a is not perpendicular to $\mathbf{b}$

(f) $\mathbf{a} \cdot \mathbf{b} = 0$ a is also perpendicular to $\mathbf{b}$

5. $\cos \alpha = \dfrac{\mathbf{a} \cdot \mathbf{b}}{ab}$

(a) $\cos \alpha = -1$ thus $\alpha = \pi$

(b) $\cos \alpha = \frac{2}{3}$ thus $\alpha \approx 48°$

6. $U = \mathbf{F} \cdot \mathbf{s}$

(a) $15\,\mathrm{N\,m}$ (b) $5\,\mathrm{N\,m}$ (c) 0

7.   (a) $\mathbf{c}$ is parallel to the z-axis
      (b) $\mathbf{c}$ is parallel to the x-axis

8. $|\mathbf{a} \times \mathbf{b}| = ab \sin \alpha$

(a) $2 \times 3 \times \dfrac{\sqrt{3}}{2} \approx 5.196$ (b) 0

(c) 6

9.   (a) $-\frac{8}{3}\mathbf{c}$ (b) $\frac{3}{2}\mathbf{b}$ (c) $-\frac{3}{2}\mathbf{b}$
      (d) $-6\mathbf{a}$ (e) 0 (f) $6\mathbf{a}$

10. $\mathbf{c} = (a_y b_z - a_z b_y,\ a_z b_x - a_x b_z,\ a_x b_y - a_y b_x)$

(a) $\mathbf{c} = (10, -9, 7)$
(b) $\mathbf{c} = (3, 6, -9)$

Chapter 5

1.   (a) 0 (b) $\frac{1}{2}$ (c) -1
      (d) 1 (e) $\frac{1}{2}$ (f) 2
      (g) 6

2.   (a) -1 (b) $\frac{1}{2}$ (c) 5
      (d) 0 (e) 1 (f) 0

3.   (a) At $x = 0$ the function is continuous but not differentiable. It is shown in figure 5.40.

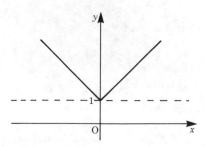

Figure 5.40

(b) The function is shown in figure 5.41. It is discontinuous at the points $x = 1, 2, 3, \ldots$

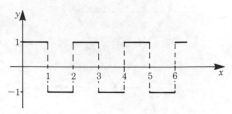

Figure 5.41

(c) The function is discontinuous at the points $x = 2, 4, 6, 8, \ldots$

4.   (a) $S_5 = 2 + \frac{3}{2} + \frac{4}{3} + \frac{5}{4} + \frac{6}{5} = 7\frac{17}{60} \approx 7.28$

(b) $S_{10} = 3 \times \dfrac{1 - (1/2)^8}{1/2} = 6 \times \frac{511}{512} \approx 5.988$

(c) $S = 3 \times \dfrac{1}{1/2} = 6$

5.   (a) End points of the secant: $P_1(1, -1)$, $P_2(\frac{3}{2}, \frac{3}{8})$

Slope of the secant: $m_s = \dfrac{\Delta y}{\Delta x} = 2.75$

Slope of the tangent: $m_T = y'(1) = 1$

(b) $v(t) = \dfrac{ds}{dt} = 6t - 8$, $v(3) = 10\,\mathrm{m/s}$

(c) (i) $dy = (2x + 7)\,dx$
    (ii) $dy = (5x^4 - 8x^3)\,dx$
    (iii) $dy = 4x\,dx$

6.   (a) $15x^4$ (b) 8
      (c) $\frac{7}{3}x^{4/3}$ (d) $21x^2 - 6\sqrt{x}$
      (e) $\dfrac{x^2 + 2}{5x^2}$

7.   (a) $6x^2$ (b) $\dfrac{1}{3\sqrt[3]{x^2}}$

(c) $-\dfrac{2}{x^3}$ (d) $\dfrac{8}{(4 + x)^2}$

(e) $6x(x^2 + 2)^2$ (f) $4x^3 - \dfrac{1}{x^2}$

(g) $\dfrac{x}{\sqrt{1 + x^2}}$ (h) $\dfrac{3b}{x^2}\left(a - \dfrac{b}{x}\right)^2$

8.   (a) $-18 \sin 6x$ (b) $8\pi \cos(2\pi x)$
      (c) $A e^{-x}\left[2\pi \cos(2\pi x) - \sin(2\pi x)\right]$

(d) $\dfrac{1}{x + 1}$ (e) $\cos^2 x - \sin^2 x$

(f) $2x \cos x^2$ (g) $12x(3x^2 + 2)$

(h) $ab \cos(bx + c)$ (i) $6x^2 e^{(2x^2 - 4)}$

9. (a) $y' = -\dfrac{c}{\sqrt{1-(cx)^2}}$

(b) $y' = \dfrac{A}{1+(x+2)^2}$

(c) $y' = \dfrac{2x}{\sqrt{1-x^4}}$

(d) $y' = \dfrac{1}{2\sqrt{x}\,(1+x)}$

10. (a) $y' = 0.1\,C\cosh(0.1x)$

(b) $u' = \pi[1-\tanh^2(v+1)]$

(c) $\eta' = \tanh\xi$

(d) $\dot{s} = \tanh t$

(e) $y' = 0$ (since $y = -1 = $ constant)

(f) $y' = 2\coth x\,(1-x\coth x)$

11. (a) $y' = \dfrac{10A}{\sqrt{1+100x^2}}$

(b) $u' = -\dfrac{C}{v^2+2v}$

(c) $\eta' = \dfrac{1}{\cos\xi}$

(d) $y' = \dfrac{1}{\sqrt{x^4+x^2(x-1)^2}}$

12. (a) $a\cos\phi + \dfrac{1}{\cos^2\phi}$

(b) $e^u(2+u)$

(c) $-\dfrac{1}{x^2}$

(d) $120x$

13. (a) $x_1 = 2$ $(\sqrt{2}, -8)$, minimum

$x_2 = -2$ $(0, 0)$, maximum

$x_3 = x_4 = 0$ $(\sqrt{2}, -8)$, minimum

(b) $x = k\pi$,

$x = \pm\dfrac{\pi}{2}, \pm\dfrac{5\pi}{2}, \pm\dfrac{9\pi}{2}, \dots$ maximum

$(k = 0, \pm1, \pm2, \dots)$,

$x = \pm\dfrac{3\pi}{2}, \pm\dfrac{7\pi}{2}, \pm\dfrac{11\pi}{2}, \dots$ minimum

(c) $x = 2\pi k$,

$x = \pm\pi, \pm5\pi, \pm9\pi, \dots$ maximum

$(k = 0, \pm1, \pm2, \dots)$,

$x = \pm3\pi, \pm7\pi, \pm11\pi, \dots$ minimum

(d) $x = \sqrt[3]{-4}$, none

(e) $x = (2k+1)\dfrac{\pi}{2} - 2$,

$x = 2k\pi - 2$, maximum

$(k = 0, \pm1, \pm2, \dots)$,

$x = (2k+1)\pi - 2$, minimum

(f) $x_1 = 4.85$, $(-1, 3\tfrac{1}{3})$, maximum

$x_2 = 1.85$, $(3, -18)$, minimum

$x_3 = 0$,

14. (a) $y = 11$ (b) $y = -8.89$

15.

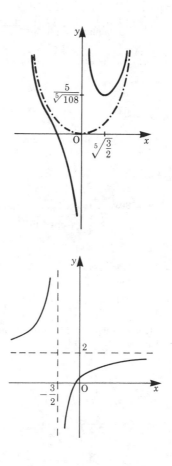

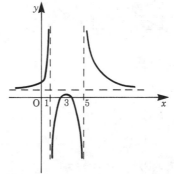

Figure 5.42

16. (a) 2.44%, 1.63%   (b) 12.95 m

17. (a) 5.27

(b) 14.14 in magnitude

18.   (a) 1   (b) 0   (c) 1

   (d) $-\frac{1}{2}$

   Hint: $\dfrac{1}{x}\left(\dfrac{1}{\sinh x}-\dfrac{1}{\tanh x}\right)=\dfrac{1-\cosh x}{x\sinh x}$

   (e) $e^{-2/\pi}$

   Hint: $x^{\tan(\pi/2)x}=e^{(\ln x)/\cot(\pi/2)x}$

   (f) 0   (g) $\dfrac{b^2-a^2}{2}$

19.   (a) $y'=\dfrac{-2x}{3y}$

   (b) $y'=-\dfrac{9x^2y^2+\cos y}{6x^3y-x\sin y}$

   (c) $y'=-\dfrac{2(x+y)+2}{2(x+y)+1}=-2$

20.   (a) $y'=30x-29$

   (b) $y'=x^{\sin x}\left(\dfrac{\sin x}{x}+\cos x\,\ln x\right)$

   (c) $y'=x^x(1+\ln x)$

21.   (a) $y'=(vu-gx)/u^2$

   (b) $y'=\tan t$

22. $x(t)=R\cos 3\pi t$
   $y(t)=R\sin 3\pi t$

23.   (a) It is a straight line in three-dimensional space (figure 5.43)

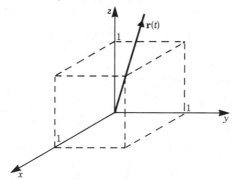

Figure 5.43

   (b) The curve is an ellipse with axes of lengths $2a$ and $2b$, respectively.
   $$b^2x^2+a^2y^2=a^2b^2$$
   $$\dfrac{x^2}{a^2}+\dfrac{y^2}{b^2}=1$$

24.   (a) $a_x(t)=-\omega v_0\cos\omega t$
   $a_y(t)=-\omega v_0\sin\omega t$ or
   $\mathbf{a}(t)=(-\omega v_0\cos\omega t,\ -\omega v_0\sin\omega t)$

   (b) $\mathbf{v}(t)=(-R\omega\sin\omega t,\ R\omega\cos\omega t,\ 1)$
   $\mathbf{v}\left(\dfrac{2\pi}{\omega}\right)=(0,\ R\omega,\ 1)$

   (c) $\mathbf{v}(t)=(v_0,\ 0,\ -gt)$

Chapter 6

1.   (a) $F(x)=\frac{3}{2}x^2+C;$ $C=\frac{1}{2},$
   $F(x)=\frac{3}{2}x^2+\frac{1}{2}$

   (b) $F(x)=x^2+3x+C;$ $C=-4,$
   $F(x)=x^2+3x-4$

2.   (a) 3   (b) 6   (c) 0

3.   (a) $\left[\dfrac{x^2}{2}-2x\right]_{-2}^{0}=-6,$ $F=|-6|=6$

   (b) $F=|-2|=2$

   (c) $F=\left[\dfrac{x^2}{2}-2x\right]_{0}^{2}+\left[\dfrac{x^2}{2}-2x\right]_{2}^{4}=4$

4.   (a) $\dfrac{d}{dx}\left(\dfrac{x-1}{x+1}\right)=\dfrac{2}{(x+1)^2}$

   (b) $\dfrac{d}{dx}\left\{x-\dfrac{1}{8}\sin(8x-2)\right\}$
   $=1-\cos\{2(4x-1)\}$
   $=1-\cos^2(4x-1)+\sin^2(4x-1)$
   $=\sin^2(4x-1)+\sin^2(4x-1)$
   $=2\sin^2(4x-1)$

   (c) $\dfrac{d}{dx}\left(\dfrac{x}{1+x^2}\right)=\dfrac{1-x^2}{(1+x^2)^2}$

5.   (a) $\ln(x-a)+C$

   (b) $\tan x+C$

   (c) $a\ln(x+\sqrt{x^2+a^2})+C$

   (d) $\frac{1}{2}(\alpha-\sin\alpha\cos\alpha)+C$

   (e) $\dfrac{a^t}{\ln a}+C$

   (f) $\dfrac{3}{10}x^{10/3}=\dfrac{3}{10}\sqrt[3]{x^{10}}$

   (g) $\dfrac{5}{3}x^3+\dfrac{5}{4}x^4+C$

   (h) $\dfrac{3}{8}t^4+2t^2$

6.   (a) $\dfrac{x^2}{4}(2\ln x-1)+C$

   (b) $x^2\sin x+2x\cos x-2\sin x+C$

   (c) $\dfrac{x^3}{9}(3\ln x-1)+C$

   (d) $a\sinh\dfrac{x}{a}(x^2+2a^2)-2ax\cosh\dfrac{x}{a}+C$

   (e) $\dfrac{1}{n}\sin x\cos^{n-1}x+\dfrac{n-1}{n}\int\cos^{n-2}x\,dx$

   (f) $\dfrac{x^{n+1}}{(n+1)}[(n+1)\ln x-1]+C$

7.   (a) $-\dfrac{1}{\pi}\cos(\pi x)+C$

   (b) $e^{3x-6}+C$

   (c) $\dfrac{1}{2}\ln|2x+a|+C$

   (d) $\dfrac{1}{6a}(ax+b)^6+C$

8. (a) $\ln\sqrt{|\sin 2x|} + C$

    (b) $\ln|a + x^2| + C$

    (c) $\dfrac{1}{40}\ln|x^{40} + 21| + C$

    (d) $-\dfrac{1}{\cosh u} + C$

9. (a) $\dfrac{1}{5}\sin^5 x + 2\sin^4 x + \dfrac{1}{2}\sin^2 x + C$

    (b) $\dfrac{2}{45}(3x^5 - 1)^{3/2} + C$

    (c) $\sqrt{a - x^2} + C$

    (d) $\dfrac{1}{2}\sin(x^2) + C$

10. (a) $\ln|e^x + 1| + C$

    (b) $\sin\left(x - \dfrac{\pi}{2}\right) + C = \cos x + C$

    (c) $\sin x - \dfrac{1}{3}\sin^3 x + C$

    (d) $\ln|\ln x| + C$

    (e) $\ln|x^3 - x| + C$

    (f) $\ln|\tan^{-1} x| + C$

11. (a) $\ln\left|\left(\dfrac{x + 2}{1 - x}\right)^{1/3}\right| + C$

    (b) $\ln\left|\dfrac{(x - 1)^{5/3}}{x^{3/2}(x + 2)^{1/6}}\right| + C$

    (c) $\ln\left|\dfrac{(x - 1)^{1/2}(x - 3)^{9/2}}{(x - 2)^4}\right| + C$

    (d) $\ln\left|\left(\dfrac{x^2 - 2}{x^2 + 1}\right)^{1/6}\right| + C$

    (e) $\dfrac{1}{3(x + 2)} + \ln\left|\left(\dfrac{x - 1}{x + 2}\right)^{1/9}\right|$

    (f) $\ln\left|\left(\dfrac{x^2 - x + 1}{x^2 + x + 1}\right)^{1/2}\right|$

    (g) $\ln\left|\dfrac{(x - 1)^2}{(x^2 + 2x + 5)^{1/2}}\right| - 2\tan^{-1}\left[\dfrac{1}{2}(x + 1)\right]$

    Note: $\dfrac{x^2 + 15}{(x - 1)(x^2 + 2x + 1)}$

    $= \dfrac{2}{x - 1} - \dfrac{x + 5}{x^2 + 2x + 5}$

12. (a) -32.5 (b) $\ln 2$ (c) 1.416
    (d) 75

13. (a) $\dfrac{255}{64}$ (b) 4.5 (c) $\dfrac{27}{16}$

14. (a) $\dfrac{1}{4}$ (b) ∞ (c) $\gamma\dfrac{1}{r_0}$

    (d) ∞ (e) $\dfrac{1}{2}$ (f) ∞

    (g) 1 (h) ∞

15. $U = \{2(x_0 + 2 - x_0) + 6(y_0 - y_0)$
    $+ 1(z_0 - z_0)\}\,\text{N m} = 4\,\text{N m}$

16. $U = \displaystyle\int_0^P \mathbf{F}\cdot d\mathbf{r} = \int_0^5 x\,dx\ \text{N m} = \dfrac{25}{2}\,\text{N m}$

17. Force and path element are perpendicular. Thus the scalar product $\mathbf{F}\cdot d\mathbf{r}$ vanishes. The line integral is zero.

18. The path element:
    $$d\mathbf{r}(t) = \left(-\sqrt{2}\sin t,\ -2\sin 2t,\ \dfrac{2}{\pi}\right)dt$$

    $$\mathbf{F}(t) = \left(0,\ -\dfrac{2t}{\pi},\ \cos 2t\right)$$

    $$U = \int_0^{\pi/2} \mathbf{F}(t)\,d\mathbf{r}(t)$$

    $$= \int_0^{\pi/2}\left(\dfrac{4t}{\pi}\sin 2t + \dfrac{2}{\pi}\cos 2t\right)dt$$

    Since $\displaystyle\int t\sin 2t\,dt = \dfrac{\sin 2t}{4} - \dfrac{t\cos 2t}{2} + C$

    $$U = \dfrac{4}{\pi}\left[\dfrac{\sin 2t}{4} - \dfrac{t\cos 2t}{2}\right]_0^{\pi/2} + \dfrac{1}{\pi}\left[\sin 2t\right]_0^{\pi/2}$$

    $$= \dfrac{4}{\pi}\,\dfrac{\pi}{4} = 1$$

Chapter 7

1. 720 square units

2. 9π square units

3. 2.14 square units

4. 0.693 square units

5. 100 square units

6. 2.97 square units

7. 19.24 square units

8. (a) 8 square units
    (b) 10.67 square units
    (c) 36 square units

9. (1) 98.12 units
    (2) 9.42 units
    (3) 4.064 units

10. 0.1109 square units

11. (a) 20.47 square units
    (b) 8.34 cubic units

12. (a) 67.02 square units
    (b) 49.35 cubic units

13. 314.16 cubic units

14. $\bar{x} = \dfrac{4a}{3\pi},\ \bar{y} = \dfrac{4b}{3\pi}$

15. 236.81 mm

16. 5/6 from the vertex

17. 46.875 mm

18. (a) MR^2

   (b) $M\left(\dfrac{R^2}{2} + \dfrac{L^2}{3}\right)$

   (c) $\dfrac{M}{2}\left(R^2 + \dfrac{L^2}{6}\right)$

19. (a) $32.98\ \text{kg/m}^2$

   (b) $131.9\ \text{kg/m}^2$

20. $I = 1.52\ \text{kg/m}^2$

21. $I = 76.04 \times 10^6\ \text{mm}^4$, $k = 74.36\ \text{mm}$

   Note that the centroid is 77.27 mm from the bottom.

22. 230 kN, 6.66 m below the surface.

23. $5.23 \times 10^5\ \text{N}$, 4 m

Chapter 8

1. (a) $f(x) = \sqrt{1-x}$

$$= 1 - \frac{1}{2}x - \frac{1}{2^2}\frac{x^2}{2!} - \frac{3}{2^3}\frac{x^3}{3!} - \cdots$$

   These are two ways of arriving at a solution:

   (i) with the help of equation 8.2;
   (ii) by using the binomial series $(n = \frac{1}{2}; a = 1)$.

   (b) $f(t) = \sin(\omega t + \pi) = -\omega t + \dfrac{\omega^3 t^3}{3!}$

$$-\frac{\omega^5 t^5}{5!} + \frac{\omega^7 t^7}{7!} - \cdots$$

   (c) $f(x) = \ln[(1+x)^5]$

$$= 5x - \frac{5}{2}x^2 + \frac{5}{3}x^3 - \frac{5}{4}x^4 + \cdots$$

Derivatives	Values
$f'(x) = \dfrac{\mathrm{d}}{\mathrm{d}x}[\ln(1+x)^5]$	$f'(0) = 5$
$\quad = \dfrac{5}{1+x}$	
$f''(x) = -5(1+x)^{-2}$	$f''(0) = -5$
$f'''(x) = 5 \times 2(1+x)^{-3}$	$f'''(0) = 5 \times 2$
$f^{(4)}(x) = -5 \times 2$ $\times 3(1+x)^{-4}$	$f^{(4)}(0) = -5 \times 3 \times 2$

   (d) $f(x) = \cos x = 1 - \dfrac{x^2}{2!} + \dfrac{x^4}{4!} - \dfrac{x^6}{6!} + - \cdots$

   (e) $f(x) = \tan x = x + \dfrac{1}{3}x^3 + \dfrac{2}{15}x^5$

$$+ \frac{17}{315}x^7 + \cdots$$

Derivatives	Values
$f(x) = \tan x$	$f(0) = 0$
$f'(x) = \dfrac{1}{\cos^2 x}$	$f'(0) = 1$
$f''(x) = \dfrac{2\sin x}{\cos x}\dfrac{1}{\cos^2 x} = 2ff'$	$f''(0) = 0$
$f'''(x) = 2(ff'' + f'^2)$	$f'''(0) = 2$
$f^{(4)}(x) = 2(ff''' + f'f'' + 2f'f'')$ $= 2(ff''' + 3f'f'')$	$f^{(4)}(0) = 0$
$f^{(5)}(x) = 2(ff^{(4)} + 4f'f''' + f''^2)$	$f^{(5)}(0) = 16$

   (f) $f(x) = \cosh x = 1 + \dfrac{x^2}{2!} + \dfrac{x^6}{6!} + \cdots$

2. (a) The formula 8.7a cannot be applied because every other coefficient vanishes ($a_n = 0$ if n is even). Instead we use the formula 8.7b:

$$R = \frac{1}{\lim\limits_{n \to \infty} \sqrt[n]{|a_n|}}.$$

   Since $a_n = \dfrac{1}{n!}$, if n is odd we find

$$R = \lim_{n \to \infty} \sqrt[n]{n!} = \infty$$

   (b) $\left|\dfrac{a_n}{a_{n+1}}\right| = \dfrac{3^n}{3^{n+1}} = \dfrac{1}{3}$

   Therefore $R = \lim\limits_{n \to \infty}\left|\dfrac{a_n}{a_{n+1}}\right| = \dfrac{1}{3}$

3. (a) $P_1(x) = x$

   $P_2(x) = 0$

   $P_3(x) = x + \dfrac{1}{3}x^3$

   For details of the solution see 1(e).

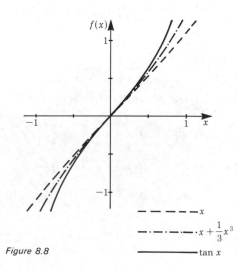

- - - - - - x

- $\cdot$ - $\cdot$ - $\cdot$ $\ x + \dfrac{1}{3}x^3$

——————— $\tan x$

Figure 8.8

(b) $y = \dfrac{x}{4-x} = \dfrac{4}{4-x} - 1 = \dfrac{1}{1-x/4} - 1$

(geometric series)

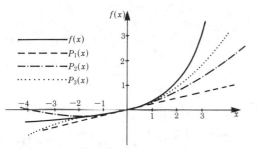

$P_1(x) = \dfrac{1}{4}x, \; P_2(x) = \dfrac{1}{4}x + \dfrac{1}{16}x^2,$

$P_3(x) = \dfrac{1}{4}x + \dfrac{1}{16}x^2 + \dfrac{1}{64}x^3$

P_3 is a parabola of the third degree with a point of inflection at $(-\tfrac{4}{3}, -\tfrac{7}{27})$.

Figure 8.9

4. (a) $y = \sin x$

$= -(x - \pi) + \dfrac{(x-\pi)^3}{3!} - \dfrac{(x-\pi)^5}{5!} + \dots$

Derivatives	Values
$y' = \cos x$	$y'(\pi) = -1$
$y'' = -\sin x$	$y''(\pi) = 0$
$y''' = -\cos x$	$y'''(\pi) = 1$
$y^{(4)} = \sin x$	$y^{(4)}(\pi) = 0$
$y^{(5)} = \cos x$	$y^{(5)}(\pi) = -1$

(b) $y = \cos x = -1 + \dfrac{1}{2!}(x-\pi)^2$

$\qquad - \dfrac{1}{4!}(x-\pi)^4 + \dots$

5. $f(x) = \ln x$

$= (x-1) - \dfrac{(x-1)^2}{2} + \dfrac{(x-1)^3}{3} - + \dots$

Derivatives	Values
$f'(x) = x^{-1}$	$f'(1) = 1$
$f''(x) = (-1)x^{-2}$	$f''(1) = -1$
$f'''(x) = 2x^{-3}$	$f'''(1) = 2$

6. $f(x) = \dfrac{4}{1-3x} = -\dfrac{4}{5} + \dfrac{12}{25}(x-2) - \dfrac{36}{125}$

$\qquad (x-2)^2 + \dfrac{108}{625}(x-2)^3 - + \dots$

Derivatives	Values
$f'(x) = -\dfrac{4 \times 3}{(1-3x)^2}$	$f'(2) = \dfrac{12}{25}$
$f''(x) = \dfrac{72}{(1-3x)^3}$	$f''(2) = -\dfrac{72}{125}$
$f'''(x) = \dfrac{648}{(1-3x)^4}$	$f'''(2) = \dfrac{648}{625}$

7. An intersection at $(1, \tfrac{5}{3})$

$f_1(x) = e^x - 1 \approx x + \dfrac{x^2}{2} + \dfrac{x^3}{6},$

$f_2(x) = 2 \sin x \approx 2x - \dfrac{x^3}{3}$

$x + \dfrac{x^2}{2} + \dfrac{x^3}{6} = 2x - \dfrac{x^3}{3}$

$x^3 + x^2 - 2x = 0$

$x_1 = 0, \qquad y_1 = 0$

$x_2 = 1, \qquad y_2 = \tfrac{5}{3}$

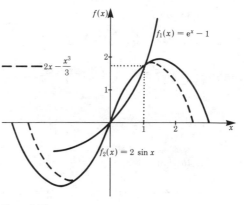

Figure 8.10

8. $\sqrt{42} \approx 6.4807$

Solution

$\sqrt{36 + 6} = \sqrt{36\left(1 + \dfrac{6}{36}\right)}$

$\qquad = 6 \times \sqrt{1 + \dfrac{1}{6}}$

We set $x = \tfrac{1}{6}$ and apply the expansion into a binomial series for

$\sqrt{1+x} \approx 1 + \dfrac{x}{2} - \dfrac{x^2}{8} + \dfrac{x^3}{16} - \dfrac{5x^4}{128} + - \dots$

(c.f. exercise 1(a))

Hence

$$6 \times \sqrt{1 + \frac{1}{6}} \approx 6\left(1 + \frac{1}{2} \times \frac{1}{6} - \frac{1}{8} \times \frac{1}{36}\right.$$
$$\left. + \frac{1}{16} \times \frac{1}{216} - \frac{5}{128} \times \frac{1}{1296} + - \ldots\right)$$

$$= 6(1 + 0.083\,33 - 0.003\,47$$
$$+ 0.000\,29 - 0.000\,03)$$
$$\approx 6.4807$$

Note that an alternative approach is to use $\sqrt{7^2 - 7} = 7\sqrt{1 - \frac{1}{7}}$.

9. (a) $\ln(1 + x) \approx x - \dfrac{x^2}{2}$

(b) $\dfrac{1}{\sqrt{1 + x}} \approx 1 - \dfrac{x}{2} + \dfrac{3}{8}x^2$

10. (a) $e^{0.25} \approx 1 + 0.25 = 1.25$

(b) $\ln 1.25 \approx \dfrac{1}{4} - \dfrac{1}{2}\left(\dfrac{1}{4}\right)^2 = \dfrac{7}{32} \approx 0.219$

(c) $\sqrt{1.25} \approx 1 + \dfrac{1}{2} \times \dfrac{1}{4} = \dfrac{9}{8} = 1.125$

11. (a) $\displaystyle\int \dfrac{dx}{1 + x} = x - \dfrac{x^2}{2} + \dfrac{x^3}{3} - \dfrac{x^4}{4} + \ldots$

$$= \ln(1 + x) + C$$

Solution:

$$\int \frac{dx}{1 + x} = \int (1 - x + x^2 - x^3 + x^4 \ldots)\,dx$$

$$= x - \frac{x^2}{2} + \frac{x^3}{3} - \frac{x^4}{4} + \frac{x^5}{5} \ldots + C$$

$$= \ln(1 + x) + C$$

(b) $\displaystyle\int \cos x\, dx = x - \dfrac{x^3}{3!} + \dfrac{x^5}{5!} - \dfrac{x^7}{7!} \cdots$

$$= \sin x + C$$

Solution:

$$\int \cos x\, dx = \int \left(1 - \frac{x^2}{2!} + \frac{x^4}{4!} - \frac{x^6}{6!} + \ldots\right) dx$$

$$= x - \frac{x^3}{3!} + \frac{x^5}{5!} - \frac{x^7}{7!} + \ldots + C$$

$$= \sin x + C$$

12. (a) $\displaystyle\int_0^{0.58} \sqrt{1 + x^2}\, dx \approx 0.6111$

Solution

Expand the integrand by means of the binomial series.

Put $\sqrt{1 + x^2} = (1 + x^2)^{1/2}$ and obtain

$$\sqrt{1 + x^2}\, dx = \int \left(1 + \frac{1}{2}x^2 - \frac{1}{8}x^4 + \frac{1}{16}x^6 - \frac{5}{128}x^8\right.$$
$$\left. - \frac{5}{128}x^8 + \ldots\right) dx$$

$$\int_0^x \sqrt{1 + t^2}\, dt = x + \frac{x^3}{6} - \frac{x^5}{40} + \frac{x^7}{112} - \frac{5x^9}{1152} + \ldots$$

$$(|x| < 1)$$

Substituting the limits $x = 0.58$ and $x = 0$, we find the value 0.6111.

(b) $\displaystyle\int_0^x \dfrac{\sin t}{t}\, dt = x - \dfrac{x^3}{3 \times 3!} + \dfrac{x^5}{5 \times 5!}$

$$- \frac{x^7}{7 \times 7!} + \ldots$$

Solution:

The integrand can be represented by the series

$$\frac{\sin t}{t} = \frac{1}{t} \sin t = \frac{1}{t}\left(t - \frac{t^3}{3!} + \frac{t^5}{5!} - \frac{t^7}{7!} + \ldots\right)$$

$$= 1 - \frac{t^2}{3!} + \frac{t^4}{5!} - \frac{t^6}{7!} + \ldots$$

$$\int_0^x \frac{\sin t}{t}\, dt = x - \frac{x^3}{3 \times 3!} + \frac{x^5}{5 \times 5!} - \frac{x^7}{7 \times 7!} + \ldots$$

13. (a) $\dfrac{dx}{\sqrt{1 - x^2}} = \sin^{-1} x = x + \dfrac{1 \times x^3}{2 \times 3}$

$$+ \frac{(1 \times 3 \times 5)x^7}{2 \times 4 \times 6 \times 7} + \frac{(1 \times 3 \times 5 \times 7)x^9}{2 \times 4 \times 6 \times 8 \times 9} + \ldots$$

$$(|x| \leqslant 1)$$

(b) $\dfrac{\pi}{2} = 1 + \dfrac{1}{6} + \dfrac{3}{40} + \dfrac{5}{112} + \dfrac{35}{1152}$

$$+ \ldots \approx 1.3167 + \ldots$$

Comparing the approximation by the first five terms with the numerical value 1.5707 ..., we find that the approximation is wrong by more than 16%! Although the sequence converges, it does so very slowly indeed. This series has no significance for numerical purposes.

Chapter 9

1. (a) $j\sqrt{3}$ (b) $12j$ (c) $\dfrac{\sqrt{5}}{2j}$
    (d) $10j$

2. (a) 1 (b) $-j$ (c) j
    (d) j

3. (a) $6j\sqrt{3}$
    (b) $(2\sqrt{3} - 2\sqrt{2} + \sqrt{0.6})j$
    (c) -3 (d) $j\sqrt{ab}$ (e) $10j$
    (f) $-j$ (g) 4 (h) j
    (i) $-2\sqrt{3}$ (j) 0
    (k) $j(a - b)$ for $a > b$; $j(b - a)$ for $b > a$
    (l) $\pm\dfrac{6j}{a}$

4. (a) 7 (b) 15

5. (a) $z* = 5 - 2j$ (b) $z* = \frac{1}{2} + j\sqrt{3}$

6. (a) $z_1 = -2 + 3j$ (b) $z_1 = -\frac{3}{4} + j$

   $z_2 = -2 - 3j$ $z_2 = -\frac{3}{4} - j$

7. (a) $10 + 3j$ (b) $\frac{3}{2}$

8. (a) $w = -1 - 5j$ (b) $w = 1 - 13j$

9. (a) $w = 2$ (b) $w = 23 + 2j$

10. (a) $4\sqrt{2} + \frac{1}{2}j$ (b) $-\frac{1}{2}\sqrt{3} - 2j$

   (c) $-0.4 + 0.7j$

   (d) $\frac{1}{2} - \frac{1}{2}j$ (e) $2j$

   (f) $8 + 4j\sqrt{3}$

11. (a) $(2x + 3jy)(2x - 3jy)$

   (b) $(\sqrt{a} + j\sqrt{b})(\sqrt{a} - j\sqrt{b})$

12.

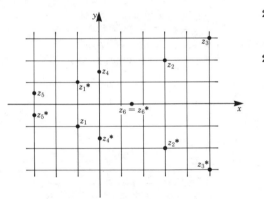

Figure 9.21

13. $z_1 = 2 + j$ $z_2 = j - 1$ $z_3 = -3$

   $z_4 = -2 - j$ $z_5 = -2j$ $z_6 = 2 - j$

14. (a) $z = \sqrt{2}\left(\cos\frac{3\pi}{4} + j\sin\frac{3\pi}{4}\right)$

   (b) $z = \sqrt{2}\left(\cos\frac{5\pi}{4} + j\sin\frac{5\pi}{4}\right)$

15. (a) $z = \frac{5}{2} - \frac{5}{2}j\sqrt{3}$

   (b) $z = -2\sqrt{2} - 2j\sqrt{2}$

16. (a) $z = 6(\cos 60° + j\sin 60°) = 3 + 3j\sqrt{3}$

   (b) $z = 5(\cos 120° + j\sin 120°) = -\frac{5}{2}$

     $+ \frac{5}{2}j\sqrt{3}$

17. (a) $z = \cos 45° + j\sin 45° = \frac{1}{2}\sqrt{2} + \frac{1}{2}j\sqrt{2}$

   (b) $z = \cos 330° + j\sin 330° = \frac{1}{2}\sqrt{3} - \frac{1}{2}j$

18. Multiplication by $-j = j^3$ means an anticlockwise rotation of 270°.

   Division by $-j = j^3$ means a clockwise rotation of 270°.

19. (a) $\sqrt{2}(-4 + 4j)$ (b) -1

20. (a) $(\cos 50° - j\sin 50°)^4$

   $= [\cos(-50°) + j\sin(-50°)]^4$

   $= \cos(-200°) + j\sin(-200°)$

   $= \cos 200° - j\sin 200°$

   (b) $(\cos\alpha + j\sin\alpha)^n = \cos(n\alpha) + j\sin(n\alpha)$

21. (a) $z_1 = 2 + 3j$

   $z_2 = -2 - 3j$

   (b) $z_1 = 0.966 + 0.259j$

   $z_2 = -0.259 + 0.966j$

   $z_3 = -0.966 - 0.259j$

   $z_4 = 0.259 - 0.966j$

22. (a) $e^{j\pi/2} = \cos\frac{\pi}{2} + j\sin\frac{\pi}{2} = j$

   (b) $\frac{1}{2} + \frac{1}{2}j\sqrt{3}$

23. (a) $\cos\alpha = 1, \quad \alpha = 0$

   $\sin\alpha = 0$

   (b) $\cos\alpha = -1, \quad \alpha = \pi$

   $\sin\alpha = 0$

   (c) $\cos\alpha = 0, \quad \alpha = -\frac{\pi}{2}$

   $\sin\alpha = -1$

   (d) $\cos\alpha = \frac{1}{2}\sqrt{3}, \quad \alpha = \frac{\pi}{6}$

   $\sin\alpha = \frac{1}{2}$

24. (a) $r = e^3$ (b) $r = e^2$

   $\alpha = 2$ $\alpha = -\frac{1}{2}$

25. (a) $w = e^z = -\sqrt{e}$

   $(r = \sqrt{e}, \alpha = \pi)$

   (b) $w = -\sqrt{e^3}$

   $(r = \sqrt{e^3}, \alpha = -\pi)$

   (c) $w = \frac{1}{e}j$

   $\left(r = \frac{1}{e}, \alpha = \frac{\pi}{2}\right)$

   (d) $w = e^3(\cos 1 - j\sin 1)$

   $\approx e^3 \times 0.54 - e^3 \times 0.841j$

26. (a) (i) $\mathrm{Re}[w(t)] = e^{-t}\cos 2\pi t$

   (ii) Period $= 1$

   (iii) Amplitude $= e^{-2} \times 1 = \frac{1}{e^2} \approx 0.135$

   (b) (i) $\mathrm{Re}[w(t)] = e^{2t}\cos\left(-\frac{3}{2}t\right)$

   $= e^{2t}\cos\left(\frac{3}{2}t\right)$

(ii) Period $= \frac{4}{3}\pi$

(iii) Amplitude $= e^4 \cos 3$

$\approx e^4(-0.99)$

≈ -54.0

27. (a) $e^{j\pi} = -1$ (b) $\frac{3}{4}e^{-j\pi/2} = -\frac{3}{4}j$

28. (a) $z_1{}^*/z_2 = 4\,e^{-j\pi} = -4$
    (b) $\frac{1}{3}e^{j\pi/2} = \frac{j}{3}$

29. (a) $z^5 = 32\,e^{j\pi} = -32$
    (b) $z^3 = \frac{1}{8}e^{j3\pi/4}$

30. (a) $z^{1/5} = 2\,e^{j2\pi} = 2$
    (b) $z^{1/4} = \frac{1}{2}\,e^{j3\pi/2} = -\frac{j}{2}$

31. (a) $z = 3e^{j\pi}$ (b) $z = \frac{1}{2}e^{j2\pi/3}$

32. (a) $z = 7.0711\,e^{-0.7854j}$
    or $z = 7.0711\,e^{-j45°}$
    (b) $z = 19.8494\,e^{-0.71413j}$
    or $z = 19.8494\,e^{-j40°55'}$

33. (a) $z = 1.8134 + 1.7209j$
    (b) $z = 0.4384 + 0.8988j$

34. (a) $x = \dfrac{-1}{\sqrt{2}},\ y = \dfrac{1}{\sqrt{2}}$
    (b) $z = (\cos 135° + j\sin 135°)$
    $(= e^{j3\pi/4})$

35. (a) $-\sqrt{3}+j$ (b) $2\,e^{j5/6\pi}$

Chapter 10

1. The linear first- and second-order DEs
   with constant coefficients are (b), (e) and
   (f).

2. (a) non-homogeneous, second order
   (b) homogeneous, second order
   (c) homogeneous, first order
   (d) non-homogeneous, second order
   (e) homogeneous, second order

3. (a) $y = C_1 e^{5x} + C_2 e^x$
   Auxiliary equation: $2r^2 - 12r + 10 = 0$
   Roots: $r_1 = 5,\ \ r_2 = 1$
   (b) $y = e^{1.5x}(C_1 + C_2 x)$
   Auxiliary equation: $4y^2 - 12r + 9 = 0$
   Roots: $r_1 = r_2 = 1.5$
   (c) Complex solution:
   $y = e^{-x}[(C_1 + C_2)\cos 2x$
   $+ j(C_1 - C_2)\sin 2x]$
   Auxiliary equation: $r^2 + 2r + 5 = 0$
   Roots: $r_1 = -1 + 2j,\ \ r_2 = -1 - 2j$
   Real solution:
   $y = e^{-x}(A\cos 2x + B\sin 2x)$

(d) Complex solution:
$y = e^{0.25x}[(C_1 + C_2)\cos 0.75x$
$+ j(C_1 - C_2)\sin 0.75x]$
Real solution:
$y = e^{0.25x}(A\cos 0.75x + B\sin 0.75x)$
(e) $y = C_1 e^{2x} + C_2 e^{-4x}$
(f) Complex solution:
$y = e^{0.2x}(C'\cos 0.4x + jC''\sin 0.4x)$
Real solution:
$y = e^{0.2x}(A\cos 0.4x + B\sin 0.4x)$

4. (a) $y(x) = C\,e^{-4x}$ (b) $y(x) = C\,e^{30x}$
   (c) $y(x) = C\,e^{2x}$

5. (a) $S(t) = \dfrac{t^3}{3} + C_1 t + C_2$
   (b) $x(t) = \cos\omega t + C_1 t + C_2$

6. (a) $y_p = 2x + 1$ (b) $y_p = x - \frac{1}{2}$

7. (a) $y = C_1 e^x + C_2 e^{-3x/7} - 2$
   (b) $y = C_1 e^{9x} + C_2 e^x + x + \frac{10}{9}$
   (c) $y = C_1 e^{4x/3} + C_2 e^{-x} - \frac{1}{4}x^2 + \frac{1}{8}x - \frac{13}{32}$
   (d) $y = e^{-x}(A\cos 2x + B\sin 2x)$
   $+ \frac{1}{17}(\cos 2x + 4\sin 2x)$

8. $y(x) = C_1 e^{5x} + C_2 e^x + \frac{3}{10}x^2 + \frac{18}{25}x + \frac{43}{125}$

9. (a) $y = 3\,e^{-4x}$ (b) $y = e^{-21}\,e^{2.1x}$

10. (a) $C = 0,$ $y(x) = 0$
    (b) $C = -2,$ $y(x) = -2\,e^{2x}$
    (c) $C = e^2,$ $y(x) = e^2\,e^{2x}$
    (d) $C = 1,$ $y(x) = e^{2x}$

11. $y(x) = C_1 \cos 2x + C_2 \sin 2x$
    (a) $y(x) = +\sin 2x$
    (b) $y(x) = \cos 2x - \frac{1}{2}\sin 2x$
    (c) $y(x) = \frac{1}{2}\sin 2x$
    (d) $y(x) = \dfrac{-b}{4}\cos 2x + a\sin 2x$

12. $y = e^x - x\,e^x = e^x(1 - x)$

13. (a) $y = x + Cx^2$
    (b) $y = x^2 + Cx$
    (c) $y = -2\cos^2 x + C\cos x$
    (d) $y = e^{-x}\left(\dfrac{x}{2} + \dfrac{C}{x}\right)$

14. (a) $u = y^{-2},\ \ u' - 2xu = -2x,$
    $y = \dfrac{1}{\sqrt{1 + C\,e^{x^2}}}$
    (b) $u = y^{-1},\ \ u' + \dfrac{2}{x^2 - 1}u = 1,$
    $y = \dfrac{x-1}{x+1}\dfrac{1}{x - \ln(x+1)^2 + C}$
    (c) $u = y^3,\ \ x^2 u' + 3xu = 3,$
    $y = \sqrt[3]{\dfrac{3}{2x} + \dfrac{C}{x^3}}$

(d) $u = y^2$, $u' + \dfrac{2}{x}u = -2(x+1)$,

$$y = \frac{1}{x}\sqrt{C - \frac{x^4}{2} - \frac{2x^3}{3}}$$

15.   (a) $y = \frac{1}{2}\ln|2\,e^x + C|$

(b) $y^2 = \ln|C(x+1)^2| - 2x$

(c) $y = \dfrac{2}{(\ln|x|)^2 + C}$

(d) $y_1 = -x + C$
$y_2 = x + e^x(1-x) + C$

16.   (a) $\dfrac{\partial}{\partial x}\left(\dfrac{2y}{x}\right) = -\dfrac{2y}{x^2} = \dfrac{\partial}{\partial y}\left(4 - \dfrac{y^2}{x^2}\right)$,

$F = \dfrac{y^2}{x} + 4x$, $y^2 = Cx - 4x^2$

(b) $\dfrac{\partial}{\partial x}(1 - x\,e^{-y}) = -e^{-y} = \dfrac{\partial}{\partial y}\,e^{-y}$,

$F = y + x\,e^{-y}$, $y + x\,e^{-y} = C$

(c) $\dfrac{\partial}{\partial x}(2y - x^2\sin 2y) = -2x\sin 2y$

$= \dfrac{\partial}{\partial y}(2x\cos^2 y)$, $F = y^2 + x^2\cos^2 y$,

$y^2 + x^2\cos^2 y = C$

(d) $\dfrac{\partial}{\partial x}(2x - 3) = 2 = \dfrac{\partial}{\partial y}(3x^2 + 2y)$,

$F = x^3 + 2xy - 3y$, $y = \dfrac{x^3 + C}{3 - 2x}$

17.   (a) Special case 1: $\mu = e^x$,
$e^{3x} - 3e^x\cos y = C$

(b) Special case 2: $\mu = e^{-y}$,
$y + x\,e^{-y} = C$

18.   (a) $x = e^{6t}(A\cos t + B\sin t)$
$y = e^{6t}[(A-B)\cos t + (A+B)\sin t]$

(b) $x = A\cos t + B\sin t$
$y = \frac{1}{2}(B - 3A)\cos t - \frac{1}{2}(A + 3B)\sin t$

(c) $x = (A + Bt)e^t + (E + Ft)e^{-t}$
$y = \frac{1}{2}(B - A - Bt)e^t - \frac{1}{2}(E + F + Ft)e^{-t}$

Chapter 11

1.   (a) $\dfrac{3}{2s^4}$ (b) $\dfrac{5}{s+2}$

(c) $\dfrac{4s}{s^2+9}$ (d) $\dfrac{2}{s(s^2+4)}$

2.   (a) $\frac{1}{2}\sin\frac{1}{2}t$ (b) $\frac{1}{4}(1 - e^{-4t})$

(c) $\frac{2}{9}(1 - \cos 3t)$ (d) $-6\sinh t$

(e) $t - \sin t$ (f) $\frac{1}{2}e^{4t} - e^{2t} + \frac{1}{2}$

3.   (a) $y = \frac{2}{3}(e^{-t} - e^{-4t})$

(b) $y = \frac{1}{5}\sin 2t - \frac{7}{15}\sin 3t + \cos 3t$

(c) $y = \frac{1}{5}\sin t + \frac{2}{5}\cos t + \frac{3}{5}e^{-2t}$

4. $y = 2 - 3\,e^x + 3\,e^{2x}$

5. $y = \frac{1}{2}e^t - \frac{1}{2}e^{-t} + \frac{1}{6}t^3 - t$

6. $y = 25 - 9\,e^t + 5t\,e^t - 16\,e^{-t/4}$

7. $y = e^{-t/6} - e^{-t/2}$, $x = 1 - \frac{1}{2}(e^{-t/6} + e^{-t/2})$

8. $Q = 4 \times 10^{-4}(1.12\sin 447t - \sin 500t)$

Chapter 12

1.

x \ y	-2	-1	0	1
-2	-2	4	6	4
-1	2	5	6	5
0	6	6	6	6
1	10	7	6	7
2	14	8	6	8

2.   (a) The function $z = -x - 2y + 2$
represents a plane. The intersecting
curves of the surface are

(1) with the x-y plane: $y = -\dfrac{x}{2} + 1$

(2) with the x-z plane: $z = -x + 2$

(3) with the y-z plane: $z = -2y + 2$

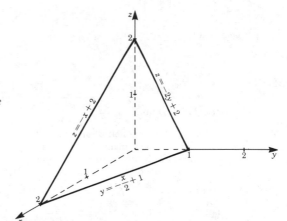

Figure 12.31

(b) The function $z = x^2 + y^2$ represents
a hyperboloid of revolution about the
z-axis. Intersecting curves with planes
parallel to the z-axis are parabolas.
Intersecting curves with planes
parallel to the x-y plane are circles.

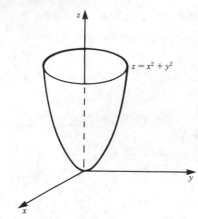

$z = x^2 + y^2$

Figure 12.32

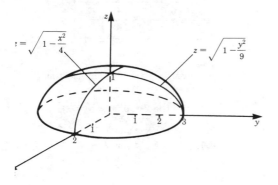

$z = \sqrt{1 - \dfrac{x^2}{4}}$

$z = \sqrt{1 - \dfrac{y^2}{9}}$

Figure 12.33

(c) The function $z = \sqrt{1 - \dfrac{x^2}{4} - \dfrac{y^2}{9}}$

represents one half of an ellipsoid above the x–y plane. The intersecting curves with the x–z plane and the y–z plane are semi-ellipses.

3. (a) $f_x = \cos x,$ $f_y = -\sin y$

(b) $f_x = 2x\sqrt{1 - y^2},$ $f_y = \dfrac{-x^2 y}{\sqrt{1 - y^2}}$

(c) $f_x = -2x\, e^{-(x^2 + y^2)},$
$f_y = -2y\, e^{-(x^2 + y^2)}$

(d) $f_x = yz + y,$ $f_y = xz + x,$
$f_z = xy + 1$

(e) $f_x = e^x \ln y,$ $f_y = \dfrac{e^x}{y},$ $f_z = 4z^3$

(f) $f_x = e^{\sin x} \cos x - e^{\cos(x+y)} \sin(x + y),$
$f_y = -e^{\cos(x+y)} \sin(x + y)$

4. Tangent in x-direction: $2x$
Slope in x-direction at point P: 0
Tangent in y-direction $2y$
Slope in y-direction at point P: 2

5. $f_{xx} = -2$ $f_{yx} = 0$
$f_{xy} = 0$ $f_{yy} = -2$

6. With $f_x = \dfrac{2x}{y^2}\, e^{(x/y)^2}$ and $f_y = -\dfrac{2x^2}{y^3}\, e^{(x/y)^2}$

the statement follows.

7. (a) $dz = \dfrac{-x\, dx}{\sqrt{1 - x^2 - y^2}} - \dfrac{y\, dx}{\sqrt{1 - x^2 - y^2}}$

(b) $dz = 2x\, dx + 2y\, dy$

(c) $dz = \dfrac{-2}{(x^2 + y^2 + z^2)^2}\,(x\, dx + y\, dy + z\, dz)$

8. (a) 7.6%   (b) 5.13%

9. (a) The contour lines are straight lines.
$y = -\dfrac{x}{2} + 1 + C$

grad $f = (-1, -2)$

(b) The contour lines are ellipses.
$\dfrac{x^2}{4} + \dfrac{y^2}{9} = 1 + C^2$

$\mathrm{grad}\, f = \dfrac{-1}{\sqrt{1 - x^2/4 - y^2/9}}\left(\dfrac{x}{4}, \dfrac{y}{9}\right)$

(c) The contour lines are circles centered at the origin.

$\mathrm{grad}\, f = \dfrac{-10}{\sqrt{(x^2 + y^2)^3}}\,(x, y)$

10. (a) The surfaces of constant function value are planes.

$z = \dfrac{x}{3} + \dfrac{y}{3} + C$

grad $f = (1, 1, -3)$

(b) The surfaces of constant function value are cylinders centered at the origin.
$C = x^2 + y^2$
grad $f = (2x, 2y, 0)$

(c) The surfaces of constant function value are spheres centered at the origin.
$C = x^2 + y^2 + z^2$
$\mathrm{grad}\, f = 3(x^2 + y^2 + z^2)^{1/2}(x, y, z)$

11. (a) $\sin 2t - 3 \cos 2t$ (b) $3t^2 + 4 + \dfrac{3}{t^2}$

12. (a) $2x + 2ay$ (b) 1

(c) $-\dfrac{\pi}{8}$ (d) 0

13. (a) Minimum at $x = 1$, $y = 1$

(b) Maximum at $x = 10$, $y = 8$

(c) $\theta = 30°$, $h = 0.32\,\text{m}$, $l = 0.20\,\text{m}$

14. Cable (a)

$$f(x, t) = 0.5\cos\left(2\pi \times 5 \times t - \frac{2\pi x}{1.2} + \phi\right)\text{m}$$

or $f(x, t) = 0.5\cos 2\pi\left(5t - \frac{x}{1.2} + \phi_1\right)\text{m}$

or $f(x, t) = 0.5\cos\frac{2\pi}{1.2}(12.5t - x + \phi_2)\,\text{m}$

Cable (b)

$$f(x, t) = 0.2\cos 2\pi\left(0.8t - \frac{x}{4.0} + \phi\right)$$

The wave velocities are unequal:

$c_a = 6\,\text{m/s}$ $c_b = 3.2\,\text{m/s}$

15. $\dfrac{\partial t}{\partial t} = -2v(vt - x)e^{-(vt-x)^2}$

$\dfrac{\partial^2 f}{\partial t^2} = -2v^2 e^{-(vt-x)^2}$
$\qquad + (2v)^2(vt - x)^2 e^{-(vt-x)^2}$

$\dfrac{\partial^2 f}{\partial x^2} = -2e^{-(vt-x)^2} + 4(vt - x)^2 e^{-(vt-x)^2}$

Consequently

$$\frac{\partial^2 f}{\partial t^2} = v^2 \frac{\partial^2 f}{\partial x^2}$$

Chapter 13

1. (a) ab (b) $\frac{2}{3}$

(c) 4   (d) 12

(e) 4   (f) $\dfrac{1}{a}(e^{az_1} - e^{az_0})(y_1 - y_0)$

2. (a) $10\frac{2}{3}$ (b) $\frac{4}{3}$

(c) $ab\pi$; $\left(\dfrac{4a}{3\pi}, 0\right)$

3. (a) $r = 3\sqrt{2}$, $\phi = \dfrac{\pi}{4}$

(b) $R^2 = x^2 + y^2$, $r = R$, in polar coordinates

(c) $r = \dfrac{\phi}{2\pi}$ (d) $\dfrac{a^3}{3\sqrt{2}}$

4. (a) $V = \pi h(R_2^2 - R_1^2)$

(b) $V = \frac{1}{3}\pi R^2 h$
$\quad I = \frac{3}{10}MR^2$ (M = total mass = ρV)

5. $I = \frac{2}{5}MR^2$, $M = \frac{4}{3}\pi R^3 \rho$

Chapter 14

1. $x' = x$, $y' = y$, $z' = z - 2$

2. The transformations are

$x = x' - 2$, $y = y' + 3$

Substitution in the equation gives

$y' = -3x' + 8$

3. $\mathbf{r}' = \begin{pmatrix} 1 + 2\sqrt{3} \\ -\sqrt{3} + 2 \end{pmatrix}$

4. The transformation equations are

$$x = x'\frac{\sqrt{3}}{2} - y'\frac{1}{2}, \quad y = x'\frac{1}{2} + y'\frac{\sqrt{3}}{2}$$

Substitution in the equation gives

$$y' = \frac{6}{\sqrt{3}} - \frac{3}{\sqrt{3}}x'$$

5. The transformation equations are

$x' = 3\cos 30° + 3\sin 30°$
$\quad = 4.0981$ to 4 d.p.

$y' = -3\sin 30° + 3\cos 30°$
$\quad = 1.0981$ to 4 d.p.

$z' = 3$

Hence $\mathbf{r}' = (4.0981, 1.0981, 3)$

6. (a) $\mathbf{A} + \mathbf{B} = \begin{vmatrix} 3 & 3 \\ 1 & 8 \\ -1 & 9 \end{vmatrix}$

(b) $\mathbf{A} - \mathbf{B} = \begin{vmatrix} -1 & 3 \\ 3 & 2 \\ 1 & 5 \end{vmatrix}$

7. (a) $6\mathbf{A} = \begin{vmatrix} 12 & 42 \\ 18 & 0 \\ 54 & -6 \end{vmatrix}$

(b) $\mathbf{AB} = \begin{vmatrix} 25 & -8 & 28 \\ 27 & 9 & 0 \\ 80 & 29 & -4 \end{vmatrix}$

$\mathbf{BA} = \begin{pmatrix} 27 & 63 \\ 32 & 3 \end{pmatrix}$

Hence $\mathbf{AB} \neq \mathbf{BA}$.

8. No matrix multiplication is possible in this case.

9. $\begin{pmatrix} x' \\ y' \end{pmatrix} = \begin{pmatrix} x - 2y \\ 5x + 7y \end{pmatrix}$

10. (a) $\mathbf{A}^{\mathrm{T}} = \begin{pmatrix} 1 & 4 & 3 \\ 2 & -3 & 0 \end{pmatrix}$

(b) $(\mathbf{A}^{\mathrm{T}})^{\mathrm{T}} = \begin{vmatrix} 1 & 2 \\ 4 & -3 \\ 3 & 0 \end{vmatrix} = \mathbf{A}$

11. 3.

12. $\begin{vmatrix} 54 & 0.5 & 4.5 \\ 0.5 & 26 & 52 \\ 4.5 & 52 & 9 \end{vmatrix} + \begin{vmatrix} 0 & 0.5 & -3.5 \\ -0.5 & 0 & -32 \\ 3.5 & 32 & 0 \end{vmatrix}$

13.

$$\mathbf{AA}^{-1} = \frac{1}{13} \begin{vmatrix} -8+0+21 & 6+0-6 \\ -16+9+7 & 12+3-2 \\ -8-6+14 & 6-2-4 \end{vmatrix}$$

$$\begin{matrix} 9+0-9 \\ 18-15-3 \\ 9+10-6 \end{matrix}$$

$$= \frac{1}{13} \begin{vmatrix} 13 & 0 & 0 \\ 0 & 13 & 0 \\ 0 & 0 & 13 \end{vmatrix} = \mathbf{I}$$

Similarly, $\mathbf{A}^{-1}\mathbf{A} = \mathbf{I}$.

Chapter 15

1. (a) $x_1 = -1$, $x_2 = 6$, $x_3 = -5$

(b) The second and third equation are linearly dependent. Thus the solution contains z as parameter free to

$$x = \frac{21-10z}{25}, \quad y = \frac{-79+65z}{25}$$

(c) $x_1 = 13$, $x_2 = 15$, $x_3 = -20$

(d) $x = \frac{0.42-1.5z}{0.12}$, $y = \frac{0.24-1.8z}{0.12}$

The first and third equations are linearly dependent.

2. (a) $\begin{vmatrix} \frac{1}{4} & \frac{1}{2} & -\frac{1}{4} \\ \frac{1}{2} & -1 & \frac{1}{2} \\ \frac{3}{8} & -\frac{3}{4} & \frac{1}{8} \end{vmatrix}$ (b) $\frac{1}{20}\begin{pmatrix} 7 & -8 \\ -6 & -4 \end{pmatrix}$

3. (a) $x_1 = x_2 = x_3 = 0$

(b) $x = -\frac{7z}{20}$, $y = \frac{z}{10}$

4. (a) -9 (b) 0 (c) -322
    (d) -186 (e) 22

5. (a) $r = 2$ (b) $r = 3$

6. (a) $\det \mathbf{A} = -104 \neq 0$; hence unique solution.

(b) $\det \mathbf{A} = 0$; no unique solution exists.

(c) $\det \mathbf{A} \neq 0$; unique solution exists.

(d) $\det \mathbf{A} = 0$; first and third equation are dependent.

(e) $\det \mathbf{A} = 0$; third equation is a linear combination of the first two equations.

Chapter 16

1. (a) The characteristic equation is

$$\det \begin{pmatrix} 4-\lambda & 2 \\ 1 & 3-\lambda \end{pmatrix} = (4-\lambda)(3-2\lambda)-2$$
$$= \lambda^2 - 7\lambda + 10 = 0$$

$\lambda_1 = 2$, $\lambda_2 = 5$

For $\lambda = 2$, solve

$$\begin{pmatrix} 2 & 2 \\ 1 & 1 \end{pmatrix}\begin{pmatrix} x_1 \\ y_1 \end{pmatrix} = 0,$$

i.e. $\begin{matrix} 2x_1 + 2y_1 = 0 \\ 1x_1 + 1y_1 = 0 \end{matrix}$

This reduces to $x_1 + y_1 = 0$. A convenient solution is

$$\mathbf{r}_1 = \begin{pmatrix} 1 \\ -1 \end{pmatrix}$$

For $\lambda_2 = 5$, solve

$$\begin{pmatrix} -1 & 2 \\ 1 & -2 \end{pmatrix}\begin{pmatrix} x_2 \\ y_2 \end{pmatrix} = 0,$$

i.e. $\begin{matrix} -1x_2 + 2y_2 = 0 \\ 1x_2 - 2y_2 = 0 \end{matrix}$

This reduces to $x_2 - 2y_2 = 0$. A convenient solution is

$$\mathbf{r}_2 = \begin{pmatrix} 2 \\ -1 \end{pmatrix}$$

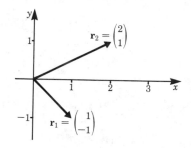

Figure 16.4

2. No. The characteristic equation is a real polynomial equation of degree $2z$. We know from algebra that if z is a complex root then z^* is a root as well, i.e. this characteristic equation has either two complex roots or two real roots.

3. The characteristic equation is

$$(3-\lambda)(1-\lambda) + 4 = \lambda^2 - 4\lambda + 7 = 0$$

There are no real roots, since $\lambda_{1,2} = 2 \pm \sqrt{4-7}$ are complex numbers.

4. (a) The characteristic equation is

$$\det\begin{pmatrix} -1-\lambda & -1 & 1 \\ -4 & 2-\lambda & 4 \\ -1 & 1 & 5-\lambda \end{pmatrix}$$

$$= -\lambda^3 + 6\lambda^2 + 4\lambda - 24 = 0$$

If λ is an integral root, then it must divide into 24, the last coefficient. $\lambda_1 = 2$, $\lambda_2 = -2$, $\lambda_3 = 6$.

(b) For $\lambda_1 = 2$; solve

$$-3x_1 - y_1 + z_1 = 0$$
$$-4x_1 + 4z_1 = 0$$
$$-x_1 + y_1 + 3z_1 = 0$$

This reduces to $x_1 = z_1$ and $y_1 = -2x_1$. $x_1 = 1$ gives the particular solution:

$$r_1 = \begin{pmatrix} 1 \\ -2 \\ 1 \end{pmatrix}$$

For $\lambda_2 = -2$, solve

$$x_2 - y_2 + z_2 = 0$$
$$-4x_2 + 4y_2 + 4z_2 = 0$$
$$-x_2 + y_2 + 7z_2 = 0$$

This reduces to $x_2 - y_2 + z_2 = 0$ and $x_2 - y_2 - z_2 = 0$. Hence $x_2 = y_2$ and $z_2 = 0$.

Choosing $x_2 = 1$ gives the particular solution:

$$r_2 = \begin{pmatrix} 1 \\ 1 \\ 0 \end{pmatrix}$$

For $\lambda_3 = 6$, solve

$$-7x_3 - y_3 + z_3 = 0$$
$$-4x_3 - 4y_3 + 4z_3 = 0$$
$$-x_3 + y_3 - z_3 = 0$$

This reduces to $x_3 + y_3 - z_3 = 0$ and $x_3 - y_3 + z_3 = 0$.

Hence $x_3 = 0$ and $y_3 = z_3$. Choosing $y_3 = 1$ gives the particular solution:

$$r_3 = \begin{pmatrix} 0 \\ 1 \\ 1 \end{pmatrix}$$

5. $\lambda_1 = 1$, $\lambda_2 = 1$.

For the first eigenvalue an eigenvector can be quickly found:

$$r_1 = \begin{pmatrix} 1 \\ 0 \end{pmatrix}$$

But for λ_2 we should like to have another eigenvector which is truly different (i.e. not merely a multiple of r_1). Unfortunately, no such vector exists.

Chapter 17

1. (a) $x_0 = 1.316\,9579$
    (b) $x_0 = 0.657\,298\,10$
    (c) $x_0 = 1.895\,494\,27$

2. (a) Integration of $\dfrac{1}{x^2+1}$

s	Trapezoidal rule	Simpson's rule	$\tan^{-1} s$
0.1	0.099 63	0.099 67	0.099 66
0.2	0.197 32	0.197 40	0.197 39
0.3	0.291 35	0.291 46	0.291 45
0.4	0.380 38	0.380 51	0.380 50
0.5	0.463 51	0.463 65	0.463 64
0.6	0.540 28	0.540 42	0.540 41
0.7	0.610 59	0.610 73	0.610 72
0.8	0.674 62	0.674 74	0.674 74
0.9	0.732 70	0.732 82	0.732 81
1.0	0.785 29	0.785 40	0.785 39
1.1	0.832 89	0.832 98	0.832 98
1.2	0.875 97	0.876 06	0.876 05
1.3	0.915 03	0.915 10	0.915 10
1.4	0.950 48	0.950 55	0.950 54
1.5	0.982 73	0.982 79	0.982 79
1.6	1.012 14	1.012 20	1.012 19
1.7	1.039 03	1.039 07	1.039 07
1.8	1.063 66	1.063 70	1.063 69
1.9	1.086 28	1.086 32	1.086 31
2.0	1.107 12	1.107 15	1.107 14

(b) Integration of $\dfrac{1}{x}$

s	Trapezoidal rule	Simpson's rule	$\ln s$
2.0	0.693 19	0.693 15	0.693 14
3.0	1.098 66	1.098 61	1.098 61
4.0	1.386 34	1.386 29	1.386 29
5.0	1.609 49	1.609 44	1.609 43

(c) Integration of e^{-x^2}

d	Trapezoidal rule	Simpson's rule
1.0	0.886 318 602	0.836 214 302
0.5	0.886 226 926	0.886 196 367
0.1	0.886 226 926	0.886 226 924

True value $\dfrac{\sqrt{\pi}}{2} = 0.886\,226\,926\ldots$

3. (a) $y = \dfrac{100e^{x/10}}{e^{x/10}-\frac{1}{2}}$ (b) $y = \dfrac{1}{\sqrt[3]{3x^2-x^3}}$

x	Euler's method	Improved Euler's method	Correct value
0.0	200.00	200.00	200.00
0.4	192.24	192.46	192.45
0.8	185.34	185.73	185.72
1.2	179.19	179.69	179.68
1.6	173.66	174.25	174.24
2.0	168.66	169.32	169.31
2.4	164.14	164.84	164.83
2.8	160.02	160.75	160.74
3.2	156.27	157.01	157.00
3.6	152.82	153.58	153.57
4.0	149.66	150.42	150.41
4.4	146.75	147.51	147.50
4.8	144.05	144.81	144.80
5.2	141.56	142.31	142.30
5.6	139.25	139.99	139.98
6.0	137.11	137.83	137.82
6.4	135.11	135.81	135.80
6.8	133.24	133.93	133.92
7.2	131.50	132.17	132.17
7.6	129.87	130.53	130.52
8.0	128.34	128.98	128.98
8.4	126.91	127.53	127.53
8.8	125.56	126.17	126.17
9.2	124.30	124.89	124.88
9.6	123.11	123.68	123.68
10.0	121.99	122.55	122.54
10.4	120.93	121.47	121.47
10.8	119.93	120.46	120.45
11.2	118.99	119.50	119.49
11.6	118.10	118.59	118.59
12.0	117.25	117.73	117.73
12.4	116.46	116.92	116.92
12.8	115.70	116.15	116.15
13.2	114.98	115.42	115.42
13.6	114.30	114.73	114.72
14.0	113.66	114.07	114.06
14.4	113.04	113.44	113.44
14.8	112.46	112.85	112.84
15.2	111.91	112.28	112.28
15.6	111.38	111.74	111.74
16.0	110.88	111.23	111.23
16.4	110.40	110.74	110.74
16.8	109.95	110.28	110.28
17.2	109.52	109.84	109.83
17.6	109.11	109.42	109.41
18.0	108.71	109.01	109.01
18.4	108.34	108.63	108.63
18.8	107;98	108.26	108.26
19.2	107.64	107.91	107.91
19.6	107.32	107.58	107.58
20.0	107.01	107.26	107.26

x	Euler's method	Improved Euler's method	Correct value
0.333	1.500	1.500	1.500
0.4	1.326	1.329	1.340
0.5	1.150	1.158	1.170
0.6	1.026	1.037	1.050
0.7	0.934	0.947	0.961
0.8	0.862	0.876	0.892
0.9	0.804	0.820	0.838
1.0	0.756	0.775	0.794
1.1	0.716	0.737	0.758
1.2	0.683	0.705	0.728
1.3	0.654	0.678	0.703
1.4	0.629	0.655	0.683
1.5	0.607	0.636	0.667
1.6	0.588	0.620	0.653
1.7	0.572	0.606	0.643
1.8	0.557	0.594	0.636
1.9	0.544	0.584	0.631
2.0	0.533	0.577	0.630
2.1	0.523	0.571	0.632
2.2	0.514	0.567	0.637
2.3	0.507	0.565	0.646
2.4	0.500	0.565	0.661
2.5	0.495	0.567	0.684
2.6	0.491	0.571	0.718
2.7	0.487	0.578	0.770
2.8	0.485	0.589	0.861
2.9	0.483	0.604	1.059

Please note that the numerical values may depend on the internal accuracy with which real numbers are represented by the computing device. The figures given were obtained on a personal computer.

4. (a)

$$\mathbf{A}^{-1} = \begin{pmatrix} -1 & 2 & 1 & 3 \\ 0 & 4 & 1 & 8 \\ 0 & 1 & 0 & 2 \\ -2 & 10 & 2 & 19 \end{pmatrix}, \qquad \mathbf{x} = \begin{pmatrix} 6 \\ 1 \\ 7 \\ 2 \end{pmatrix}$$

(b)

$$\mathbf{A}^{-1} = \begin{pmatrix} 1 & 2 & 0 & -1 & -2 \\ 4 & -3 & 1 & -2 & 2 \\ -8 & 4 & -1 & 4 & -2 \\ 3 & -2 & 0 & -1 & 1 \\ -5 & 3 & -1 & 2 & -1 \end{pmatrix},$$

$$\mathbf{x} = \begin{pmatrix} 2 \\ 2 \\ 6 \\ 2 \\ 7 \end{pmatrix}$$

5. $$\mathbf{A}^{-1} = \begin{pmatrix} 10 & -99 \\ -1 & 10 \end{pmatrix},$$

$$\mathbf{A}'^{-1} = \begin{pmatrix} 5 & -49.5 \\ -0.5 & 5.05 \end{pmatrix}$$

The matrix $\mathbf{A}$ is ill conditioned.

6. $$\mathbf{A}^{-1} = \begin{vmatrix} 5 & 7 & 6 & 5 \\ 5 & 7 & 9 & 10 \\ 10 & 14 & 14 & 13 \\ 2 & 3 & 2 & 2 \end{vmatrix},$$

$$\mathbf{A}'^{-1} = \begin{vmatrix} 2.5 & 3.5 & 3 & 2.5 \\ 2.5 & 3.5 & 6 & 7.5 \\ 5 & 7 & 8 & 8 \\ 1 & 1.6 & 0.8 & 1 \end{vmatrix}$$

The matrix $\mathbf{A}$ is ill conditioned.

Chapter 18

1. Since $f(x)$ is an even function, the coefficients b_n vanish.

$$a_0 = \frac{1}{\pi} \int_{-\pi}^{\pi} f(x)\, dx = 1$$

$$a_n = \frac{1}{\pi} \int_{-\pi/2}^{\pi/2} \cos nx\, dx = \frac{1}{\pi n} [\sin nx]_{-\pi/2}^{\pi/2}$$

$$= \frac{1}{\pi n} \left(\sin \frac{n\pi}{2} - \sin\left(-\frac{n\pi}{2}\right) \right) = \frac{2}{\pi n} \sin \frac{n\pi}{2}$$

For n even a_n is zero. Thus the Fourier series is

$$f(x) = \frac{1}{2} + \frac{2}{\pi} \sum_{n=1}^{\infty} \frac{(-1)^{n-1}}{2n-1} \cos(2n-1)x$$

2. The function is odd. Thus all coefficients a_n are zero.

$$b_n = \frac{1}{\pi} \int_{-\pi}^{\pi} f(x) \sin nx\, dx$$

$$= \frac{1}{\pi} \int_{-\pi}^{0} \sin nx\, dx - \frac{1}{\pi} \int_{0}^{\pi} \sin nx\, dx$$

$$= \frac{-1}{\pi n} [1 - \cos(-n\pi)] + \frac{1}{\pi n} [\cos n\pi - 1]$$

$$= \frac{-2}{\pi n} + 2\frac{(-1)^n}{\pi n}$$

For n even the coefficients b_n are zero. Thus we obtain

$$f(x) = -\frac{4}{\pi} \sum_{n=0}^{\infty} \frac{1}{2n+1} \sin(2n+1)x$$

3. $$a_0 = \frac{1}{\pi} \left(-\int_{-\pi}^{0} \sin x + \int_{0}^{\pi} \sin x \right) = \frac{4}{\pi}$$

$$a_n = \frac{1}{\pi} \left(-\int_{-\pi}^{0} \sin x \cos nx\, dx \right.$$

$$\left. + \int_{0}^{\pi} \sin x \cos nx\, dx \right)$$

$$= \frac{1}{\pi} \times \frac{1}{2} \left[\left[\frac{\cos(n+1)x}{(n+1)} - \frac{\cos(n-1)x}{(n-1)} \right]_{-\pi}^{0} \right.$$

$$\left. + \left[-\frac{\cos(n+1)x}{(n+1)} + \frac{\cos(n-1)x}{(n-1)} \right]_{0}^{\pi} \right]$$

$$a_n = \begin{cases} \dfrac{4}{\pi(n+1)(n-1)}, & n \text{ even} \\ 0, & n \text{ odd} \end{cases}$$

$$b_n = \frac{1}{\pi} \left(\int_{-\pi}^{0} -\sin x \sin nx\, dx \right.$$

$$\left. + \int_{0}^{\pi} \sin x \sin nx\, dx \right) = 0$$

The Fourier series of the rectified waveform

$$f(x) = \frac{2}{\pi} - \frac{4}{\pi} \frac{1}{1 \times 3} \cos 2x - \frac{4}{\pi} \frac{1}{3 \times 5} \cos 4x$$

$$- \frac{4}{\pi} \frac{1}{5 \times 7} \cos 6x - \dots$$

4. A similar function, with the period 2π, has been treated in the example on p. 432.

$$f(x) = \frac{1}{2} + \frac{2}{\pi} \sum_{n=1}^{\infty} \frac{(-1)^{n-1}}{2n-1} \cos\frac{(2n-1)}{2}x$$

Chapter 19

1. Let A, B, C, D, E be the five dishes. The sample space consists of the following sets of possible pairs:

{ AB, AC, AD, AE, BC, BD, BE, CD, CE, DE }

2. $P = \frac{1}{2}$

3. $P = \frac{1}{3}$

4. $h_A = \frac{1}{30}$

5. $P(\text{blue}) = 0.8$, $P(\text{green}) = 0.2$
    Compound probability $= 0.16$

6. $P = \dfrac{1}{36} \times \dfrac{1}{9} = \dfrac{1}{324}$

7. $P = \dfrac{1}{36} + \dfrac{2}{36} + \dfrac{3}{36} = \dfrac{1}{6}$

8. $N_p = 5! = 120$

9. $N = \begin{pmatrix} 15 \\ 3 \end{pmatrix} = \dfrac{15!}{3!(15-3)!} = 455$

Chapter 20

1. The probability distribution for the random variable 'sum of the number of spots' was given in section 20.1.1.

Random variable	2	3	4	5	6	7	8	9	10	11	12
Probability	$\frac{1}{36}$	$\frac{2}{36}$	$\frac{3}{36}$	$\frac{4}{36}$	$\frac{5}{36}$	$\frac{6}{36}$	$\frac{5}{36}$	$\frac{4}{36}$	$\frac{3}{36}$	$\frac{2}{36}$	$\frac{1}{36}$

The mean value is

$$\bar{x} = 2 \times \tfrac{1}{36} + 3 \times \tfrac{2}{36} + 4 \times \tfrac{3}{36} + 5 \times \tfrac{4}{36} + 6 \times \tfrac{5}{36}$$
$$+ 7 \times \tfrac{6}{36} + 8 \times \tfrac{5}{36} + 9 \times \tfrac{4}{36} + 10 \times \tfrac{3}{36}$$
$$+ 11 \times \tfrac{2}{36} + 12 \times \tfrac{1}{36}. \text{ Thus } \bar{x} = 7.$$

2. $\bar{x} = \displaystyle\int_{-\infty}^{+\infty} x f(x)\, dx = \int_0^2 x \frac{x}{2}\, dx = \left[\frac{x^3}{6}\right]_0^2$

$= \dfrac{4}{3}$

3. $P = \dbinom{10}{8}(0.6)^8 (0.4)^2 = 45 \times 0.016 \times 0.16$

$= 0.12$

4. The random variable which is distributed according to

$$f(x) = \frac{1}{\sigma\sqrt{2\pi}}\, e^{-[(x-\mu)/\sigma]^2/2}$$

has the mean value μ. Hence it follows that
(a) $\bar{x} = 2$,   (b) $\bar{x} = -4$.

Chapter 21

1. (a1) S   (a2) R   (b) S   (c) S

2. (a)

ρ_i (g/cm^3)	$\rho_i - \bar{\rho}$ (g/cm^3)	$(\rho_i - \bar{\rho})^2$ (g/cm^3)^2
3.6	0.4	0.16
3.3	0.1	0.01
3.2	0	0
3.0	−0.2	0.04
3.2	0	0
3.1	−0.1	0.01
3.0	−0.2	0.04
3.1	−0.1	0.01
3.3	0.1	0.01
Sum 28.8	0	0.28

$\sigma^2 = \dfrac{0.28}{8} = 0.035\,(\text{g/cm}^3)^2$

$\sigma = 0.19\,\text{g/cm}^3$

$\rho = 3.2\,\text{g/cm}^3$

(b) Mean value:

$\bar{v} = \dfrac{\sum v_i}{n} = \dfrac{12.80}{10}\,\text{m/s} = 1.28\,\text{m/s}$

Variance:

$\sigma^2 = \dfrac{\sum (v_i - \bar{v})^2}{N-1} = \dfrac{0.011}{9}\,(\text{m/s})^2$

$= 0.00122\,(\text{m/s})^2$

Standard deviation:

$\sigma = 0.035\,\text{m/s}$

3. Mean value:

$\mu = \displaystyle\int_0^1 x\, dx = \dfrac{1}{2}$

Variance:

$\sigma^2 = \displaystyle\int_0^1 \left(x - \dfrac{1}{2}\right)^2 dx = \dfrac{1}{12}$

4. (a) $\sigma_M = \dfrac{\sigma}{\sqrt{N}} = \dfrac{0.19\,\text{g/cm}^3}{3} = 0.06\,\text{g/cm}^3$

Confidence intervals:
$3.14\,\text{g/cm}^3 \leqslant \rho \leqslant 3.26\,\text{g/cm}^3$
$3.08\,\text{g/cm}^3 \leqslant \rho \leqslant 3.32\,\text{g/cm}^3$

(b) $\sigma_M = \dfrac{0.035\,\text{m/s}}{3.16} = 0.01\,\text{m/s}$

Confidence intervals:
$1.27\,\text{m/s} \leqslant v \leqslant 1.29\,\text{m/s}$
$1.26\,\text{m/s} \leqslant v \leqslant 1.30\,\text{m/s}$

5. 16%

6. (a) $A = \bar{x}\bar{y} = 120 \times 90\,\text{cm}^2 = 10\,800\,\text{cm}^2$

Calculation of σ_{MA} using Gaussian error propagation law:

$A_x = \dfrac{\partial}{\partial x}(xy) = y,$

$A_y = \dfrac{\partial}{\partial y}(xy) = x$

$A_x(\bar{x}, \bar{y}) = 90\,\text{cm}$
$A_y(\bar{x}, \bar{y}) = 120\,\text{cm}$
$\sigma_{MA}^2 = A_x^2 \sigma_x^2 + A_y^2 \sigma_y^2$
$\quad = 90^2 (0.2)^2\,\text{cm}^4 + 120^2 (0.1)^2\,\text{cm}^4$
$\quad = 468\,\text{cm}^4$
$\sigma_{MA} = 21.63\,\text{cm}^2$
$A = (10\,800 \pm 21.63)\,\text{cm}^2$

(b) $V = \dfrac{4}{3}\pi\left(\dfrac{D}{2}\right)^3 = 124.79\,\text{cm}^3$

$\rho = \dfrac{M}{V} = \dfrac{1000}{124.79}\,\text{g/cm}^3$

$\quad = 8.014\,\text{g/cm}^3$

Calculation of σ_{MV} using Gaussian error propagation law:

$\dfrac{\partial}{\partial M}\left(\dfrac{M}{V}\right) = \dfrac{1}{V} = \dfrac{1}{124.79\,\text{cm}^3}$

$\quad = 0.008\,\dfrac{1}{\text{cm}^3}$

$$\frac{\partial}{\partial D}\left(\frac{M}{V}\right) = \frac{\partial}{\partial D}\left(\frac{6M}{\pi D^3}\right) = \frac{-18m}{\pi D^4}$$

$$= 3.877$$

$$\sigma_M{}^2 = (0.008)^2(0.1)^2\left(\frac{g}{cm^3}\right)^2$$

$$+ (3.88)^2(0.01)^2\left(\frac{g}{cm^3}\right)^2$$

$$= 0.001\ 5041\left(\frac{g}{cm^3}\right)$$

$$\sigma_M = 0.039\frac{g}{cm^3}$$

$$= (8.014 \pm 0.039)\frac{g}{cm^3}$$

7.

	m (g)	m^2 (g²)	S (cm)	mS (g cm)
1	2	4	1.6	3.2
2	3	9	2.7	8.1
3	4	16	3.2	12.8
4	5	25	3.5	17.5
5	6	36	4.0	24
$\sum$	20	90	15	65.6

$$\bar{m} = 4\ g \qquad \bar{S} = 3\ cm$$

$$a = \frac{\sum m_i S_i - n\bar{m}\bar{S}}{\sum m_i{}^2 - nm^2} = \frac{65.6 - 5 \times 4 \times 3}{90 - 5 \times 4^2}$$

$$= \frac{5.6}{10} = 0.56$$

$$b = \bar{S} - a\bar{m} = 3 - 0.56 \times 4 = 0.76$$

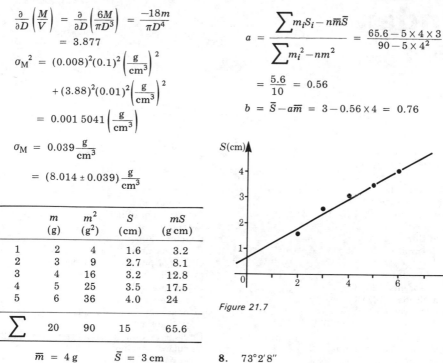

Figure 21.7

8. $73°2'8''$

Index